LIQUID
SEMICONDUCTORS

MONOGRAPHS IN SEMICONDUCTOR PHYSICS

LIQUID
SEMICONDUCTORS

Vasilii M. Glazov
Moscow Institute for Steel and Alloys

Svetlana N. Chizhevskaya
and
Natal'ya N. Glagoleva
Baikov Institute of Metallurgy
Academy of Sciences of the USSR, Moscow

Translated from Russian by
Albin Tybulewicz
Editor, *Soviet Physics — Semiconductors*

℗ Springer Science+Business Media, LLC 1969

Vasilii Mikhailovich Glazov was born in Krasnodar in 1931 and is a graduate of the M. I. Kalinin Institute for Nonferrous Metals and Gold in Moscow. Between 1954 and 1963 Glazov was employed at the Baikov Institute of Metallurgy, first as a junior and then as a senior member of the scientific staff, and attained his doctorate in 1966. He has been employed at the Moscow Institute for Steel and Alloys since 1963, and is presently a professor in the Department of Physico-chemical Properties of Semiconductors. His investigations have been concerned mainly with the physicochemical analysis of semiconductors in the liquid state, the doping of semiconductors, phase diagrams of metal systems, and the structure and crystallization of metal alloys. He has published three books and more than 100 articles on these subjects.

Svetlana Nikolaevna Chizhevskaya was born in Moscow in 1933. Graduated from the M. I. Kalinin Institute for Nonferrous Metals and Gold in Moscow in 1956, she has been employed since then first as a junior and now as a senior member of the scientific staff of the Laboratory of Semiconducting Materials at the Baikov Institute of Metallurgy. She was awarded the degree of Candidate of Technical Sciences in 1963, her dissertation being concerned with the subject of liquid semiconductors.

Natal'ya Nikolaevna Glagoleva was born in Moscow in 1934. Graduated from the M. I. Kalinin Institute for Nonferrous Metals and Gold in Moscow in 1956, she was a member of the faculty of the Metallography Department at this Institute from then until 1961. Since 1962 she has been a junior member of the staff in the Laboratory for Semiconducting Materials of the Baikov Institute of Metallurgy, and was awarded the degree of Candidate of Technical Sciences in 1966. Her dissertation dealt with liquid semiconductors, and her other interests include phase diagrams of metallic alloys, crystallization processes, and the structure of alloys.

Library of Congress Catalog Card Number 68-31237

ISBN 978-1-4899-6220-1   ISBN 978-1-4899-6451-9 (eBook)
DOI 10.1007/978-1-4899-6451-9

The original Russian text, published by Nauka Press in Moscow in 1967, has been corrected by the authors for this English edition.

Глазов Василий Михайлович, Чижевская Светлана Николаевна,
Глаголева Наталья Николаевна
Жидкие полупроводники
Zhidkie Poluprovodniki

Preface to the American Edition

Recently interest in liquid semiconductors has increased significantly. This is due to the following two factors:

1. potential practical applications of liquid semiconductors in the solution of the problem of direct conversion of thermal energy into electrical power, and

2. potential improvements in the technologically important processes of preparation of semiconducting materials with specified parameters on the basis of the knowledge of their properties in the molten state (the processes concerned include synthesis, strong purification by crystallization methods, preparation of perfect single crystals, doping, etc.).

Investigations of liquid semiconductors are also important in the development of the physics of the liquid state, which at present lags considerably behind the physics of solids and gases.

The present book is the first monograph on the subject of liquid semiconductors and it collects the available experimental data on changes in the physical properties of semiconductors at the melting point and on the temperature dependences of various properties in a wide range of temperatures below and above the melting point. A considerable fraction of the monograph is based on the authors' own studies of the physical properties of semiconductors at the melting point and in the liquid state. A special feature of the approach of the authors to the behavior of semiconducting elements and compounds in the liquid state has been the extensive use of the principles of the physicochemical analysis enunciated by Academician N. S. Kurnakov and concerned with certain features of the phase diagrams.

This approach seems very fruitful and other investigators are now using it to study the properties of melts (cf., for example, Dancy [590]). We have also used extensively the modern theory of the solid state and of the chemical binding. For this reason each chapter dealing with a group of semiconductors with a particular structure begins with the section giving full information on the crystal structure and on the nature of chemical binding in these semiconductors. The present monograph probably gives the fullest information on the structure and chemical binding of semiconductors collected within the covers of one book.

General problems of the physics and physical chemistry of the liquid state are discussed in the final chapters. Some of the views put forward in this book are controversial. However, the reported experimental data should be valuable irrespective of these views.

The chapter describing methods for investigating the properties of semiconductors consists almost entirely of descriptions of the methods used by the present authors. This is justified by the limited number of available methods and by the content of the monograph, which reports results obtained solely by these methods.

The American edition differs little from the Russian original. The authors have simply corrected errors and inaccuracies which were unavoidable in the first edition and which have been noticed only after the appearance of the book.

The authors welcome the appearance of their monograph in the USA and are hopeful that their efforts will be received favorably by the American readers.

The authors will be grateful to learn of any errors and faults of the book.

V. Glazov

Preface to the Russian Edition

The rapid development of semiconductor technology has stimulated comprehensive investigations of the physicochemical properties of semiconducting materials over a wide range of temperatures, including the liquid state. Numerous investigations of molten semiconductors have been carried out. Only ten years ago, there were few original papers on the properties of molten semiconductors. Now, more than 500 papers on the subject have been published. However, no reviews have yet been published. The present monograph is intended to fill this gap.

Investigations of the properties of semiconductors in the liquid state have taken many man-years and have involved tens of substances. The results have been published in various journals. It seemed desirable to review this work, to compare the results obtained by different authors, and to present them from one particular point of view.

This point of view is rooted, on the one hand, in the basic principles of physicochemical analysis developed by Academician N. S. Kurnakov, and, on the other hand, in the modern concepts of chemical binding in solids.

In spite of the great interest in the physicochemical properties of molten semiconductors, many aspects of the subject have not yet been investigated. Therefore, the present book includes a special section (Chapter 1, §3), where the principal scientific and practical problems concerning the melts are formulated. This section not only indicates the possible directions of further in-

vestigations but also discusses the possible applications of the data on the properties of melts.

Tables at the end of the book give numerical experimental data on the viscosity, density, electrical conductivity, and thermoelectric power. These data (cf. Appendix) can be used for reference purposes, in contrast to the graphical data in the main part of the book, which are presented only to show the general nature of the temperature and concentration dependences of the various properties.

The title of the book is somewhat imprecise. We shall discuss substances which are metals in the liquid phase (for example, germanium, silicon, and some compounds) as well as substances which are true liquid semiconductors, i.e., which retain their predominantly covalent type of binding in their transition to the liquid state.

The monograph does not claim to be comprehensive. In particular, the results of the structure and thermodynamic investigations are practically ignored. This is due partly to the absence of really reliable data.

The authors are aware that there must be deficiencies in a first review of this kind. They will be very grateful for any comments and suggestions from the readers.

Contents

Chapter 1

Introduction—
Importance of the Physicochemical
Analysis of Liquid Semiconductors

§1. Some General Problems

At the beginning of the present century, Nikolai Semenovich Kurnakov, a leading Russian chemist, developed a new branch of chemistry known as the physicochemical analysis.

Kurnakov described it thus [1]: "The physicochemical analysis is a quantitative investigation of the equilibrium properties of systems, formed by two or more components, as a function of their composition. In the simplest case of two components, we can plot the composition along the abscissa and the measured values of the investigated property along the ordinate. We thus obtain a chemical composition–property diagram which consists of a number of lines or surfaces whose positions define the state of the system. We can obtain, for various equilibria, the numerical data on the composition and the conditions of formation of new phases, without actually producing these phases."

Quantitative determinations of such chemical composition–property dependences yield information on various phases, from which the nature of chemical compounds and other species can be deduced.

The analysis of composition–property diagrams allows us, in some cases, to find singular points [1]. This analysis of the

1

diagrams is based on two general principles of continuity and correlation, which govern the relationships between the composition
and properties in these diagrams.

These principles were formulated by N. S. Kurnakov. In the
case of physicochemical analysis, they can be expressed as follows.

1. The Continuity Principle. The thermodynamic
equilibrium parameters (the temperature t, the pressure P, and
the concentrations of the components $x_1, \ldots, x_n$) and the measured properties of the various phases (for example, the viscosity
η, the electrical conductivity σ, the density d, etc.) are related by
continuous functions which determine the equilibrium state of a
system

$$f(t, P, x_1, \ldots, x_n, \eta),$$
$$f_1(t_1, P_1 x_1 \ldots, x_n, \sigma),$$
$$f_2(t_2, P_1, x_1, \ldots, x_n, d).$$

When the composition of a liquid or solid solution (their
similarity has been pointed out by Van t'Hoff) is altered gradually,
the properties of such a solution (η, σ, d, etc.) vary continuously.
As pointed out by S. N. Kurnakov, the principle of continuity can
be used to find the relationships and mutual transitions even in
those cases when apparent discontinuities are observed in the geometrical diagrams. The classical investigations of van der Waals
have established the continuity of the gaseous and liquid states in
systems consisting of one or two components. These investigations
have accounted for such phenomena as the mutual solubility of liquids, the existence of a critical solution temperature, etc. [1]. In
our opinion, the conclusion of Frenkel' [2] that there are no basic
differences between the solid and liquid phases near the melting
point is also based on the principle of continuity.

Kurnakov stressed that the idea of continuity had been the
guiding principle in investigations of chemical equilibria from the
time of Berthollet, and had been most clearly formulated by D. P.
Konovalov: "a chemical transformation is the result of interac-
tions which obey the law of continuity." However, Kurnakov was
the first to apply the principle of continuity to phase equilibria described by phase diagrams.

2. The Correlation Principle. The second basic rule used in analysis of chemical composition—property diagrams is the principle of geometrical affinity or correlation [1].

According to this principle, each phase of an equilibrium system, i.e., each physicochemical species, corresponds to a particular geometrical image in a complex property diagram [1].

The application of this principle to the analysis of chemical and phase equilibria is based on those properties of phases and compounds that are governed by their physicochemical nature, i.e., their structure and chemical binding. It should be possible to solve the converse problem: the analysis of complex geometrical property diagrams should yield information on the nature of the chemical binding and the structure of a given phase or a physicochemical species.

We must stress that these two principles have a much wider application than the analysis of composition—property diagrams. This was pointed out indirectly by Kurnakov himself, who referred to the work of van der Waals on liquid—vapor equilibria. The phase diagram of, for example, a binary system is a three-dimensional figure plotted in coordinates of temperature, pressure, and composition. The nature of a change in the properties, considered as a function of any one of these coordinates, should be governed by the principles of continuity and correlation. Consequently, when temperature or pressure are altered at a transition from one phase region to another (in particular, at the melting point) the correlation principle shows that the property diagram should have certain geometrical figures, whose relative positions can give information on the physicochemical nature of the phases and on changes taking place at the phase transition.

This idea, implicit in Kurnakov's principles, was taken up and applied in brilliant manner by Regel' [3], who carried out the first systematic investigation and analysis of the nature of transitions from the solid-to-liquid state for a large number of elements and compounds with different structures and types of chemical binding. Regel' concluded that the analysis of singular points in the property—temperature diagrams can yield information on the structure and on the nature of the binding of a substance in the liquid phase.

The classical physicochemical method of analysis of the liquid state was developed by Kurnakov, who carried a large number of experimental investigations of binary liquid systems [4], investigating their viscosity, electrical conductivity, density, refractive index, etc., at various temperatures. On the basis of these investigations, Kurnakov proposed a classification of systems in accordance with the nature of their composition—property diagrams in the liquid state. According to this classification, the composition—property diagrams of compounds which do not dissociate when melted is characterized by the presence of an extremal singularity, corresponding to the ordinate of the relevant chemical compound.

The composition—property diagrams of systems forming more or less dissociating compounds ("irrational systems") are also characterized by an extremum near the composition of the chemical compound but this extremum is broad and shifted in the direction of the component having the maximum (or minimum) value of a given property. For example, in the viscosity—composition diagrams the viscosity maximum of irrational systems shifts, when the temperature is increased, in the direction of the more viscous component. Numerous investigations on metallic systems, carried out many years after Kurnakov's pioneer work, confirmed the general validity of his hypotheses on the nature of the concentration dependences of the properties of systems which include a chemical compound.

Kurnakov and later investigators have analyzed the composition—property diagrams of liquid phases, ignoring changes in the nature of the chemical binding at the melting point of a compound and of the components of which it is formed. This problem has begun to interest investigators only after the discovery of semiconductors. In this case, two aspects of physicochemical analysis are combined: the principles of continuity and correlation are used to analyze the nature of the property diagrams when the composition and temperature are varied.

§ 2. Molten Semiconductors

The considerable successes in the physics and chemistry of semiconductors have led to the development of semiconductor technology. Many semiconducting materials (germanium, silicon,

indium antimonide, etc.) are now manufactured on an industrial scale. A great variety of semiconductor devices is made from these materials. Semiconductors are promising in the solution of the problem of direct energy conversion from the thermal to the electrical form.

Much work has already been done on new semiconductors satisfying the technological requirements. The interest has spread from crystalline materials to organic, amorphous, and liquid semiconductors.

It is also necessary to improve further the technology of preparation of semiconducting materials and their quality; this requires a comprehensive study of the physicochemical properties of semiconducting materials, including liquid semiconductors.

We must distinguish clearly the concepts of "liquid semiconductors" and "molten semiconductors" or "semiconductor melts." In the former case, we mean that a particular substance has semiconducting properties in the liquid state, i.e., the substance is a true semiconductor. The second concept of molten semiconductors or semiconductor melts is wider and includes the first concept. In this case, we mean the liquid phase of substances which have semiconducting properties in the solid state but not necessarily in the liquid state.

The first systematic and comprehensive investigation of the relationships governing changes in the physicochemical properties of semiconductors at the melting point and in the liquid state was carried out by Regel' et al. [3, 5-18]. They investigated the electrical conductivity and certain other properties and demonstrated conclusively that some liquids can have semiconducting properties. They also found that some semiconductors, such as germanium and silicon, become metallic after melting.

Regel' has investigated a large number of substances and has suggested a classification based on the change in the electrical properties at the melting point. According to this classification, semiconductors can exhibit two types of solid-to-liquid transition:†

---

† It is reported in [24, 25] that one further type of solid-to-liquid transition may be exhibited by semiconductors: semiconductor → ionic conductor (this happens in the case of cuprous iodide).

semiconductor → semiconductor, or

semiconductor → metal.

Moreover, we can have also an intermediate case:

semiconductor → semiconductor → metal.

In the last case, the transition to the metallic state takes place in the liquid phase after a further, sometimes quite considerable, increase in the temperature. This is the most frequent case, because many substances pass through an intermediate state, and often only a very slight increase in the temperature is required to go over from the intermediate to the metallic state [19, 20]. This follows from investigations of the viscosity and from calculations of the thermodynamic parameters of the viscous flow of such substances as germanium and $A^{III}B^V$ compounds [21-23].

A transition of the first type is typical of the compounds Bi_2S_3, Sb_2S_3, Cu_2S, CdTe, ZnTe, and some others (cf. [3, 5-7, 26-30]. These substances are the true liquid semiconductors. The existence of liquid semiconductors, i.e., substances which retain semiconducting properties in the absence of long-range order is remarkable in itself and important consequences follow from this observation. One of these consequences, established by Regel' himself, is the limited validity of the band theory of solids, since this theory is based on the assumption of a regular crystalline structure of matter.

Transitions of the second type are observed when germanium, silicon, $A^{III}B^V$ compounds and certain other substances are melted. The most typical example of the third (intermediate) type of transition from the solid-to-liquid state is provided by tellurium.

The retention of semiconducting properties in the liquid phase has been attributed by Regel' to the conservation − in the molten state − of the short-range order observed in the solid state. The transition to the metallic state is explained by the destruction of the three-dimensional system of covalent bonds and a basic rearrangement of the short-range order structure. The experimental investigations of Regel' et al. have thus proved conclusively the validity of Ioffe's principle [31, 32], that the generation and motion of free electrons is governed mainly by the short-range order in the distribution of atoms, i.e., by the number and chemi-

cal nature of the nearest neighbors in the first coordination sphere, by the geometry of the distribution of these neighbors, and by the absolute values of the interatomic spacings.

§ 3. General Scientific and Practical Problems Encountered in the Investigations of the Physicochemical Properties of Molten Semiconductors

Bearing in mind the two basically different types of solid-to-liquid transition, we shall attempt to formulate some general scientific and practical problems encountered in the investigations of the physicochemical properties of molten semiconductors.

These investigations are important to the development of the physics and physical chemistry of the liquid state. The liquid state has been investigated less than the solid or gaseous state. There is as yet no rigorous theory which would describe sufficiently fully and reliably the physicochemical nature and properties of substances in the liquid state. The available theories of the liquid state are based on models which are far removed from real conditions (cf., for example, reviews [33, 34]).

Therefore, further investigations of melts, in particular semiconductor melts, are necessary to enable us to develop sufficiently clear ideas about the nature of liquids and to construct a theoretical description of the liquid state, such as that already available for the solid and gaseous states.

There are other unsolved problems associated with changes in the short-range order and in the nature of chemical binding at the melting point and during further heating of melts. Analysis of the investigations reported in [5-30] shows that it would be best to study these problems in semiconductors because the nature of the chemical binding in solid semiconductors may vary from the purely covalent to the almost ionic or from the purely covalent to the almost metallic.

Investigations of the physicochemical properties of semiconductors at their melting points and in the liquid phase should answer the following questions.

1. What is the influence of the nature of the chemical binding in the solid state on the nature of the changes at the solid-to-liquid transition and during further heating of the melt ?

2. What is the role of the initial solid-state structure in the formation of the short-range order in the liquid ?

3. Is the change of the structure of a solid phase to the final structure of a liquid an isothermal process or does it extend over a range of temperatures below and above the melting point ?

4. What is the relationship between the nature of the composition−property diagram in the liquid state and the form of the phase equilibrium diagram ?†

In §2 we have mentioned two main directions of investigations of semiconducting materials. Therefore, we should consider two corresponding aspects of investigations of the physicochemical properties of semiconductor melts.

A. Preparation of Liquid Semiconductors. The discovery of liquid semiconductors raises the question of their practical applications. At present, it is difficult to see what these applications could be in the future, but there have been papers suggesting liquid thermoelectric energy converters [36, 37].

Some substances (for example, complex copper chalcogenides) have very high values of the thermoelectric parameters in the liquid state. Bearing in mind that large temperature gradients can be established in liquids, it follows that we can increase considerably the efficiency of the thermal-to-electrical energy conversion by the use of liquids. The use of liquid semiconductors in the direct conversion of thermal to electrical energy promises more efficient use of the heat evolved in nuclear reactors or collected by solar concentrators. It would seem desirable to search for new liquid semiconductors with suitable electrical (including thermoelectric) properties.

B. Investigations of Physicochemical Properties (in connection with the development and improvement of

---

† The currently held views are contradictory (cf. the summary of a discussion on the structure and properties of liquid metals, which took place at the A. A. Baikov Metallurgy Institute in 1961 [35]).

the technology of preparation of semiconducting materials). A number of problems in the preparation of semiconducting materials can be solved by investigating the physicochemical properties of semiconductors in the liquid state. The technological processes used in the preparation of the semiconducting materials include the extraction of a substance or its synthesis (in the case of compounds), purification, preparation of single crystals, and doping. The various stages of these processes involve the melting of a substance and its crystallization under specified conditions. Consequently, the knowledge of the properties of the liquid phase is very desirable.

The main problems encountered in the preparation of semiconductors are as follows.

In many cases, the synthesis of semiconducting compounds involves the fusion of components in their stoichiometric ratios. It is essential to ensure that the chemical interaction between components in the melt is as complete as possible because incompleteness affects the properties of a semiconductor. The best example of this is aluminum antimonide, the formation of which in the liquid phase is complete only after a fairly long treatment at high temperatures. If the treatment is too short, three phases (aluminum, antimony, and aluminum antimonide) are found in the crystallized melt [38-43].

Studies of the synthesis of some compounds have established that a two-phase structure is obtained when some stoichiometric melts are crystallized. A one-phase alloy is obtained only if there is an excess of one of the components [44-50]. In such systems, it is essential to know the nature of the chemical interaction in the liquid phase because a deviation from stoichiometry, necessary for the formation of one phase, may decrease when the temperature is increased and may disappear altogether in the liquid phase [48]; this makes it possible to prepare homogeneous metastable phases using high cooling rates, resulting in diffusionless crystallization [52].

Some compounds begin to dissociate in the liquid phase above a certain temperature. Consequently, the synthesis and purification by crystallization must be carried out at temperatures within the thermal stability limits of the given compound in the liquid

phase. The thermal stability limits of compounds in the liquid phase can be determined by physicochemical analysis, plotting the chemical composition—property diagrams [1-4] obtained from investigations of the concentration dependences of the physicochemical properties.

In many cases, semiconducting materials must be prepared in the form of single crystals of nearly perfect structure. Consequently, we must investigate carefully the precrystallization phenomena in the liquid phase, because the state and structure of the melt affects the structure and properties of crystals formed from it. This applies particularly strongly to semiconductors, since the structures of many semiconducting elements and compounds differ considerably in the solid and liquid states.

Semiconducting materials are currently prepared in the form of dendritic ribbons. This process takes place at high pulling rates of a solid from a strongly supercooled melt. It follows that we must know the physicochemical properties of supercooled melts and the relationships governing strong supercooling.

Investigations of the physicochemical properties of complex (doped) solutions are important in connection with the doping and the interaction between dopants. Once again, we can use the physicochemical analysis method in order to obtain information on the state of dopants in liquid and solid solutions based on such semiconductors as germanium or silicon.

Finally, in the design of apparatus and processes used in the preparation of semiconducting materials, it is necessary to have reliable numerical data on such important properties as the electrical conductivity, magnetic susceptibility, viscosity, density, etc. Consequently, it is necessary to investigate systematically these properties so as to be able to design scientifically the apparatus and to control the processes in which the material passes, at some stage, through the liquid phase.

Our discussion shows that investigations of the physicochemical properties of molten semiconductors are important both scientifically and from the practical point of view. We shall now review the most important results obtained on molten semiconductors.

Chapter 2

Methods for Investigating the Physicochemical Properties of Molten Semiconductors

§1. Selection of Properties and Investigation Methods

The electrical conductivity, magnetic susceptibility, thermoelectric power, density, and viscosity are among the widely investigated physicochemical properties. Sometimes also the thermal conductivity, Hall effect, and other properties are studied. These properties are very sensitive to changes in the structure and in the nature of the chemical binding. Therefore, analysis of changes in these properties during melting and further heating can yield information on the structural transitions and changes in the nature of the chemical binding. This applies particularly to combined investigations of several properties of the same substance. We shall review these properties briefly and discuss the methods used in investigating them.

The electrical conductivity is one of the most sensitive indicators of changes in the nature of the chemical binding. In general, the electrical conductivity is governed by the product of the carrier density and the carrier mobility. In the presence of carriers of one kind:

$$\sigma = neu, \tag{1}$$

where n is the number of carriers, u is the mobility, and e is the electron charge.

A change in the nature of the chemical binding alters primarily the carrier density and this changes the electrical conductivity. Structural changes are accompanied by changes in the carrier mobility, which are also reflected in the electrical conductivity. Consequently, the electrical conductivity is also a structure-sensitive property. (The very early investigations of metals
showed that the conductivity decreases approximately by a factor
of 2 at the melting point) The behavior of antimony, bismuth, and
gallium — whose electrical conductivity increases after melting —
has been regarded as anomalous; it was attributed by Bridgman to
an increase in the close packing of the structure in the transition
from the solid to liquid state [53]. Later, Perlitz [54] related the
nature of the changes in the electrical conductivity of metals at
the melting point to their structure in the solid state and formulated
a rule, according to which the electrical conductivity of closely
packed metals doubles after melting, that of metals with the bcc
lattice increases by a factor of 1.5, while the conductivity of semimetals falls to half the value in the solid state. Mott [55] assumed
that the short-range order of a solid is retained in the molten
state and deduced a dependence which relates the value of the
change in the electrical conductivity at the melting point to the
heat of fusion

$$\frac{\sigma_s}{\sigma_l} = -\exp(80 L_f / T_{mp}), \qquad (2)$$

where σ_s is the electrical conductivity of a solid at the melting
point; σ_l is the same property in the liquid phase; L_f is the heat
of fusion; T_{mp} is the melting point.

However, Eq. (2) is not obeyed by semimetals because, as
correctly pointed out by Mott, of the changes in their structure at
the melting point. Thus, changes in the structure and chemical
binding are reflected in the electrical conductivity. The absolute
value of the electrical conductivity and the sign of its temperature
coefficient can be used to classify various solids into semiconductors, metals, and dielectrics and to draw at least qualitative conclusions about the nature of the chemical binding.

The magnetic susceptibility is a complex property
which depends on the state of all electrons (free, valence, inner).
Naturally, when the nature of the chemical binding changes and

there is a change in the state of electrons, for example, at the melting point, the magnetic susceptibility should also change. Unfortunately, such changes cannot be interpreted quantitatively because of the absence of a rigorous theory. However, we can draw qualitative conclusions, particularly when a study of the susceptibility is combined with investigations of other properties, such as the electrical conductivity. In the opinion of Professor Ya. G. Dorfman (private communication, 1963), investigations of the magnetic susceptibility at the melting point and in the liquid state are of interest in investigations of compounds which are chemical analogs: in this case, the nature of the chemical binding varies gradually along a series of such compounds.

The magnetic susceptibility of the melts is usually measured by the Faraday method [57], which is most convenient for the melts, particularly near the melting point [58].

The density is an integral structure-sensitive property, which depends on the coordination number and on the interatomic spacing, i.e., it is directly related to the short-range order. Volume changes at the melting point are governed completely by changes in these two quantities and therefore, if one of them is known (the interatomic distance or the coordination number), we can calculate the other property if we know the density. Density (volume) changes at various transitions (including fusion) reflect basic internal changes in the state of a system or a material. Therefore, a change in the volume, accompanied by certain thermal effects, represents a change in one of the principal thermodynamic parameters of state [59]. Analysis of the volume changes gives information on changes in the structure and nature of the chemical binding and yields other important data. In particular, we can deduce the pressure dependence of the melting-point temperature [59, 60]. The density of liquids is usually measured by the following two methods: determination of the volume in a calibrated quartz pycnometer (also known as the "thermometer" method) and hydrostatic weighing. Our use of these two methods has been dictated by the chemical properties of the melts (we shall discuss this point in detail in connection with the results obtained). Changes in the density of a solid can be determined conveniently by the dilatometric method [61].

The viscosity yields information on changes in the
short-range order when the melts of compounds are heated. This
is one of the principal methods in physicochemical analysis and –
according to N. S. Kurnakov and A. I. Bachinskii – it is very sensitive to structure [1, 4, 62]. A close relationship between the
viscosity and the structure of a liquid was first pointed out by
Bachinskii, who found that the free volume plays a decisive role
in the process of viscous flow [62, 63]. The relationship between
the viscosity and the structure or the nature of the chemical binding has been pointed out in many papers. In particular, Stewart
[64] has shown that the viscosity of a liquid falls when the degree
of order in the distribution of molecules decreases. Bernal [65]
has shown that a liquid with directional covalent bonds should have
a low coordination number and a high viscosity.

The majority of modern theories of viscous flow are based
on the assumption of a direct relationship between the viscosity
and the structure of a liquid (cf. a review of theories of viscosity
[66]). Ravikovich [67] has compared the viscosity with the dimensions of the structural units in a liquid and has shown that the viscosity is directly proportional to these dimensions. Golik and
Karlikov [68] have demonstrated the relationship between the viscosity and the structure of a liquid by a direct comparison of the
diffraction patterns of melts of the same viscosity. They have
found that the isoviscous liquids have identical diffraction patterns
and consequently identical short-range order.

These examples demonstrate conclusively the high sensitivity of the viscosity to changes in the structure of a liquid. Moreover, the viscosity approach has been found to be useful in the
physicochemical analysis of systems including semiconducting
chemical compounds [19, 24-29, 69].

The viscosity of the melts at high temperatures is often determined by the Poiseuille method, based on the measurement of
the velocity of flow of a liquid through a capillary; other extensively used methods are those based on the oscillations of a solid of
revolution, immersed in the investigated liquid (external hydrodynamic problem) and on damped oscillations of a hollow sphere
or cylinder filled with the investigated liquid and suspended by an
elastic filament (internal hydrodynamic problem). Detailed descriptions of these methods can be found in many reviews [70-74].

The damped oscillations method is most convenient and reliable in measurements of the viscosity of melts at high temperatures. Details of this method can be found in a monograph by Shvidkov-skii [74]. This method was suggested at the end of the nineteenth century by Meyer [75].

The thermoelectric power and thermal conductivity are used to study melts for the following reasons. First, the analysis of changes in the thermoelectric power at the melting point, combined with the data on changes in the electrical conductivity and density, can yield additional information on changes in the short-range order and in the nature of the chemical binding. Moreover, the sign of the thermoelectric power can be used to determine the type of conduction, including conduction in liquids.

Finally, both these properties are fundamental thermoelectric characteristics. The thermoelectric efficiency of a semiconductor is given by the expression

$$z = \frac{\alpha^2 \sigma}{\varkappa},$$ (3)

where α is the thermoelectric power, σ is the electrical conductivity, and $\varkappa$ is the thermal conductivity.

Therefore, the thermoelectric power and the thermal conductivity are among important properties of melts.

§ 2. Methods and Apparatus for
Investigating Physicochemical
Properties of Melts

Experimental investigations of the electrical conductivity, thermoelectric power, magnetic susceptibility, density, viscosity, and thermal conductivity of molten substances, which may have a considerable chemical activity at high temperatures, are fairly difficult. However, suitable apparatus helps to overcome the difficulties.

We shall now consider briefly the basic principles of the methods, the apparatus, and the particular methods used in the measurements of these properties.

1. Electrical Conductivity

The electrical conductivity of melts can be conveniently determined by an electrodeless method using a rotating magnetic field, based on the application of induction currents. A review of the work on the use of induction currents in the measurement of the electrical conductivity has been given by Regel' [3, 76]. The use of this method avoids the difficulties associated with the selection of suitable contact materials and it makes it possible to seal hermetically the ampoule containing the substance under investigation, which is particularly important in the case of many volatile semiconductors. This method is particularly suitable for high-conductivity substances. It is based on the measurement of a torque acting on a conductor placed in a rotating magnetic field. The system is, in principle, similar to the asynchronous motor in which the sample being investigated is the rotor and the stator is iron-free in order to obtain a more homogeneous magnetic field. The torque acting on the sample is easily determined from the angle of twist of the suspension. A calculation of the moment acting on a conducting spherical solid placed in a rotating magnetic field was given in the general form by Hertz at the beginning of the nineteenth century. Regel' [76] used the Hertz solution to derive calculation formulas, which can be employed to carry out absolute measurements of the electrical conductivity by the rotating magnetic field method. The work of Regel' [76] has played a decisive role in the development and wide application of this method, particularly in investigations of the electrical conductivity of molten metals and semiconductors. Therefore, we shall briefly consider the fundamentals of this method.

A uniform rotating magnetic field can be represented by

$$H_x = H_0 \cos \omega t; \quad H_y = H_0 \sin \omega t; \quad H_z = 0. \tag{4}$$

Under steady-state conditions, when a sphere rotates at a constant angular velocity ω, the energies of the magnetic and electric fields of the induced currents are constant, and consequently all the work done by the external forces in maintaining the rotation at a constant angular velocity ω is used up in the maintenance of the current, i.e., it is equal to the Joule heat W, evolved in a sphere in one second:

$$W = \omega M. \tag{5}$$

Consequently, having calculated W, we can find from Eq. (5) the value of the torque acting on a fixed sphere placed in a rotating magnetic field:

$$M = \frac{W}{\omega}. \tag{6}$$

Appropriate calculations give the following results:

$$W = \frac{2\pi}{15} \cdot \frac{\omega H_0^2 R^5 \sigma}{c^2}. \tag{7}$$

Allowing for the self-induction, we obtain the following expression, which is valid at all frequencies:

$$W = \frac{3}{4} \omega R^3 H_0^2 N(t), \tag{8}$$

$$t = \frac{R}{c} \sqrt{2\pi\omega\sigma}, \tag{9}$$

where H_0 is the intensity of the uniform rotating magnetic field, in Oe; ω is the angular frequency of rotation of the field, rad/sec; σ is the electrical conductivity of the material of the sphere, in cgs esu units; R is the radius of the sphere, in cm; and c is the velocity of light, in cm/sec.

In the limiting cases when t is either very small or very large, we easily find from Eq. (7) that

when $t \ll 1$,

$$W_1 = \frac{2\pi}{15\,c^2}(\sigma\omega^2 H_0^2 R^5), \tag{10}$$

and when $t \gg 1$,

$$W_2 = \frac{3}{4\sqrt{2\pi\sigma}}(c H_0^2 R^2 \sqrt{\omega}). \tag{11}$$

The expressions for N(t) are found from Eqs. (6), (10), and (11). The values of M, R, and H_0 are measured. Having calculated the function N(t) and knowing ω, we can find the electrical conductivity σ. If an accuracy of about 1% is acceptable, we can show that

when $t \leq 0.85$,

$$N_1(t) = 8.83\,(8) \times 10^{-2}\,t^2, \tag{12}$$

when $t \geq 2.5$,

$$N_2(t) = \frac{t-1}{t^2}. \tag{13}$$

Formula (8) can be used in the absolute measurements of the electrical conductivity by the rotating magnetic-field method, provided the values of M, R, H_0, and ω are measured accurately.

It is difficult to produce a perfectly spherical sample. Consequently, Regel' has carried out a special investigation of the influence of departure from the spherical shape. This investigation, carried out on mercury, has established that the sample need not be perfectly spherical.

It follows directly from the principle of the rotating magnetic-field method that we can carry out relative measurements. If we have a sample made of a material of known electrical conductivity, it is sufficient to determine the angle through which the sample is twisted ($\Delta\varphi$) from the equilibrium position as a function of the average current in the stator which is producing the rotating magnetic field. Then, knowing $\Delta\varphi$, σ, and i_{av} [$i_{av} = (i_1 + i_2 + i_3)/3$, where i_1, i_2, and i_3 are the currents in the three parts of the stator], we can determine the instrument constant k:

$$k = \sigma i_{av}^2 / \Delta\varphi. \tag{14}$$

Knowing k, we can determine the electrical conductivity of other samples of the same dimensions and shape. The conditions, and particularly the dimensions of the sample, should be such that the self-induction effect can be neglected. Regel' has shown that the self-induction can be neglected in samples of 1-1.5 cm^3 volume even if these samples are made of copper; for materials of lower electrical conductivity the self-induction can be neglected with even greater justification. Changes in the volume of a sample due to the slight expansion caused by heating can be neglected provided there is no phase transition. However, in the case of considerable volume changes, particularly at phase transitions (for example, at the melting point), it is necessary to make a correction for the change in the volume [15]:

$$k_1 = k_0 \, (V_0/V_1)^{2/3}, \tag{15}$$

where V_0 is the volume of a sample at the melting point in the solid phase; V_1 is the volume of the same sample at the melting point in the liquid phase.

To measure the electrical conductivity by this method, we have developed and constructed apparatus [16, 77] in accordance with the principles established by Regel' [76]; the same apparatus has been used also in measurements of the kinematic viscosity by the Meyer–Shvidkovskii method.

Figure 1 shows schematically the apparatus mounted on a Textolite base. The base is supported by brackets attached to a load-bearing wall of a building. The main units are: a casing, a furnace, a suspension system, a stator, a device for measuring the angle of twist of the suspension, and a control panel with a power supply and with measuring instruments.

The casing is a water-cooled quartz chamber 1, to which a molybdenum glass tube 3, containing a suspension system 11, is connected by a ground-glass joint 2.

The furnace consists of a heater 4, in the form of a tube of molybdenum sheet 0.15 mm thick and 230 mm long. One end of the heater is attached to a water-cooled flange 6 and the other end is attached to a refractory stainless-steel flange, which is connected by a massive stainless-steel cylinder 9 to a water-cooled flange 5. The current is supplied to the flanges 5 and 6, which are insulated from each other by a rubber spacer 7; this spacer also acts as a vacuum seal. The massive steel cylinder 9 is used as an efficient heat screen (its specific heat is high); moreover, it compensates the electric field of the heater by forming a bifilar configuration with it. The working space in which the sample is placed is screened by a system of molybdenum disks 10 in order to reduce the radiation losses. The temperature in the furnace is measured with a platinum–platinorhodium thermocouple 8, which enters the furnace through the lower flange and is connected to a millivoltmeter. This construction makes for easy and rapid assembly and dismantling. This is particularly important in the case of semiconducting materials, since the crystallization of many of them is accompanied by an increase in their volume; ampoules containing the investigated substance fracture, and their contents fall into the furnace.

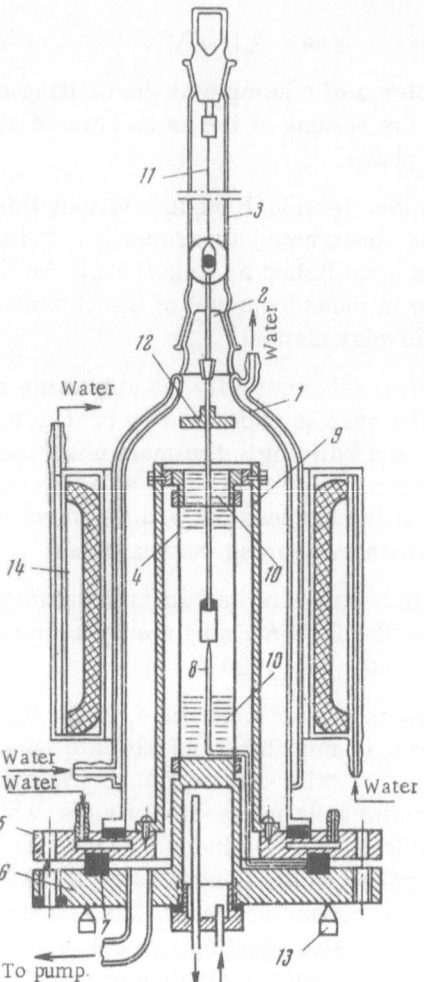

Fig. 1. Apparatus for the measurement of the electrical
conductivity by the electrodeless method in a rotating
magnetic field.

After such accidents, it is necessary to take the furnace apart rapidly and to remove the samples being investigated, as well as the fragments of the ampoule.

The suspension system is one of the most important parts of the apparatus. A tungsten filament 11 is attached to a ground-glass stopper at the top and to a large clamp at the bottom; this clamp holds the sample and carries a small mirror. The diameter

of the suspension filament depends on the dimensions of the sample and on its electrical conductivity. In investigations of low-conductivity samples, we use thin filaments, 30-40 μ in diameter, in order to obtain high sensitivity. The suspension system is made perfectly vertical by adjusting three screws 13, on which the apparatus is resting. After centering the suspension system, the correct position is set by means of special screws.

The stator, which produces the rotating magnetic field, consists of three pairs of coils and each pair consists of 400 turns; these coils are placed at angles of 120° to one another and are joined in a star configuration. The windings are made of enamelled wire, 0.75 mm in diameter, and are placed in a water-cooled cylinder 14, filled with transformer oil. The current in the various parts of the stator is controlled by three autotransformers and measured with three astatic ammeters. The stator surrounds the casing but makes no contact with it and is placed on a base which is attached by means of two brackets to the wall of the building.

The scale is in the form of a graduated rule 0.5 m long, placed at a distance of 1.2 m from the mirror attached to the suspension system. A beam of light is reflected by this mirror onto the calibrated rule. The position of the rule and the angle of rotation of the mirror can be adjusted. This arrangement is very simple and gives accurate values of the angle of rotation of the suspension system.

The control panel is mounted separately on a special stand. The general power supply of the apparatus is taken from the standard alternating current city supply, which is connected to a distributor supplying the furnace (through TNN-40 and OSU-40 transformers), the stator (through LATR-1 autotransformers), and vacuum pumps. Each unit has its own switch and all the switches are placed on the control panel, which also carries the instruments used to measure the currents in the coils of the stator, the current supplying the furnace, the temperature of the sample, and vacuum in the furnace. This arrangement makes it possible to carry out measurements in vacuum down to 10^{-4} mm Hg, at temperatures up to 1600°C.

Table 1. Electrical Conductivity and Density
of Standards Used in Measurements of the
Electrical Conductivity by the Contactless
Relative Method [78-80]

Standard	t, °C	$\sigma \cdot 10^{-4}$, Ω^{-1} cm^{-1}	d, g/cm^3
In	300	2.92	6.916
Bi	400	0.745	9.91
Bi	600	0.688	9.66
Sn	700	1.66	6.64

The electrical conductivity of semiconductors near the melting point is best measured using cylindrical samples, 8 mm in
diameter and 8 mm high, sealed into quartz ampoules evacuated
to 10^{-3} mm Hg, and suspended by a tungsten filament of 46-μ
diameter. Substances which react chemically with quartz can be
placed in corundum crucibles with ground stoppers, and these
crucibles can then be sealed into evacuated quartz ampoules. The
weight of a crucible and a stopper should not differ by more than
0.01 g from the weight of a similar crucible in which a standard
sample is placed. In this case, the diameter of the tungsten filament should not be less than 60 μ. In all cases, the quartz ampoules should be selected so their diameters (inner and outer) do
not differ by more than 0.1 mm from the diameters of the ampoule
containing the standard substance; the lengths of the ampoules
should also be the same. Before making the measurements at the
melting point, it is necessary to apply the requisite temperature
for 1.5 h in order to melt the sample completely; at other temperatures, the measurements can be made after 20 min. At each
temperature, 6-7 measurements must be carried out. The relative method is most convenient. Standard samples can be made of
molten indium, tin, or bismuth.

Table 1 lists the values of the electrical conductivity and
density of these standard materials.

Measurements using this apparatus show that the relative
error in the determination of the instrument constant k, using different standard materials, does not exceed 1.5%. The error in the
determination of the absolute values of the electrical conductivity

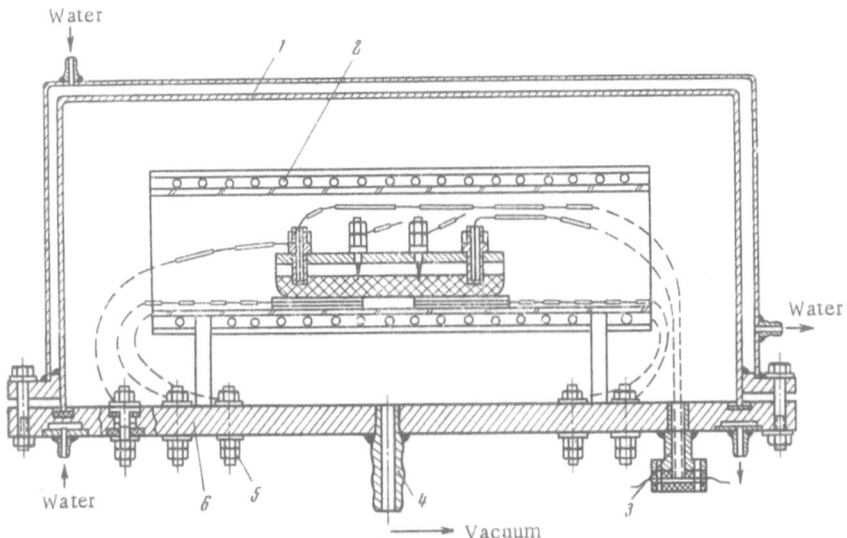

Fig. 2. Apparatus for the measurement of the electrical conductivity by the contact method.

is governed by the accuracy of the determination of the instrument constant, the angle of twist of the suspension filament, and the current in the three branches of the stator [cf. Eq. (14)]:

$$\frac{\Delta\sigma}{\sigma} = \frac{\Delta k}{k} + \frac{\Delta(\Delta\varphi)}{\Delta\varphi} + \frac{2\Delta i_{av}}{i_{av}}. \tag{16}$$

The total error in the measurement of the electrical conductivity by this method is of the order of 3.5%. However, the error in the determination of the temperature dependence of the electrical conductivity is somewhat less. Since the electrical conductivity varies strongly with temperature and concentration [5-30], such accuracy can be regarded as fully satisfactory. Special features of the measurements associated with particular properties of various materials will be mentioned in the sections reporting on the experimental results. This contactless method is not recommended for high-resistivity materials because the angle of twist is small and the relative error in the determination of this angle is very considerable. In this case, it is better to measure the electrical conductivity by a contact method, which also has the valuable advantage that one can measure directly the temperature

of the melt rather than the temperature of the interior of the furnace. The apparatus for the measurement of the electrical conductivity of semiconductors by a contact method is described in
[81].

Figure 2 shows the general arrangement used. The main
parts of the apparatus are: a casing, a furnace, a measuring unit,
and a measuring circuit. The casing consists of a cover 1 and a
base 6. The cover is welded, made of stainless steel, and cooled
with water. The base 6 is also made of stainless steel and also
cooled with water at the point where the cover is resting on the
base; the joint between the cover and the base is made vacuum-
tight. Vacuum-tight leads 5, supplying the power to the furnace
and connecting with the measuring unit, pass through the base. A
connecting pipe with flanges 3 is used for thermocouple leads. The
apparatus is pumped out and filled with an inert gas through another pipe 4. The required temperature is established in a furnace 2 (this furnace has a bifilar winding on a quartz tube 50 mm
in diameter; the heater is made of molybdenum or Kh25Yu5 alloy;
the length of the furnace is 300 mm; the heater is screened by a
series of quartz and molybdenum screens). The measuring unit
is a quartz boat 100 mm long, filled with the substance being investigated. Two molybdenum sleeves, coated (with the exception
of the end surface) with a thin layer of Al_2O_3, are placed at the
edges of the boat; they pass through a ceramic cover into the substance being investigated. Calibrated thermocouple junctions
are attached to the bottom of each sleeve by capacitor-discharge
welding. The boat with its cover is clamped in a sheet of niobium
or zirconium, 0.5-0.7 mm thick, which also acts as a getter. The
electrical conductivity is measured by a two-probe method for two
directions of a constant current, using either an R 2/1 potentiometer or a high-impedance N373-1 millivoltmeter. The electrical
conductivity of semiconductors in the liquid and solid state can be
measured in this apparatus up to ~1200°C. The maximum error
in the measurement of the electrical conductivity is ±5%. The
same apparatus can be used to measure the thermoelectric power
of the melts. The next subsection describes the method for measuring the thermoelectric power and thermal conductivity and the
construction of the apparatus developed for this purpose.

2. Thermoelectric Power and Thermal Conductivity

A detailed analysis of the modern methods for investigating the thermoelectric power and thermal conductivity is given in [82]. Bearing in mind the special requirements in the case of molten semiconductors, we have developed and constructed apparatus for high-temperature measurements of the thermoelectric power [83]. This method has been used to carry out the majority of our measurements on semiconductor melts.

The method used differs from those employed earlier [84, 85] in that the measurements of the thermoelectric power can be carried out over a wider range of temperatures and an excess inert-gas pressure can be used in investigations of volatile substances.

The apparatus is shown schematically in Fig. 3. A two-section tubular furnace 3, closed at its two ends by heat-insulation screens 4, is attached by means of two molybdenum rods 2 to a rotatable water-cooled base 1. The upper part of the furnace has two apertures 5 for the introduction of thermocouples 6. The furnace can be made of quartz for measurements up to 1200°C. At higher temperatures, we can use a corundum tube, of 35 mm internal diameter, grooved on the outer surface. The grooves carry a molybdenum wire which is covered by a layer of aluminum oxide and a heat-insulating layer. A boat 7 with the investigated substance is placed in the furnace. The boat is covered with a special cover 8, which has two apertures through which graphite rods 9 are passed. The furnace is placed in a water-cooled cylindrical casing 10, which is fixed to a four-wheel carriage 11, which can be moved on rails. The casing 10 is connected to the flange 1 by six bolts. The flange of the casing and the base 1 have a circular channel in which a rubber ring 12 is placed in order to obtain a vacuum-tight seal. The base 1 has apertures through which current-carrying leads and thermocouples are passed through vacuum-tight seals. The base 1 is attached to a plate 13, fixed to the laboratory bench, but a ball bearing 14 allows it to rotate about its axis. To load the boat, the casing is pushed away on its rails so that the furnace, fixed to the base 1, can be rotated toward the investigator; this allows convenient loading and unloading of the boat.

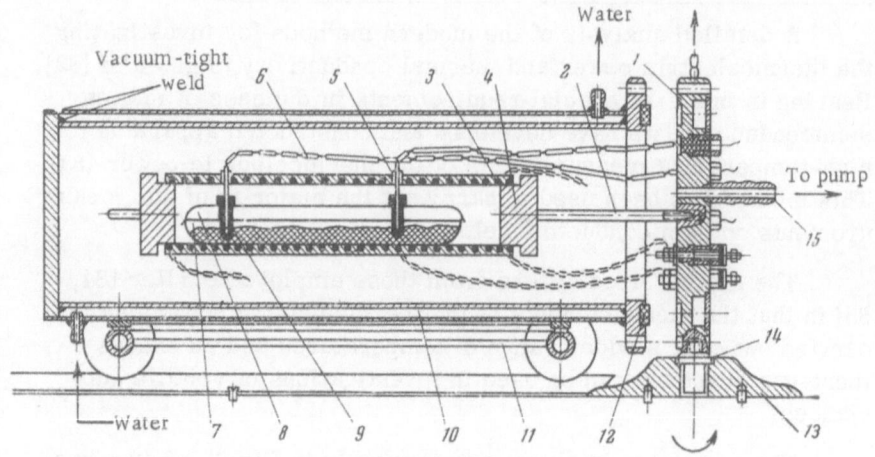

Fig. 3. Apparatus for the measurement of the thermoelectric power of liquids at high temperatures.

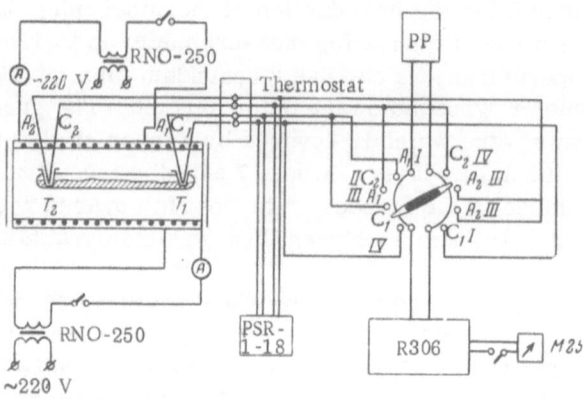

Fig. 4. Circuit used to measure the thermoelectric power of molten semiconductors.

Measurements can be carried out in this apparatus in 10^{-4} mm Hg vacuum. The apparatus is pumped with a backing-type pump VN-461 and a TsVL-100 diffusion pump, connected to the apparatus through a pipe 15. The temperature in the furnace is controlled by means of RNO-250-5 autotransformers and ammeters, connected to each section of the furnace. A recording potentiom-

eter of the PSR-1-18 type is used to record the conditions in the furnace. Chromel—Alumel thermocouples are used for the measurement of the thermoelectric power up to 1100°C; at higher temperatures, platinum—platinorhodium thermocouples (in this case, the graphite rods cannot be used) or tungsten—rhenium thermocouples are employed. The cold ends of the thermocouples are kept at room temperature. The thermoelectric power is measured using a circuit shown in Fig. 4 (this circuit is used in measurements with Chromel—Alumel thermocouples). The like and unlike ends of the thermocouples $(A_1-C_1; A_2-C_2)$ are connected to a multiposition switch. The thermoelectric power in the circuit of each of the two thermocouples and in the like branches is measured using a PP potentiometer or, more accurately, using an R306 potentiometer coupled with a small mirror galvanometer of the M25 type. In more accurate measurements full stabilization of the temperature is necessary. The temperature gradient is then recorded directly on a recorder chart.

This method allows us to determine reliably the thermoelectric power of molten materials in the range of temperatures referred to above. The values of the thermoelectric power are calculated using a method described in [86]. A fixed temperature difference is maintained between the ends of a sample:

$$\Delta T = T_1 - T_2, \qquad (17)$$

where T_1 and T_2 are the temperatures measured with thermocouples 1 and 2.

The voltages u_I and u_{II}, measured by potentiometers in the Chromel and Alumel branches of the circuit, are found from the expressions

$$u_I = \alpha_{C-M} \Delta T, \qquad (18)$$

$$u_{II} = \alpha_{A-M} \Delta T, \qquad (19)$$

where α_{C-M} and α_{A-M} are the thermoelectric powers of Chromel and Alumel, respectively, relative to the investigated material (M). Consequently,

$$u_{II} - u_I = [(u_{A-M}) T_1 + (u_{M-A}) T_2] - [(u_{C-M}) T_1 + (u_{M-C}) T_2] =$$

$$= [(u_{A-M}) - (u_{C-M})] T_1 + [(u_{M-A}) - (u_{M-C})] T_2 =$$

$$= (u_{A-C}) T_1 - (u_{A-C}) T_2. \tag{20}$$

Thus,

$$u_{II} - u_I = (u_{A-C}) T_1 - (u_{A-C}) T_2 = \alpha_{A-C} \Delta T, \tag{21}$$

where α_{A-C} is the thermoelectric power of the Alumel−Chromel pair. From Eqs. (18), (19), and (21), we obtain

$$\alpha_{C-M} = \frac{u_I}{u_{II} - u_I} \cdot \alpha_{A-C}, \tag{22}$$

$$\alpha_{A-M} = \frac{u_{II}}{u_{II} - u_I} \cdot \alpha_{A-C}. \tag{23}$$

Thus, to determine the thermoelectric power of the material being investigated, it is sufficient to measure only two voltages u_I and u_{II}, as well as ΔT. This method is superior to the usually employed determination of ΔT by a separate measurement of T_1 and T_2, because the errors are smaller. First, the values of the thermoelectric power of bismuth and antimony are determined between room temperature and 1100°C, and then these materials are used to check the operation of the apparatus.

Since this method has been used to investigate the thermo-electric power of molten semiconducting materials, we shall describe briefly our investigation of the thermoelectric powers of bismuth and antimony in the indicated range of temperatures.

A crushed substance (20-30 g) was placed in the quartz boat. The graphite rods passing through the cover of the boat were immersed in the material to a known depth and were kept at this depth by means of collars attached to the outer surface of the cover. The side surfaces of the graphite rods were coated with an insulating layer of aluminum oxide, so that only the end surface of each rod was in contact with the melt. This reduced somewhat the scatter of the results. The thermoelectric power of bismuth and antimony was measured in an atmosphere of spectroscopically pure argon, using samples of technical and semiconductor purity grades. In the latter case, bismuth of the V000 grade and antimony of the SU000 "extra" grade, with a total impurity content $<10^{-4}\%$,

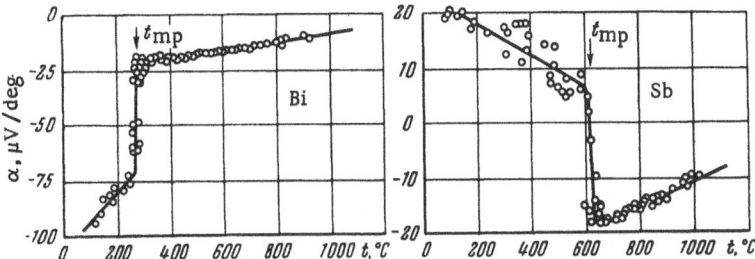

Fig. 5. Dependence of the thermoelectric power of antimony and bismuth on the average temperature in the solid and liquid states (relative to Chromel).

were employed. The measurements were begun in the liquid state, when the temperature of the cooler end of the boat was higher than the melting point of the investigated material. An approximately constant temperature difference ΔT (of the order of 40°) was maintained during the measurements. The measurements of the thermoelectric power in the solid state were begun after lowering the temperature to a level at which the hotter end of the boat was below the melting point of the investigated material. Measurements in the intermediate range of temperatures, when the boat contained both the liquid and solid phases, yielded results with a considerable scatter; this could be due to changes in the proportions of the solid and liquid phases caused by even relatively small temperature fluctuations. After crystallization and solidification of the material around the contacts, the sample could be heated again and the experiments repeated in order to check the reproducibility of the results for a given sample. The boat did not fracture because its bottom and its ends were rounded. The measurements carried out on one sample were reproducible. The scatter of the results did not exceed 2.5%.

Figure 5 shows the dependences of the thermoelectric powers of bismuth and antimony on the average temperatures; these were obtained on several samples of the purity indicated above, using a Chromel–Alumel couple. In the liquid state, the purity of the material had practically no influence on the absolute values of the thermoelectric power. The results obtained for different samples were in good agreement. Calculations showed that the maximum error in the measurements of the relative thermoelectric power in the liquid state did not exceed 2%.

The method described here made it possible to carry out measurements of the thermoelectric power of a wide range of materials, which was important in an investigation of the relationships governing the changes in the short-range order and in the nature of the chemical binding, which took place during melting and crystallization. The method for measuring the thermal conductivity was developed by Fedorova [87], with the advice of the present authors on the choice of graphite for the rods placed in the measuring unit; this choice solved the main problem in the measurements of the thermal conductivity of molten semiconductors at temperatures up to ~1200°C [88].

The apparatus used in the determination of the thermal conductivity of molten semiconductors by this method is shown in Fig. 6.

The measuring unit consists of two coaxial cylinders 1 and 2, made of spectroscopically pure graphite of M-3 grade. Graphite is used because it has a sufficiently high thermal conductivity, does not react with the molten compounds being investigated, and is easily machined. A liquid semiconductor is poured into the cylindrical gap 3 between two graphite cylinders. The width of the gap between the cylinders is 3 mm; this ensures the absence of convection (Gr · Pr ≪ 1000). The gap between the cylinders is maintained by spacers 4 and 5, placed in the upper and lower parts of the inner cylinder outside the limits of the region where measurements are carried out. An additional cavity 6 in the upper part of the outer cylinder is filled before experiment with a finely crushed solid. The cavity 6 is covered by a graphite cover 7. The inner cylinder carries, at a depth of 180 mm, an aperture of 10 mm in diameter, into which an inner heater 8 is placed; this heater is made of a molybdenum wire, 0.5 mm in diameter, wound on an alundum cylinder 7 mm in diameter. An insulating layer of aluminum oxide is smeared on top of the heating element. The working region is 60 mm high and is located in the central part of the cylinders; here, we have molybdenum wire leads of 0.3-mm diameter, which are employed to record the voltage. The temperature in the apparatus is maintained by an external three-section furnace 9, made of molybdenum wire 0.8 mm in diameter, wound on an alundum tube of 65-mm outer diameter. The height of this furnace is 230 mm. The upper and lower sections of the outer heater can be

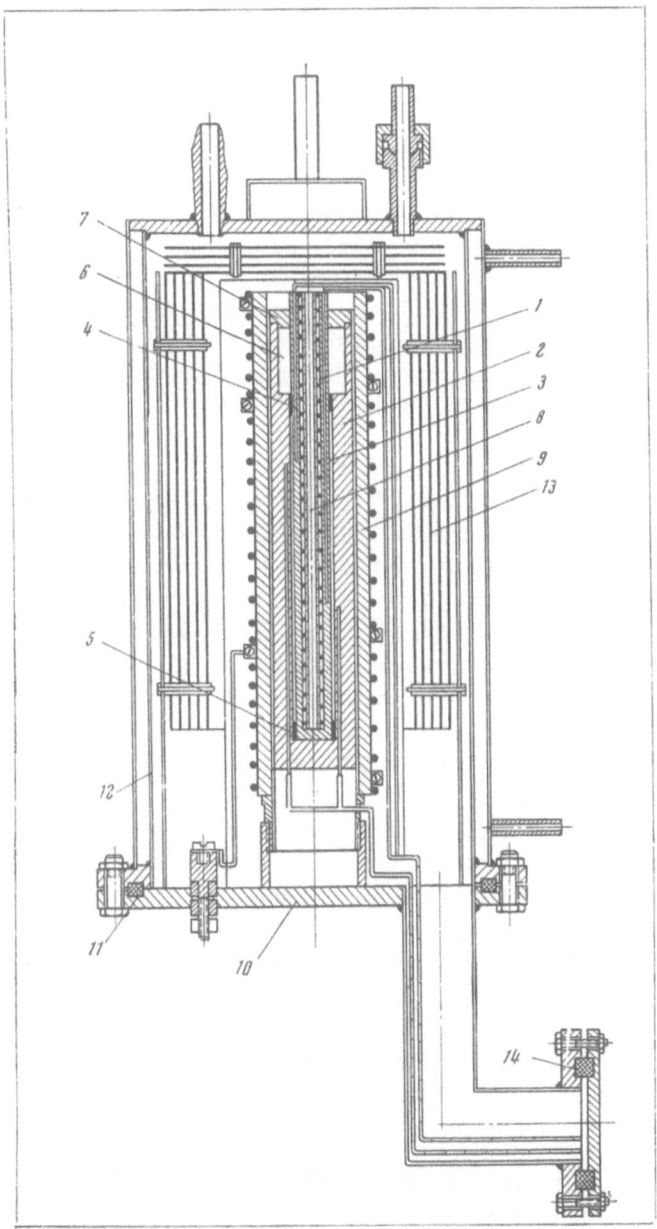

Fig. 6. Apparatus for the measurement of the thermal conductivity of molten semiconductors.

used to vary the temperature field along the height of the cylinders.
The inner heater establishes the necessary temperature drop in
the semiconductor layer being investigated. The measuring unit
and the external heater are fixed to a supporting flange 10, to
which a stainless-steel water-cooled casing 12 is attached by bolts
passing through the vacuum seals 11. A system of radiation
screens 13, reducing the heat losses from the heater, is placed between the measuring units and the casing. The leads supplying
the current to the various sections of the outer heater are made
of molybdenum wire of 1.5 mm diameter. Thermocouples are introduced through rubber seals through a tube 14 connecting the
apparatus to a vacuum pump. Tungsten—rhenium thermocouples,
0.3-0.4 mm in diameter, are used.

Both heaters are supplied with 220 V ac through a voltage
stabilizer of the S-3S type. The power supplied to the heaters is
controlled by means of RNO 250/5 autotransformers. The current
in the inner heater is measured with an ac ammeter of the D-57
type and 0.1 class, which can be used for currents up to 5 and 10 A.
The voltage in the working part is measured with a D-523 voltmeter of the 0.5 class, capable of measuring voltages up to 7.5,
15, 30, or 60 V. The thermoelectric power of the thermocouples
is measured with a semiautomatic potentiometer of the R 2/1 type.
Four thermocouples are placed in the inner cylinder: two control
couples, one 80 mm above the other, and two other couples in the
center of the working region at the same level but at diametrically
opposite positions. One of these thermocouples is used for the
measurement and the other for control purposes. Thus, temperatures are measured over a distance of 80 mm but the working part
is only 60 mm. This ensures that the problem is one-dimensional.
The maximum vertical misalignment at the height of 80 mm is
1.5-2°. The outer cylinder carries two thermocouples: one to
take measurements and the other for control purposes. These
thermocouples are also placed at the same level but at diametrically opposite positions.

After the preliminary pumping and heating, the apparatus is
filled with spectroscopically pure argon. After the establishment
of steady-state conditions (temperature drop 8-12°), the readings
of the ammeter and the voltmeter, indicating the power consumed
in the working region, as well as the readings of the thermocouples,

are recorded. The thermal conductivity $\varkappa$ is calculated using the formula for a cylindrical layer

$$\varkappa = \frac{Q \ln{(d_2/d_1)}}{2\pi l \, (t_1 - t_2)}, \qquad (24)$$

where Q is the amount of heat evolved by the heater per unit time, l is the length of the cylinder, d_1 and d_2 are the inner and outer diameters of the cylindrical liquid semiconductor layers, and t_1 and t_2 are the temperatures of the inner and outer surfaces of the semiconductor layer being investigated.

A preliminary calibration of the apparatus, carried out using water and liquid bismuth, gave results which differed by not more than 5% from those tabulated in [89]. The maximum error in the determination of the thermal conductivity by this method was ±12%. The main contribution to this error was made by the determination of the temperature.

3. Magnetic Susceptibility

In the determination of the magnetic susceptibility by the Faraday method, we measure the ponderomotive force F, acting on a sample in a nonuniform magnetic field,

$$F = \chi H m \frac{dH}{dz}, \qquad (25)$$

where m is the mass of the sample, H is the magnetic-field intensity, dH/dz is the magnetic-field gradient along the z direction, and χ is the magnetic susceptibility of the sample.

If the sample is small, the field gradient can be assumed to be constant and therefore the ponderomotive force can be taken to be proportional to the magnetic susceptibility if the mass of the sample is constant. This allows us to measure the magnetic susceptibility by a relative method:

$$F = k\chi, \qquad (26)$$

where k is the instrument constant, determined using a sample of known magnetic susceptibility.

Figure 7 shows schematically the apparatus used in the measurement of the magnetic susceptibility at high temperatures. This apparatus was developed by Vertman et al. [58].

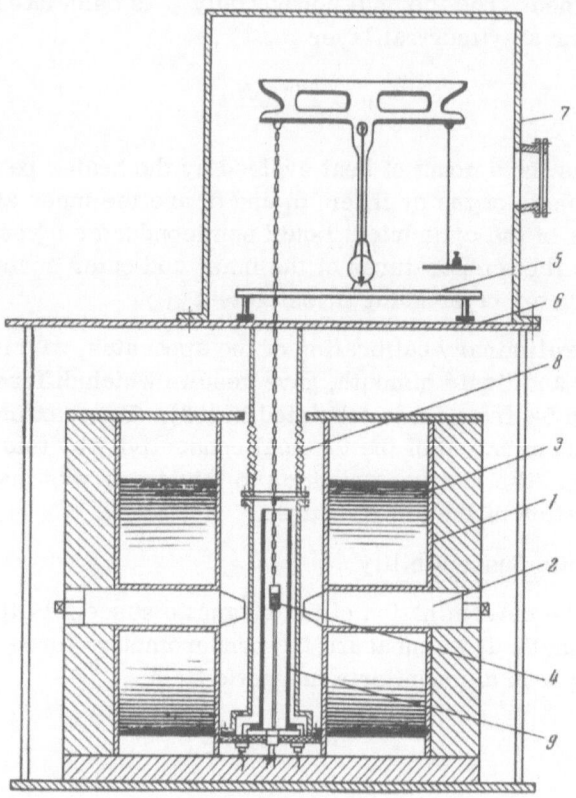

Fig. 7. Apparatus for the measurement of the magnetic susceptibility by the Faraday method.

The apparatus consists of three parts: an electromagnet, an
analytic balance, and a casing with a heater. The auxiliary apparatus includes transformers, a vacuum system, a system for the
supply of an inert gas, a water-cooling system, and measuring instruments. The yoke (of 140 × 120 mm cross section) and the
poles (120 mm in diameter) of the electromagnet 1 are made of
Armco iron. The pole-pieces 2 are cast from Permendur. The
conical ends of the pole-pieces produce a nonuniform magnetic
field. The coils 3 are 500 mm in diameter. A copper wire of
4.4 × 1.45 mm cross section is used in the coil windings. The
electromagnet is supplied from a set of accumulators. An ampoule with the substance being investigated 4 is suspended by a

quartz and molybdenum chain to one arm of an analytic balance 5.
The balance is fixed to a brass plate 6 and enclosed in a steel case
7, which has two viewing windows as well as devices for the re-
mote placing of weights on the pan of the balance. The chain
passes through a bellows 8 to a cylinder containing a heater 9. The
heater is in the form of a graphite tube, 250 mm long and of 18 mm
internal diameter. A vertical slot is made in this tube. A heater
of this shape is basically bifilar and its magnetic field along the
z axis is practically equal to zero. The heater is supplied from a
welding transformer. The temperature is controlled with a TNN-
40 autotransformer. The temperature is measured with a potentiometer, connected to a platinum—platinorhodium thermocouple,
whose hot junction is placed in the center of the heater at a distance of 3-5 mm from the sample. To determine the temperature
drop in the heater and in the sample, the thermocouple is calibrated
using pure metals, which are suspended in place of the sample being investigated (melting is usually accompanied by a sudden change
in the magnetic susceptibility). Measurements up to 2000°C can be
carried out using this apparatus.

It is convenient to measure the magnetic susceptibility using
samples of 10-mm diameter and 10-mm height, placed in evacuated (10^{-3} mm Hg) quartz ampoules. In the case of chemically
active compounds (for example, Mg_2B^{IV}), the sample under investigation should be placed in a corundum crucible with a ground
cover and this crucible must be placed in a sealed evacuated quartz
ampoule. The position of the ampoule relative to the pole-pieces
must be determined first from the maximum force tending to push

Table 2. Standards Used in Measurements
of the Relative Magnetic Susceptibility
[57, 90, 91]

Standard	t, C°	$\chi \cdot 10^6$
Sugar	20	—0.566
Copper	20	—0.0859
Copper	702	—0,0753
Germanium . .	20	—0,105

out the sample from the field. The measurements consist in weighing the sample with the magnetic field switched on and off. The magnetic susceptibility is calculated from Eq. (26). Copper, germanium of known degree of purity, and sugar can be used as standards for the determination of the instrument constant. The magnetic susceptibilities of these substances are listed in Table 2.

The absolute error in the determination of the magnetic susceptibility using Eq. (26) is due to errors in the determination of the instrument constant and in the difference between the weights of a sample with the magnetic field switched on and off, i.e.,

$$\frac{\Delta \chi}{\chi} = \frac{\Delta k}{k} + \frac{\Delta F}{F}, \tag{27}$$

$$F = \frac{P_2 - P_1}{P_1}, \tag{28}$$

where P_1 is the weight of the sample in the absence of a magnetic field and P_2 is the weight of the sample in a magnetic field.

Our investigations showed that the change in the weight due to the application of a magnetic field was relatively small and represented thousandths of a gram. Therefore, the relative error $\Delta F/F$, representing the accuracy of weighing with an analytic balance was of the order of 3.5%. The relative error in the determination of the temperature dependence of the magnetic susceptibility was of the same magnitude, because the error in the determination of the instrument constant could be regarded as systematic. The error in the determination of the absolute value of the magnetic susceptibility was about 7%.

4. Density

The density of semiconductors was determined using three methods: two for the liquid state and one for the solid state.

At high temperatures, many molten semiconductors do not react at all or react very slightly with quartz. Therefore, the "thermometer" method is most suitable for the determination of the density of such materials. In this method, we determine the volume of the investigated substance by placing it in a calibrated

pycnometer. The pycnometer is a specially shaped ampoule made
of fused quartz. The lower part of this ampoule has a fairly large
volume but the middle part is a capillary with a scale that has been
marked with a diamond. The volume at each scale division is
measured first by weighing the pycnometer when empty and when
filled with distilled water. The density is found from the expression

$$d = \frac{P}{V_i},\qquad(29)$$

where P is the weight of the investigated substance, and V_i is the
volume of the pycnometer up to the i-th mark. At elevated temperatures, a correction must be made for the expansion of quartz.

Figure 8 shows schematically the apparatus used to determine the density by this method [92].

A pycnometer containing the substance being investigated 1
is placed in a massive nickel block 2, which produces a uniform
temperature field along its height. An aperture 3 mm wide and of
height exactly equal to the calibrated length of the pycnometer
capillary, is made in the block. Moreover, three deep holes are
made in the block to carry thermocouples 3, which are used to
measure the temperature along the pycnometer height at three
points simultaneously. The block enclosing the pycnometer is
placed in a vertical heater with two windows 4, opposite the aperture in the block; the windows are closed with transparent quartz
plates. The heater consists of a massive refractory steel cylinder 5 wound in such a way as to avoid the window areas. This
stainless-steel cylinder is placed (together with its winding) in a
large-diameter asbestos–cement tube 7, and the gap between the
tube and the cylinder is filled with a heat-insulating powder 8. To
improve the heat insulation along the vertical direction, two insulating foamed fireclay blocks 9 are used; moreover, the heater is
closed by a cover 10. The whole heater is attached to an asbestos–
cement plate 11, which is placed on the laboratory bench. A vibrator 12 is attached below this plate and a rod from this vibrator
transmits vibrations to the cylinder carrying the pycnometer. A
powerful source of light 13 is placed opposite the windows in the
heater; this makes it possible to determine easily the level of the
melt in the pycnometer. This level can be determined either visually or with a cathetometer.

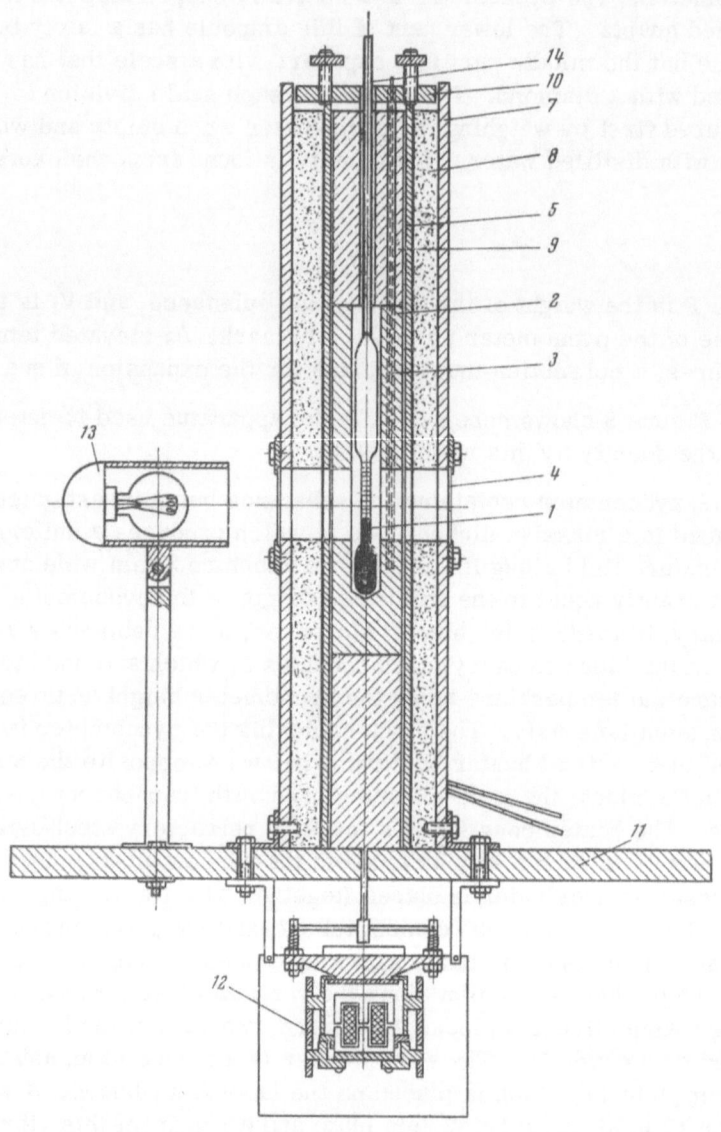

Fig. 8. Apparatus for the determination of the density of melts at high temperatures by the pycnometric method.

After loading with the sample being investigated (which is in a finely powdered form), the pycnometer is pumped down to 10^{-3} mm Hg and sealed. A quartz rod 14 is welded to the upper part of the pycnometer and this rod projects outside. During the measurements, the sample is shaken by the vibrator at a frequency of 50 cps. The vibration is necessary to remove the gas bubbles, which form in the molten substances. Moreover, the pycnometer can be shaken manually by means of the rod 14. The considerable difference between the volume of the substance in the lower part of the pycnometer and the capillary guarantees a high accuracy of the determination of the density (in our investigations, the absolute error in the determination of the density of melts by this method did not exceed 0.5%). The pycnometer method cannot be used to determine the density of substances which are highly reactive in the molten state at high temperatures, since they usually react with quartz. Therefore, the density of such substances should be measured by the hydrostatic weighing method. In this method, a standard sample of known weight and volume is weighed in the liquid under investigation. The density is then found from the expression

$$d = \frac{P_1 - P_2}{V_S} \qquad (30)$$

where P_1 is the weight of the standard sample in air, P_2 is the weight of the standard in the investigated liquid, and V_S is the volume of the standard.

The apparatus used to determine the density by this method is shown schematically in Fig. 9 [93]. A standard sample 6 is immersed in the melt 5, which is covered with a layer of flux in a corundum crucible. The crucible with the melt is placed in a Silit heater 4. The temperature in the heater is measured with a thermocouple 3. The standard is suspended, by a porcelain rod and a thin wire 2, from one arm of an analytic balance 1, which is placed at a considerable distance from the heater.

Good standard samples can be prepared from single crystals of refractory metals, such as molybdenum or tungsten; they should be in the form of cylinders with a sharp lower end (cf. Fig. 9). A porcelain capillary of 1.5 mm diameter is inserted into the upper end of the standard and this capillary is used to suspend the

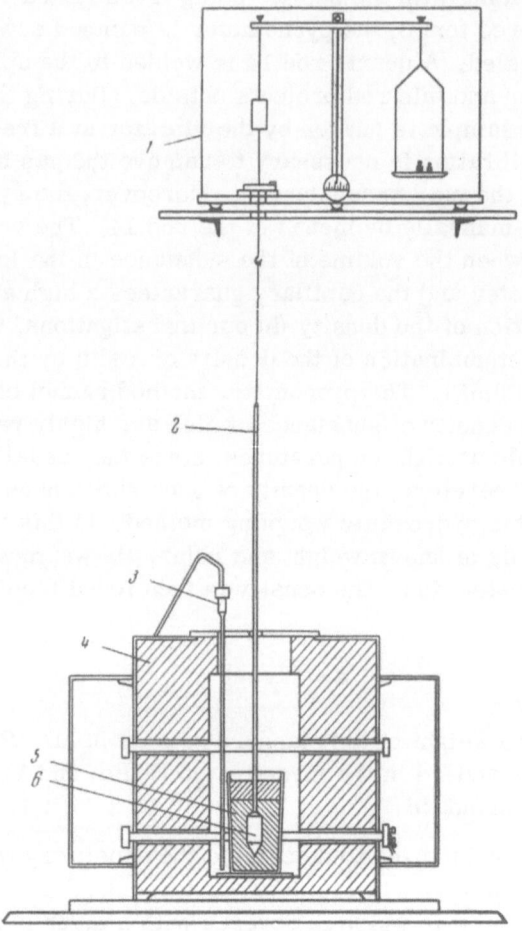

Fig. 9. Apparatus for the determination of the density of melts at
high temperatures by the hydrostatic weighing method.

sample from the balance. The volume of the standard sample is
~2 cm^3. Its volume at room temperature is determined by hydrostatic weighing in distilled water. At higher temperatures, a correction is made to allow for the thermal expansion of the material
of which the standard is made. After measurements, the standard
should be weighed again. There should be no noticeable change in
the weight, compared with the initial value in distilled water. The
absence of such a change indicates that the standard does not

interact with the melt under investigation. The density can be
measured by this method in vacuum, in an inert atmosphere, or in
air under a layer of flux. In the latter case, the error due to the
fact that part of the capillary is immersed in the flux can be ignored: the standard is weighed in water so that the capillary is
immersed to a certain mark; the level of the flux should be at the
same mark in subsequent measurements. This procedure partly
compensates for the error, which cannot be large anyway, because
the standard sample is of high density and of considerable volume.
An estimate of the error due to the immersion of the capillary
partly in the melt and partly in the flux has been obtained by measurements in water; these measurements indicate that the error is
less than 0.2%.

A formula for the calculation of the density by the hydrostatic weighing method is given in [94]; this formula allows for the
surface tension of the investigated liquid:

$$d = \frac{P_1 - P_2 + \dfrac{\sigma \pi D}{g}}{V_s}, \tag{31}$$

where P_1 is the weight of the standard in air, P_2 is the weight of
the standard in the investigated liquid, σ is the surface tension of
the investigated liquid, and D is the diameter of the capillary
which is in contact with the surface of the investigated liquid.

Before weighing a standard sample immersed in the melt, it
is necessary to maintain a given temperature for about 15-20 min
in order to allow the standard to reach this temperature. The
total error in the determination of the density by this method is of
the order of 1.5%.

The density of solids can be determined most conveniently
by the dilatometric method. This method is based on the measurement of the thermal expansion of the investigated sample and calculation of the density at a given temperature from the data on the
density at room temperature. The calculation is made using the
formula

$$d_t = \frac{d_{20}}{(1 + \Delta l_t)^3} \simeq d_{20} (1 - 3 \Delta l_t), \tag{32}$$

where d_t is the density of the investigated substance at a given temperature, d_{20} is the density of the same substance at room temperature, and Δl_t is the elongation, per 1 cm length of the sample, due to the heating from 20°C to t°C.

Figure 10 shows a dilatometer developed by us for investigations up to 1600°C in vacuum or in an inert-gas atmosphere [92]. This dilatometer differs from those described earlier by an improved construction of the heater, which is very similar to that used in the determination of the electrical conductivity by the contactless method.

A sample of the substance being investigated 1 is placed in a quartz test tube 2, which is attached rigidly to an upper flange 3. When heated, the sample expands and the elongation is transmitted through a quartz rod 5 to the foot of a displacement gauge 4 (one scale division represents 0.002 mm). The gauge is fixed rigidly to the flange 3 and is placed in a cold part of the apparatus under a hood 6. When experiments are carried out in an inert atmosphere, the hood is drawn tightly against the flange 3 by tie rods 7 and nuts 8; a vacuum sealing ring 9 is placed between the hood and the flange.

A casing 10 is tightened against the flange 3 through a vacuum rubber seal 11 and against the flange 12 through a seal 13. The temperature of the sample is measured with a thermocouple 14, placed in the direct vicinity of the sample (at a distance 1 mm from it). The thermocouple is introduced through a lower flange 15 via a vacuum seal 16, which is fixed by a nut 17. A vacuum seal is also placed between flanges 12 and 15. This construction is very convenient and reliable.

The test tube (containing the sample) and the rod which transmits the elongation to the gauge are under the same temperature conditions but they expand in opposite directions and so they compensate the elongation; therefore, a correction for the expansion of the quartz test tube should be made only in respect of the length of the sample which is in a uniform temperature field. This is checked in preliminary experiments by recording the temperature gradient along the height of the heater at various temperatures.

The thermal expansion is found by first calibrating the apparatus with substances whose thermal expansion is known well.

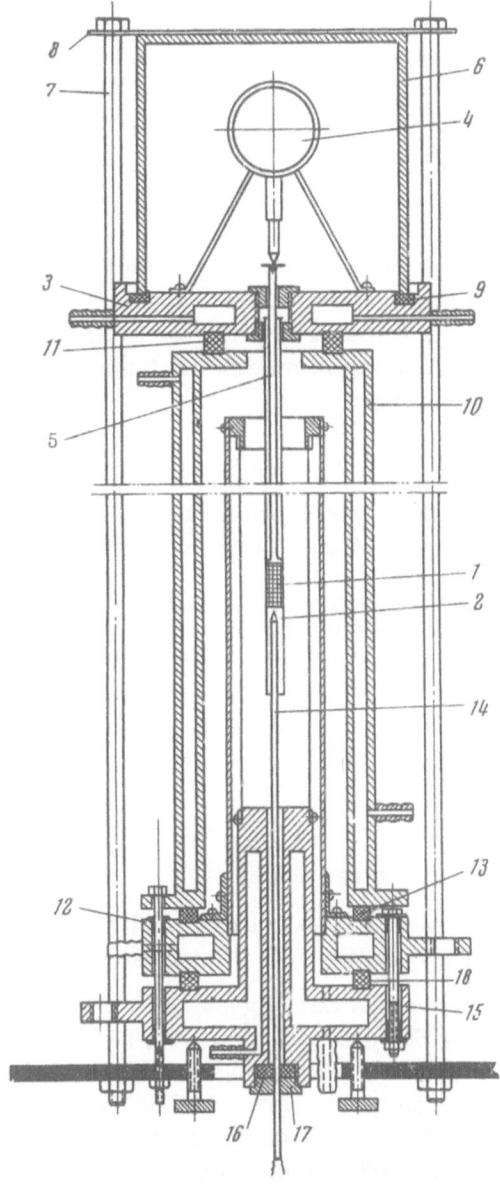

Fig. 10. Indicating dilatometer for the investigation of the thermal expansion and density of solids at high temperatures.

Such substances include germanium, copper, nickel, titanium, etc.
A comparison is made with the published values of the thermal expansion coefficients and with the temperatures of solid-state transitions (for example, in the case of nickel and titanium). The maximum error in the determination of the density by this method does not exceed 0.8%.

5. Viscosity

One of the best methods for measuring the viscosity is to determine the damped oscillations of a cylinder filled with the liquid under study and suspended by an elastic filament (the cylinder method). The force setting in motion the liquid F is balanced by the internal friction forces:

$$F = \eta \, sdv/dz, \tag{33}$$

where s is the area over which shear takes place; η is the internal friction or the viscosity of the liquid; dv/dz is the velocity gradient.

The internal friction or the viscosity of a liquid η is given by the value of the tangential force exerted by two layers of the liquid over an area of 1 cm^2, separated by 1 cm, when the velocity gradient is equal to unity. We can distinguish the dynamic viscosity η and the kinematic viscosity ν, which are related by:

$$\eta = \nu d, \tag{34}$$

where d is the density of the liquid.

The dynamic viscosity η is expressed in poises (g $\cdot$ cm^{-1} $\cdot$ sec^{-1}); the kinematic viscosity ν is expressed in stokes (cm^2/sec).

The cylinder method was suggested by Meyer [75] and developed in detail by Shvidkovskii et al. [74, 95-97]. When a cylinder filled with a liquid oscillates torsionally, the boundary layer next to the walls executes the same oscillations as the cylinder. The next layer, due to the internal friction in the liquid, is dragged by the boundary layer and also executes torsional oscillations, but of smaller amplitude. At the center of the cylinder (along the cylinder axis), the particles of the liquid should remain at rest. Consequently, an angular velocity gradient is established in the

liquid and internal friction forces are observed; the elastic deformation energy of the suspension filament is used to overcome these forces. Consequently, the oscillations of the liquid-filled cylinder are damped. The hydrodynamic problem of torsional oscillations of a liquid-filled cylinder is very complex. This problem has been solved in detail by Shvidkovskii [74]. We shall consider only the final result of his solution and give the formulas which we shall use in our calculations. When the hydrodynamic equation for the oscillations of a liquid-filled cylinder are solved, three main formulas are obtained for the calculation of the kinematic viscosity and the range of validity of each of these formulas is determined by the values of

$$\xi = R \sqrt{2\pi/\nu\tau}, \qquad (35)$$

where R is the radius of the cylinder, τ is the period of the oscillations of the suspension system, and ν is the kinematic viscosity.

For values of ξ ranging from 0 to 1.2, the viscosity is found from the equation

$$\nu = \frac{0.8225 \frac{MR^2}{2} R^2 \sigma}{\tau \left[\left(I + \frac{MR^2}{2} \right) \delta - I \frac{\tau}{\tau_0} \delta_0 \right]}, \qquad (36)$$

where M is the mass of the liquid, I is the moment of inertia of the empty cylinder, δ and δ_0 are the logarithmic decrements of the damped oscillations of the filled and empty cylinder, and τ and τ_0 are the oscillation periods of the filled and empty cylinder.

The value of σ is found from the expression

$$\sigma = 1 + 0.0250z\delta - 0.076\delta^2 - 0.1645z^2 - 0.4924 \frac{R}{2H} \qquad (37)$$

In this expression

$$z = R^2/\tau\nu^*, \qquad (38)$$

where $\nu*$ is the kinematic viscosity calculated using Eq. (36) for $\sigma = 1$, and 2H is the level of the liquid in the cylinder.

If the liquid is not in contact with the upper cover of the cylinder, the last term in Eq. (37) must be reduced by a factor of 2.

Table 3. Coefficients a, b, c, d, e for Calculation
of the Viscosity Using Eq. (40) [74]

$\nu\tau/R^2$	a	b	c	d	e
0.35	5.363	2.495	0.099	0.233	0.264
0.36	5.357	2.455	0.101	0.244	0.264
0.37	5.349	2.415	0.103	0.255	0.264
0.38	5.339	2.375	0.105	0.266	0.263
0.39	5.327	2.336	0.107	0.277	0.263
0.40	5.313	2.297	0.110	0.287	0.263
0.41	5.297	2.259	0.113	0.296	0.262
0.42	5.279	2.220	0.116	0.305	0.262
0.43	5.259	2.182	0.119	0.314	0.261
0.44	5.238	2.144	0.122	0.322	0.261
0.45	5.216	2.107	0.125	0.329	0.260
0.46	5.192	2.071	0.128	0.336	0.260
0.47	5.167	2.034	0.131	0.343	0.259
0.48	5.140	1.998	0.133	0.349	0.258
0.49	5.113	1.963	0.136	0.354	0.257
0.50	5.085	1.928	0.139	0.359	0.255
0.52	5.027	1.859	0.144	0.369	0.252
0.54	4.966	1.793	0.149	0.377	0.248
0.56	4.902	1.729	0.154	0.384	0.243
0.58	4.837	1.667	0.159	0.390	0.238
0.60	4.771	1.607	0.163	0.394	0.233
0.62	4.704	1.549	0.167	0.398	0.228
0.64	4.636	1.493	0.171	0.400	0.223
0.66	4.569	1.439	0.175	0.402	0.218
0.68	4.502	1.388	0.178	0.403	0.213
0.70	4.435	1.339	0.181	0.404	0.207
0.72	4.369	1.292	0.183	0.403	0.202
0.74	4.304	1.247	0.185	0.402	0.197
0.76	4.239	1.204	0.186	0.401	0.192
0.78	4.175	1.163	0.188	0.400	0.187
0.80	4.112	1.123	0.189	0.399	0.182
0.82	4.049	1.085	0.190	0.397	0.177
0.84	3.988	1.049	0.190	0.395	0.172
0.86	3.928	1.014	0.191	0.393	0.168
0.88	3.869	0.980	0.191	0.391	0.163
0.90	3.811	0.948	0.191	0.388	0.159
0.92	3.755	0.917	0.191	0.386	0.155
0.94	3.699	0.888	0.191	0.383	0.151
0.96	3.644	0.860	0.190	0.380	0.147
0.98	3.591	0.833	0.190	0.376	0.143
1.00	3.539	0.808	0.190	0.373	0.140
1.04	3.438	0.759	0.189	0.366	0.132
1.08	3.342	0.715	0.188	0.360	0.125
1.12	3.250	0.674	0.187	0.353	0.118
1.16	3.162	0.636	0.186	0.346	0.112
1.20	3.078	0.601	0.184	0.340	0.106
1.24	2.998	0.569	0.182	0.333	0.101
1.28	2.922	0.532	0.180	0.327	0.096
1.32	2.849	0.511	0.178	0.320	0.092

Table 3 (continued)

$\nu\tau/R^2$	a	b	c	d	e
1.36	2.779	0.485	0.176	0.314	0.088
1.40	2.712	0.461	0.173	0.308	0.084
1.44	2.648	0.439	0.171	0.302	0.080
1.48	2.586	0.418	0.168	0.297	0.076
1.52	2.527	0.399	0.166	0.291	0.073
1.56	2.470	0.381	0.163	0.285	0.070
1.60	2.416	0.363	0.161	0.280	0.067
1.64	2.364	0.347	0.158	0.275	0.065
1.68	2.314	0.332	0.156	0.270	0.062
1.72	2.266	0.319	0.153	0.265	0.060
1.76	2.220	0.306	0.151	0.260	0.057
1.80	2.175	0.293	0 148	0.255	0.055
1.84	2.132	0.281	0.146	0.251	0.053
1.88	2.091	0.270	0.143	0.246	0.051
1.92	2.051	0.260	0.141	0.242	0.049
1.96	2.012	0.250	0.139	0.238	0.047
2.00	1.975	0.241	0.137	0.234	0.046
2.05	1.930	0.230	0.135	0.229	0.044
2.10	1.887	0.220	0.132	0.224	0.042
2.15	1.847	0.210	0.130	0.220	0.041
2.20	1.808	0.201	0.127	0.215	0.039
2.25	1.770	0.193	0.125	0.211	0.037
2.30	1.734	0.185	0.123	0.207	0.036
2.35	1.700	0.177	0.121	0.203	0.034
2.40	1.666	0.170	0.119	0.199	0.033
2.45	1.634	0.164	0.117	0.196	0.032
2.50	1.603	0.158	0.115	0.192	0.030
2.60	1.545	0.146	0.112	0.186	0.028
2.70	1.490	0.136	0.109	0.179	0.026
2.80	1.439	0.127	0.105	0.173	0.024
2.90	1.391	0.119	0.102	0.168	0.023
3.00	1.347	0.111	0.099	0.163	0.021
3.10	1.305	0.104	0.096	0.158	0.020
3.20	1.265	0.098	0.093	0.153	0.019
3.30	1.228	0.092	0.090	0.149	0.018
3.40	1.193	0.087	0.088	0.145	0.017
3.50	1.160	0.082	0.086	0.141	0.016
3.60	1.129	0.078	0.083	0.138	0.015
3.70	1.099	0.074	0.081	0.134	0.014
3.80	1.071	0.070	0.079	0.131	0.013
3.90	1.044	0.067	0.077	0.127	0.013
4.00	1.019	0.063	0.075	0.124	0.012
4.20	0.971	0.057	0 072	0.119	0.011
4.40	0.927	0.052	0.069	0.114	0.010
4.60	0.887	0.048	0.066	0.109	0.009
4.80	0.851	0.044	0.063	0.104	0.008
5.00	0.818	0.041	0.060	0.100	0.008
5.20	0.787	0.038	0.058	0.096	0.007
5.40	0.758	0.035	0.056	0.092	0.006
5.60	0.731	0.033	0.054	0.089	0.006
5.80	0.706	0.030	0.053	0.087	0.006
6.00	0.683	0.028	0.051	0.084	0.006
6.20	0.661	0.026	0.050	0.081	0.005

The viscosity is calculated by the method of successive approximations. First, we find a rough value $\nu = \nu*$ for $\sigma = 1$. Then, $\nu*$ is substituted into Eq. (36) and, after the calculation of σ, a more accurate value of the viscosity is found from

$$\nu = \nu^* \sigma. \tag{39}$$

When $\xi = 1\text{-}4.2$, we can use the following formula:

$$\frac{10\,I}{MR^2}\left(\delta - \frac{\tau}{\tau_0}\,\delta_0\right) + 5\delta = a + b\delta - \frac{4R}{2H}(d + e\delta) - c\delta^2. \tag{40}$$

The coefficients a, b, c, d, and e (which are all functions of $\nu\tau/R^2$) are listed in Table 3, which was compiled by Shvidkovskii [74].

This table is used as follows. First we determine the numerical value of the quantity

$$\frac{10I}{MR^2}\left(\delta - \frac{\tau}{\tau_0}\,\delta_0\right) + 5\delta = A. \tag{41}$$

Next, we find the two nearest values A_1 and A_2 ($A_1 < A < A_2$), and then we use linear interpolation to find $\nu\tau/R^2$ corresponding to the newly found value of A. We must bear in mind that the quantity A is governed mainly by the coefficient a. Therefore, before carrying out the exact calculations, we must estimate which row in Table 3 matches approximately the calculated value of A. Usually it lies a little higher above the row with the value $A = a$.

We shall consider such a calculation reported by Shvidkovskii [74].

Let us assume that $A = 4.230$, $\delta = 0.2405$, $\tau = 9.433$ sec, $2H = 5.0$ cm, $R = 1.296$ cm, $R^2 = 1.680$ cm^2. We find in Table 3 a row with $a = 4.239 \approx A$, which corresponds to the value $\nu\tau/R^2 = 0.76$; we take little b and d from the same row. Then, $b\delta \approx 0.300$, $4Rd/2H \approx 0.400$. Consequently, to obtain the required value of A, we must take a row with a value of a which is approximately 0.100 larger than 4.239. We take the row with $a = 4.369$ and calculate $A_1 = 4.369 + 1.292\delta - 0.183\delta^2 - 4R/2H(0.403 + 0.202\delta) = 4.201$. We obtain $A_1 < A$. Now we find $A_2 > A$. For this purpose, we use a row lying a little higher and find that $A_2 = 4.435 + 1.339\delta + 0.181\delta^2 - 4R/2H(0.404 + 0.207\delta) = 4.275$. The values of A_1 and A_2 correspond to $\nu\tau/R^2$ equal to 0.72 and 0.70. Hence, we find that the required value of $\nu\tau/R^2$:

$$v \frac{\tau}{R^2} = 0.70 + 0.02 \frac{A_1 - A}{A_1 - A_2} = 0.712,$$

$$v = 0.712 \frac{R^2}{\tau} = 0.127.$$

Finally, for values of $\xi \geq 10$, we use the formulas

$$v = \frac{1}{\pi} \left(\frac{I}{MR} \right)^2 \frac{\left(\delta - \delta_0 \frac{\tau}{\tau_0} \right)^2}{\tau \sigma^2}, \tag{42}$$

$$\sigma = 1 - \frac{3}{2} x - \frac{3}{8} x^2 - a + \frac{4R}{2H} (b - cx), \tag{43}$$

$$x = \delta/2\pi. \tag{44}$$

The values of the coefficients a, b, and c, (which are functions of y) are taken from Table 4, which was also compiled by Shvidkov-skii [74]:

$$y = \xi^2 = 2\pi R^2/\tau v^*, \tag{45}$$

where v^* is the kinematic viscosity calculated using Eq. (42) with $\sigma = 1$. In order to find the coefficients a, b, and c, we plot the dependence of these coefficients on y.

Having calculated σ from Eq. (43), we find the final value of the kinematic viscosity v:

$$v = v^*/\sigma^2. \tag{46}$$

As in the preceding case, Eq. (43) for σ is obtained on the assumption that there are two flat end surfaces where the liquid is in contact with the cylinder. If the liquid is not in contact with the cover, the term 4R/2H should be reduced by a factor of 2.

Equations (37), (40), and (43) include correction terms which allow for the finite dimensions of the cylinder. These corrections are valid when the ratio between the height 2H and the radius of the liquid cylinder obeys the relationship

$$2H \geqslant 2.62R. \tag{47}$$

Table 4. Coefficients a, b, and c for Low-Viscosity Liquids

ν	a	b	c	ν	a	b	c
100	0.2121	0.0466	0.1150	1700	0.0514	0.1032	0.1683
150	0.1732	0.0587	0.1243	1800	0.0500	0.1037	0 1688
200	0.1500	0.0669	0.1312	1900	0.0487	0.1042	0.1693
250	0.1342	0.0725	0.1366	2000	0.0474	0.1047	0.1697
300	0.1224	0.0765	0.1409	2100	0.0463	0.1052	0.1701
350	0.1130	0.0798	0.1444	2200	0.0452	0.1056	0.1704
400	0.1061	0.0826	0.1472	2300	0.0442	0.1060	0.1707
450	0.1000	0.0850	0.1496	2400	0.0433	0 1064	0.1710
500	0.0947	0.0870	0.1517	2500	0.0424	0.1067	0.1713
600	0.0865	0.0901	0.1552	2600	0.0416	0.1070	0.1716
700	0.0801	0.0926	0.1579	2700	0.0408	0.1073	0.1718
800	0.0750	0.0946	0.1601	2800	0.0401	0.1076	0.1720
900	0.0708	0.0962	0.1618	2900	0.0394	0.1078	0.1722
1000	0.0671	0.0975	0.1631	3000	0.0387	0.1080	0.1724
1100	0.0639	0.0986	0.1642	3100	0.0381	0.1082	0.1725
1200	0.0612	0.0995	0.1651	3200	0.0375	0.1084	0.1726
1300	0.0588	0.1004	0.1659	3300	0.0369	0.1086	0.1727
1400	0.0567	0.1012	0.1666	3400	0.0364	0.1088	0.1728
1500	0.0547	0.1019	0.1672	3500	0.0358	0.1090	0.1729
1600	0.0530	0.1026	0.1678				

Thus, there are in fact two ranges of values of the criterion ξ: from 0 to 4.2 and from 10 to ∞, within which the viscosity can be calculated by one of the following equations: (36), (40), or (42). The first range represents liquids of high viscosity and the second those of low viscosity. The range $\xi = 4.2\text{-}10$ represents intermediate viscosity and cannot be calculated because of the very weak dependence of ν on δ. Consequently, before we design the suspension system, we must calculate the criterion ξ calculated by means of Eq. (35). To do this, we must know at least approximately the range of values of the kinematic viscosity in which measurements will be carried out. Then, assuming certain values of R and τ, we determine the corresponding extreme values of the criterion ε:

$$\xi_{min} = R \sqrt{2\pi/\tau\nu_{max}}, \tag{48}$$

$$\xi_{max} = R \sqrt{2\pi/\tau\nu_{min}}. \tag{49}$$

If the range of values of this criterion does not fit the range of validity of one of the calculation formulas (36), (40), (42), then we must alter the oscillation period and the radius of the cylinder in order to obtain such values of ξ_{max} and ξ_{min} which would enable us to use one of these formulas. In selecting the radius and the height of the cylinder, we must bear in mind Eq. (47). For a given value of R, the suspension system has a moment inertia I, which is related to the oscillation period by

$$I = \tau^2 N/4\pi^2, \tag{50}$$

where N is the rigidity of the suspension filament:

$$N = \pi D^4 G/32L \tag{51}$$

(G is the shear modulus of the material of the filament; D is the diameter of the filament; L is the length of the filament). Substituting the value of N from Eq. (51) into Eq. (50), we obtain

$$\tau = \sqrt{128\pi LI/GD^4}. \tag{52}$$

The moment of inertia of the suspension system I is found using a standard (usually a disk with an aperture) whose moment of inertia is known:

$$I = I_s \, \tau_0^2/(\tau_s^2 - \tau_0^2), \tag{53}$$

where I is the moment of inertia of the system; I_S is the moment of inertia of the standard; τ_0 is the oscillation period of the suspension system; τ_S is the oscillation period of the suspension system and the standard.

For a disk with a circular aperture at its center, we have

$$I_s = \frac{m \, (r_1^2 + r_2^2)}{2}, \tag{54}$$

where m is the mass of the standard, r_1 is the radius of the standard, and r_2 is the radius of the aperture.

Thus, the oscillation period of the system is related to the geometrical dimensions of the suspension filament and to the parameters of the material of which this filament is made. Consequently, the required value of the oscillation period of the system can be obtained by a suitable selection of the suspension filament.

The viscosity of melts can be measured using the same apparatus as that employed in the determination of the electrical conductivity by the contactless method (cf. Fig. 1). The experimental conditions are selected so that the value of the criterion ξ is greater than 10. Then, the viscosity is calculated using Eq. (42) for low-viscosity liquids. The necessary conditions are satisfied by the use of a tungsten filament 60 μ thick and 270 mm long, if the radius of the sample is about 6 mm. Standard ampoules of calibrated quartz are used in such measurements: diameters of the ampoules can differ from each other by not more than ± 0.1 mm. Moreover, the height of the ampoules should also be the same. If the investigated material reacts with quartz, the ampoules should be coated with a layer of pyrocarbon, using a method described in [98]. In some cases, a quartz ampoule can be placed in a graphite container. Then, the tungsten filament should not be thinner than 100 μ. In exceptional cases, the investigated substance must be placed in a corundum crucible, which is then located in a quartz ampoule. In this case, we again must use a filament at least 100 μ thick. More details will be given in the description of the experimental results. Before measurements are made at the melting point, the substance must be kept at the required temperature for 1.5 h in order to ensure complete melting; at other temperatures, the measurements can be begun after about 20 min. At each temperature, 5-7 measurements must be carried out. The viscosity is determined by measuring the oscillation period of an ampoule containing the substance being investigated and the logarithmic damping decrement is found from Eq. (55):

$$\delta = \frac{1}{f} \ln (A / A_f), \tag{55}$$

where f is the number of oscillations, and A_0 and A_f are the initial and final amplitude in a series of f oscillations.

In each case, we must determine first the range of values of the oscillation amplitudes in which the amplitude depends linearly on the oscillation number. Under these conditions, the deformation of the filament still obeys Hooke's law. The relative error in the determination of the logarithmic damping decrement by means of Eq. (55) is given by the following expression:

$$\frac{\Delta \delta}{\delta} = \frac{\Delta A_0 / A_0 + \Delta A_f / A_f}{\ln (A_0 / A_f)} . \tag{56}$$

Usually, the quantity $\Delta \delta / \delta$ does not exceed 1%. The errors in the determination of the oscillation period and of the moment of inertia are, respectively, 0.3 and 1.5%. The total error in the determination of the kinematic viscosity using Eq. (42)† is given by the expression

$$\frac{\Delta \nu}{\nu} = \frac{2 \Delta I}{I} + \frac{2 \Delta \delta}{\delta} + \frac{2 \Delta \tau}{\tau} . \tag{57}$$

The errors in the determination of the weight and radius of the sample as well as in the expression $\delta_0 \tau / \tau_0$, which is small compared with δ, can be neglected. Consequently, the error in the determination of the absolute values of the viscosity is about 5.6%. However, we must note that the relative error in the determination of the temperature dependence of the kinematic viscosity for a particular sample is only 2.6%, because the error in the determination of the moment of inertia is then systematic and can be ignored.

---

† The quoted values of the probable error in the determination of the viscosity apply to the formula for low-viscosity liquids [Eq. (42)], because this was the basic calculation formula [Eqs. (36) and (40) were used only in the calculation of the viscosity of selenium].

Chapter 3

Physicochemical Properties of Molten Elemental Semiconductors

§1. Silicon and Germanium

1. Crystal Structure, Nature of Chemical Binding, and Some Physicochemical Properties of Solid Silicon and Germanium

Silicon and germanium belong to Group IV of the periodic table.

They both crystallize in a diamondlike structure [99] (Fig. 11).

The unit cell of diamond is a face-centered cube in which atoms are located not only at the vertices and at the centers of the faces but also in four out of eight tetrahedral vacancies.

Thus, the unit cell contains eight atoms, each of which is surrounded by four nearest neighbors, i.e., the coordination number is four. Free unexcited silicon and germanium atoms have four electrons each in the outer electron shells in $(3s)^2 (3p)^2$ and $(4s)^2 (4p)^2$ configurations, respectively [100].

One could expect the formation of two covalent bonds, since there are only two unpaired electrons in the p-shell. However, when atoms approach each other in an assembly, electrons are redistributed between the s and p orbitals, because the $(3s)^2$ cloud of silicon and the $(4s)^2$ cloud of germanium are near the surfaces of

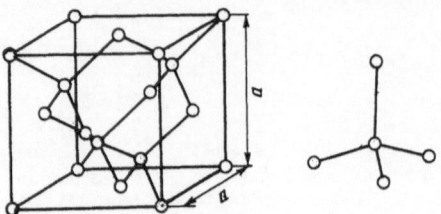

Fig. 11. Structure of diamond.

Table 5. Principal Crystal Structure Parameters of Silicon and
Germanium (diamond-type structure, K = 4 [105-108])

Element	Lattice period, Å	Interplanar distance (Å) for planes				Reticular density of planes, atoms/Å²		
		{100}	{110}	{111}^{I}	{111}^{II}	{100}	{110}	{111}
Si	5.4308	1.3577	1.9411	2.3411	0.7828	0.0675	0.0948	0.0785
Ge	5.6574	1.4143	2.0224	2.4467	0.8156	0.0625	0.0884	0.0721

Table 6. Some Physicochemical Properties of Silicon
and Germanium

Property	Si	Ge	Reference
Density, g/cm³	2.328	5.323	[109,110]
ΔE, eV (300°K)	1.09	0.665	[111,112]
u_n, cm²·V⁻¹·sec⁻¹	1550	4400	[109]
u_p, cm²·V⁻¹·sec⁻¹	480	1900	[111]
t_{mp}, °C	1420	937	[111]
Q_f, kcal/g-atom	12.1	8.35	[113—117]
Debye temperature, °K	690	400	[118, 119]
Atomization energy, kcal/mole	204	178	[120, 121]

the corresponding atoms; the s and p orbitals give rise to sp³
hybrid orbitals. Thus, each atom has four unpaired electrons and
forms four covalent bonds.

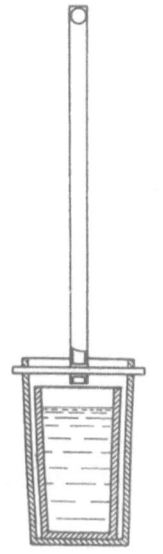

All these four bonds are equivalent and have a tetrahedral configuration.

The sp³ hybrid bonds are stronger than the s and p bonds, because their formation involves a strong overlap of the electron clouds [101]. Moreover, they are very rigid (directional). Investigations of the spatial distribution of the electron density in silicon and germanium, carried out by Sirota et al. [101-103], have shown that, in contrast to ionic compounds, the electron density along the tetrahedral directions does not decrease to zero but forms "bridges."

Fig. 12. Suspension system in measurements of the electrical conductivity of silicon.

Investigations of the microcleavage of silicon and germanium [104] indicate that the main cleavage plane is {111}.† The fracture along this plane breaks the smallest possible number of bonds, i.e., cleavage takes place along perfect covalent bonds.

The main crystal structure parameters of silicon and germanium are listed in Table 5. It is evident from this table that the interatomic spacing in germanium, whose atomic number is higher, is larger than in silicon. This is evidently due to a weaker interatomic interaction because of the stronger screening of the nuclei by a larger number of inner shells. For this reason, all the quantities which represent the strength of interatomic bonds in germanium have lower values than those in silicon. Table 6 shows that the atomization energy, the Debye temperature, the melting point, and the heat of fusion are all lower for germanium. The latter two quantities are related to the strength of bonds not only in

---

† In fact, the cleavage "plane" is a pair of {111} planes. The distance between two closely spaced planes of this kind is $a\sqrt{3}/12$; the corresponding distance between two widely spaced planes is $a\sqrt{3}/4$. Two closely spaced {111} planes are so close together and so tightly bound that they behave as a single "thick" plane (Table 5).

the solid state but also in the liquid phase; they are also associated with changes in the nature and strength of chemical bonds at
the melting point. However, we must remember that the rupture
of bonds at the melting point differs from substance to substance.
Only if the nature of the change in the chemical bonds due to melting is the same for two substances do the heat and temperature of
fusion reflect the strength of interatomic bonds. This is true of
silicon and germanium.

The weakening of the bonds of valence electrons with nuclei
reduces the activation energy of the valence electrons of germanium relative to silicon (Table 6). At the same time, the mobility
of electrons and holes is higher in germanium than in silicon.
Goryunova [109] explains this by the smoothing out of the potential
field in a crystal by the electron shells of "heavy" atoms and by
the reduction in the carrier scattering due to the lower amplitude
of thermal vibrations of "heavy" atoms.

Thus, silicon and germanium are covalent crystals in the
solid state and the "metallic component" of binding is greater for
germanium than for silicon.

2. Changes in Short-Range Order and Chemical Binding due to the Melting of Silicon and Germanium†

In order to obtain information on changes in the structure
and chemical binding at the melting point of silicon and germanium,
we must consider first the results of investigations of the electrical, magnetic, and thermoelectric properties. These properties
are related primarily to the state of valence electrons, and changes
in these properties at the melting point reflect changes in the nature of the chemical binding in the transition from the solid-to-liquid state.

We investigated silicon and germanium single crystals with
carrier densities of 10^{13}-10^{14} cm^{-3}. Germanium samples (8 mm
in diameter and 8 mm high) used in the measurements of the electrical conductivity and those (10 mm in diameter and 10 mm high)

---

† In presenting the data on the physicochemical properties at the melting point, we
shall first give our results, then compare them with the published data when such data
are available.

employed to measure the magnetic susceptibility were sealed into quartz ampoules evacuated down to 10^{-3} mm Hg. Since silicon reacts with quartz and, moreover, quartz loses its strength at the melting point of silicon, the silicon samples were placed in cylindrical corundum cans. A can with a sample was placed in a corundum crucible, which was suspended by a corundum rod (Fig. 12).

The electrical conductivity of silicon was measured in 10^{-3} mm Hg vacuum; the magnetic susceptibility was measured in a helium atmosphere. The thermoelectric power of germanium was measured in a helium or argon atmosphere (1.5 atm) in the manner described in Chapter 2. The thermoelectric power of silicon was not measured.

The results of our measurements of the electrical conductivity of silicon and germanium are given in Fig. 13, together with those obtained by other authors.

The results obtained by different workers are in fairly good agreement, particularly if we bear in mind that materials of different purity were used and the measurements were carried out using different methods: the contact method was used for the results reported in [124-126], while in our investigation and in A. R. Regel's work the electrodeless method and a rotating magnetic field were used. It is evident from Fig. 13 that at high temperatures the electrical conductivity of solid silicon and germanium increase with temperature (this indicates intrinsic conduction). Melting produces a sudden increase in the electrical conductivity.

Table 7 lists the values of the electrical conductivity of silicon and germanium at the melting point of each element.

It follows from Table 7 that the discontinuity in the electrical conductivity of germanium is less than that for silicon, because of the higher value of the conductivity of solid germanium. After melting, the electrical conductivity of germanium and silicon decreases monotonically with increasing temperature.

The electrical conductivity of silicon and germanium in the liquid state is of the order of the conductivity of liquid metals such as tin or mercury (10^4 $\Omega^{-1} \cdot cm^{-1}$). The absolute value of the electrical conductivity of molten silicon and germanium and the nega-

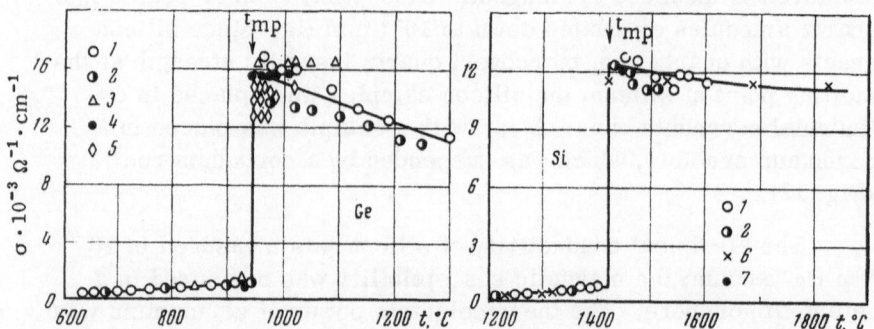

Fig. 13. Temperature dependences of the electrical conductivity of silicon and germanium in the solid and liquid state. 1) Results of the present authors; 2) Regel' [5, 6, 8, 10-13, 18]; 3) Keyes [125]; 4) Domenicali [124]; 5) Hamilton et al. [126]; 6) Baum et al. [123]; 7) Gel'd et al. [122].

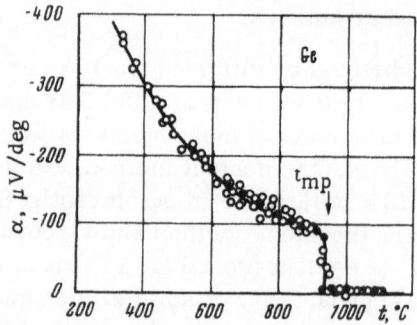

Fig. 14. Temperature dependence of the thermoelectric power of germanium in the solid and liquid states. ◯) Results of the present authors; ●) Domenicali [124].

Table 7. Electrical Conductivity of Silicon and Germanium at Their Melting Points

Element	σ_s, $\Omega^{-1} \cdot cm^{-1}$	σ_l, $\Omega^{-1} \cdot cm^{-1}$	σ_s/σ_l
Si	~580	~12000	~20
Ge	~1250	~14000	~11

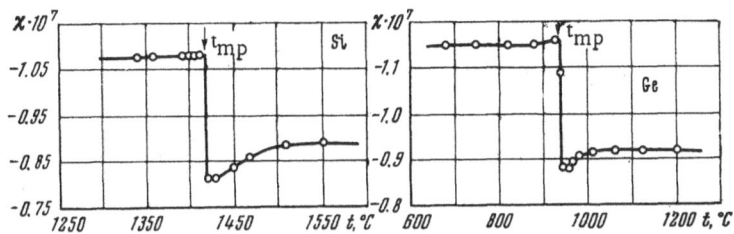

Fig. 15. Temperature dependences of the magnetic susceptibility of silicon and germanium in the solid and liquid states.

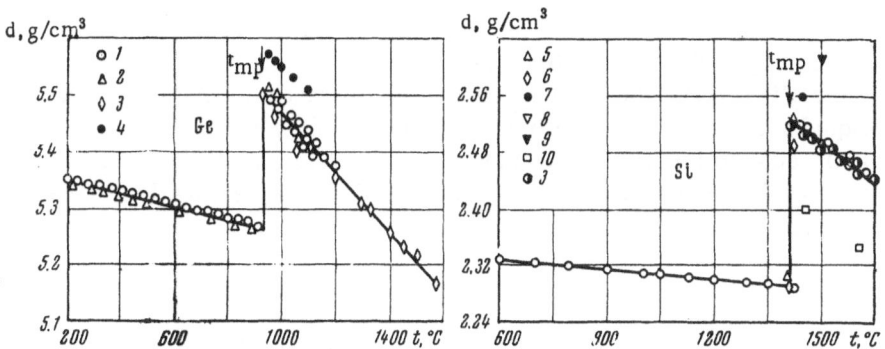

Fig. 16. Temperature dependences of the density of silicon and germanium in the solid and liquid states. 1) Results of the present authors; 2) Regel' [5, 6, 8, 12]; 3) Lucas [128, 129]; 4) Klemm [132]; 5) Wartenberg [131]; 6) Logan and Bond [134]; Vatolin and Esin [135]; 8, 9) Gel'd and Gertman [130]; 10) Koniger and Nagel [133].

Table 8. Density of Silicon and Germanium
at Their Melting Points

Element	d_s, g/cm^3	d_l, g/cm^3	$\dfrac{d_l - d_s}{d_s}$, %
Si	2.30	2.53	~10
Ge	5.26	5.51	~4.75

tive sign of the temperature coefficient of the conductivity both indicate that these substances become metallic in the molten state.

Figure 14 shows the temperature dependence of the thermoelectric power of germanium determined in our investigations;
this figure includes also the results of Domenicali [124].

It is evident from Fig. 14 that the thermoelectric power decreases at the melting point to very low values and remains practically constant at higher temperatures.

The sudden fall of the thermoelectric power of germanium
at the melting point to very low values typical of metals is evidently due to a sudden and strong increase in the carrier density.

Figure 15 shows the temperature dependences of the magnetic susceptibility of silicon and germanium [127]. It is evident
from Fig. 15 that both elements are diamagnetic in the solid and
liquid states. At the melting point, the absolute value of the magnetic susceptibility falls suddenly. In the liquid state, a small
rise in the magnetic susceptibility is followed by saturation. The
sudden fall of the total magnetic susceptibility may be attributed
to an increase in the spin paramagnetism of free electrons when
the density of these electrons increases suddenly.

Thus, we can draw the general conclusion that the melting
of silicon and germanium produces a sudden strong rise in the
conduction electron density so that the electrical conductivity
reaches values typical of liquid metals such as mercury. The sudden increase in the carrier density is attributed by Regel' to the
destruction of the three-dimensional system of the rigid sp^3 hybrid
bonds at the melting point and a transition to a metal-like state.
Thus, according to Regel's classification, the melting of silicon
and germanium is of the "semiconductor−metal" type. The change
in the nature of the chemical binding from covalent to metallic,
the latter with a high density of free electrons, should naturally be
accompanied by a change in the short-range order.

The density is an integral structure-sensitive characteristic,
which changes with the coordination number and the interatomic
spacing. Since melting usually either increases the interatomic
spacing or alters it slightly, a change in the density can be used
to deduce information on the change in the coordination number.

The temperature dependences of the density of silicon and germanium obtained by us are given in Fig. 16, where they are compared with the results of other workers [5, 6, 8, 12, 128-135].

Our results for germanium are in good agreement with those of Regel' [5, 6, 8, 12] and Lucas [128, 129]. The results of Klemm et al. [132] are somewhat higher. Our results for silicon are in good agreement with those of Gel'd and Gertman [130], Lucas [128, 129], and Logan and Bond [134] in the liquid state; the results of Vatolin and Esin [135] are somewhat higher, and those of Koniger and Nagel [133] are lower. It is evident from Fig. 16 that the density of silicon and germanium in the solid phase decreases linearly when the temperature is raised, increases suddenly at the melting point, and continues to decrease with temperature in the liquid phase.

Table 8 lists the values of the density of silicon and germanium in the solid and liquid states at their melting points.

It is evident from Table 8 that at the melting point the density of silicon increases by about 10% and that of germanium by about 4.7%. Investigations of the density show that the melting of silicon and germanium is accompanied by a change in the short-range order to a closer packing. This conclusion is in agreement with the results of an x-ray diffraction analysis of molten germanium, carried out by Hendus in 1948 [136]. Hendus analyzed the intensity curve and concluded that the covalent bonds in germanium were partly or completely destroyed by melting. The experimental intensity curve was used to calculate the atomic distribution; it was found that the coordination number of molten germanium at 1000°C was 8 and the radius of the first coordination sphere was 2.70 A (in the solid state, K = 4, r = 2.43 Å).

A. S. Lashko also investigated the structure of molten germanium by x-ray diffraction analysis. He found, by analyzing the intensity curves and the radial distribution, that the coordination number at 1000°C was 7.6.

Bearing in mind the difficulties encountered in such experiments and in the interpretation of the results obtained, we may regard the agreement between the results of Lashko and Hendus on the structure of germanium to be satisfactory. The similarity in the nature of the changes in the density of silicon and germanium,

as well as in the changes of other physical properties sensitive to
the structure and chemical binding, indicate that the melting of
silicon produces similar changes in the short-range order.

3. Model of Short-Range Order in Molten Silicon and Germanium and Changes in the Electron Mobility Due to Melting

Grigorovich [137] has suggested that the melting of germanium and silicon is accompanied by the liberation of four valence
electrons. This uncovers the $(d^4 + d^6)$ and p^6 electron shells of
germanium and silicon [138]; the outer six electrons in these shells
form six clouds elongated along the axes of orthogonal coordinates.

According to Grigorovich [137], the directional interaction
between ions should result in the formation of a structure in the
form of a body-centered cube with a coordination number equal to
8, in agreement with the x-ray diffraction data.

The assumption of the liberation of four valence electrons at
the melting point of germanium is supported by the results of investigations of the optical properties of this element, according to
which there are about four free electrons per atom in molten germanium [139-140].

To refine the model of molten germanium and silicon, we
can compare the number of atoms per 1 cm^3, calculated from the
experimental values of the density of molten germanium, with the
x-ray diffraction determination of the interatomic spacing and the
coordination number at 1000°C [141]. Simple calculations show
that for the bcc model the number of atoms per 1 cm^3 is $N_1 =
6.6 \cdot 10^{22}$. According to the measurements of the density, the num-
ber of atoms per 1 cm^3 is

$$N = \frac{d}{1.65 \cdot 10^{-24} \cdot A} , \tag{58}$$

where d is the density of the melt in g/cm^3, the quantity $1.65 \cdot 10^{-24}$
is the weight of the hydrogen atom in g, and A is the atomic weight.
For germanium at 1000°C, a calculation based on Eq. (58) gives
$N = 4.56 \cdot 10^{22}$ cm^{-3}. Comparing N_1 and N, we may conclude that
the short-range order in germanium (and in silicon) is approximately of the body-centered cubic type.

Using these results, we can easily calculate the number of carriers per 1 cm^3 of molten germanium and silicon on the assumption that n = 4N. The results of such calculations are given in Table 9.

If we assume that the electrical conductivity of molten germanium and silicon is due to electrons, then, using the experimental data on the electrical conductivity of the melts, we can estimate approximately the corresponding values of the mobility from a simple relationship

$$u = \frac{\sigma}{ne}. \tag{59}$$

It is interesting to compare the values of the density and mobility of electrons in molten germanium, calculated in this way, with the results obtained experimentally from the measurements of the Hall coefficient [142-143] and electrical conductivity. According to these results, the electron density in molten germanium is $1.16 \cdot 10^{23}$ cm^{-3} and the mobility at 1000°C is 0.54 $cm^2 \cdot V^{-1} \cdot sec^{-1}$; these values are in very good agreement with the results quoted in Tables 9 and 10, and they support Grigorovich's model of molten germanium and silicon.

We shall now consider the problem of the change in electron mobility at the melting point [141]. It has been reported [9, 144] that the carrier mobility is altered only slightly by the melting. The considerable jump in the electron mobility observed at the melting point of indium antimonide (deduced from the Hall effect [145]) is attributed to the anomalously high electron mobility of this compound in the solid state and is regarded as an exception.

To estimate the changes in the electron mobility and density at the melting point, we can use empirical formulas [111] for solid germanium

$$n^2 = 3.1 \times 10^{+32} T^3 \exp\left(-0.785/kT\right), \tag{60}$$

$$u_n = 3800 \left(\frac{300}{T}\right)^{1.66} \tag{61}$$

and for solid silicon

$$n^2 = 1.5 \times 10^{39} \cdot T^3 \exp\left(-1.21/kT\right), \tag{62}$$

Table 9. Carrier Density in Molten Silicon
and Germanium

Element	Melt temp., °C	N, cm^{-3}	n, cm^{-3}
Si	1450	$5.40 \cdot 10^{22}$	$2.16 \cdot 10^{23}$
Ge	1000	$4.56 \cdot 10^{22}$	$1.80 \cdot 10^{23}$

Table 10. Electron Mobility in Silicon
and Germanium

Element	u_s, cm$^2 \cdot$ V$^{-1} \cdot$sec^{-1}	u_l, cm$^2 \cdot$ V$^{-1} \cdot$sec^{-1}	u_s / u_l	n_s / n_l
Si	16	0.3	53	$0.5 \cdot 10^4$
Ge	370	0.4	925	$0.65 \cdot 10^4$

$$u_n = 1450 \left(\frac{300}{T}\right)^{2.6}. \tag{63}$$

The electron mobilities in solid germanium and silicon, calculated at the melting point using Eqs. (61) and (63), are given in Table 10. These values are naturally approximate, since they are obtained by the extrapolation of these formulas to the melting point. However, they give correctly the order of magnitude, because they agree (within an order of magnitude) with the values of the electrical conductivity of solid germanium and silicon calculated from Eq. (64):[†]

$$\sigma = en\,(u_n + u_p). \tag{64}$$

Thus, we may assume that these calculations give the correct orders of magnitude of n and u_n of germanium and silicon at the melting points.

[†] The values of the hole mobilities were calculated using empirical equations given in R. A. Smith's book [111]: $u_p = 1800(300/T)^{2.33}$ for germanium and $u_p = 500(300/T)^{2.3}$ for silicon.

Comparison with the corresponding values for the liquid state shows that the electron density in germanium and silicon increases by about four orders of magnitude, at the melting point, while the mobility falls sharply (Table 10). The carrier density increase is the dominant effect so that the electrical conductivity increases strongly in the molten state.

This considerable fall in the mobility of germanium and silicon in the molten state is most likely due to a strong increase in the intensity of the electron scattering by thermal vibrations and structure defects, associated with the destruction of long-range order. Such a large fall in the carrier mobility should be typical of all semiconductors whose fusion is represented by the "semiconductor – metal" transition, i.e., semiconductors whose short-range order and chemical binding are altered considerably by melting.

4. Changes in Short-Range Order during the Heating of Silicon and Germanium Melts

We shall now consider changes in the short-range order of molten silicon and germanium during heating above the melting point.

In order to follow the qualitative changes in the short-range order in molten silicon and germanium, we investigated the temperature dependence of the viscosity, which (as shown in Chapter 2) is one of the most structure-sensitive properties. The measurement method has been described in Chapter 2. We used samples 12 mm in diameter and 20 mm high, cut from very pure silicon and germanium single crystals. The germanium samples were placed in evacuated (10^{-3} mm Hg) quartz ampoules and the silicon samples were placed in corundum cans, which were enclosed in corundum crucibles and suspended by corundum rods, in the same way as in the measurement of the electrical conductivity.

Since the heats of fusion of silicon and germanium are relatively high, these elements melt very slowly at temperatures close to the melting point. To ensure the complete melting of the silicon and germanium samples, they were heated to a temperature 15-20° above the melting point and kept at this temperature for 4-5 h. During this treatment, we measured periodically the

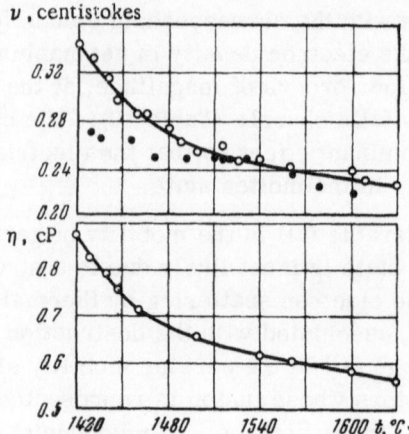

Fig. 17. Temperature dependences of the kinematic and dynamic viscosity of liquid silicon. O) Results of the present
authors; ●) Turovskii [147].

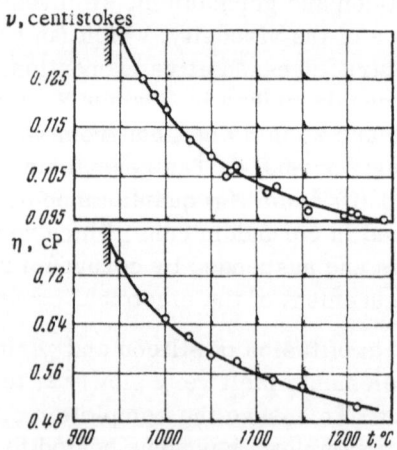

Fig. 18. Temperature dependences of the kinematic and dynamic viscosity of liquid germanium.

logarithmic decrement, whose magnitude increased with increase in the proportion of the molten substance. The samples were regarded as molten when the logarithmic decrement reached a constant value. Control samples were taken out of the apparatus and examined visually to confirm that the whole single crystal had indeed melted.

We then carried out measurements, during cooling, to the melting point and then during heating; we kept each sample at a given temperature for a long time in order to obtain reliable values. No hysteresis was observed. The results of measurements of the kinematic viscosity of silicon and germanium [21, 146] are presented in Figs. 17 and 18. Figure 17 (upper part) also includes the results of Turovskii and Lyubimov [147] for silicon.

It is evident from Fig. 17 that the viscosities found by Turovskii and Lyubimov [147] and by the present authors have similar absolute values at high temperatures but they differ considerably at lower temperatures. This may be due to the different purities of the investigated materials: Turovskii and Lyubimov [147] used technically pure silicon (~99.9% Si), while we used very pure silicon ($<10^{-6}\%$ of impurities).

The lower parts of Figs. 17 and 18 give the temperature dependences of the dynamic viscosity, calculated from the kinematic viscosity and density. The temperature dependence immediately above the melting point is fairly steep but the slope becomes less at 80-100° above the melting point. Analysis of the viscosity data on the basis of a theory of activated complexes yields information on changes in the short-range order during heating. The theory of activated complexes has been developed by Eyring [148]; according to this theory, the kinematic viscosity ν is given by the expression

$$\nu = \frac{Nh}{M} \exp\left(F_b/RT\right), \tag{65}$$

where N is Avogadro's number, h is Planck's constant, M is the molecular weight, and F_b is the isobar−isothermal potential or the free activation energy of viscous flow.

The value of F_b can be obtained from the formula

$$F_b = RT \ln \frac{M\nu}{Nh}. \tag{66}$$

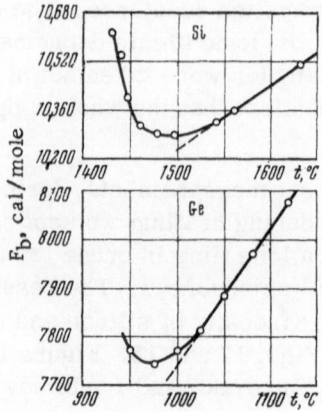

Fig. 19. Temperature dependences of the free activation
energy of the viscous flow of molten silicon and germanium.

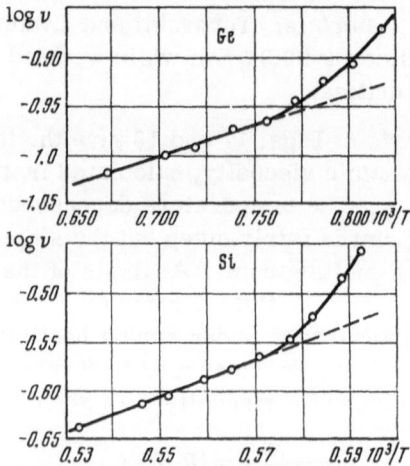

Fig. 20. Dependences of the logarithm of the kinematic
viscosity of molten silicon and germanium on the reciprocal of the absolute temperature.

The results of a calculation, in the form of the temperature dependences of the free activation energy of the viscous flow of silicon and germanium, are presented in Fig. 19. It follows from
these dependences that, at high temperatures (Δt above the melting point), the free activation energy of viscous flow increases
with increasing temperature:

$$F_b = a + bt. \tag{67}$$

When the temperature is lowered, departures from the linear dependences are observed: the values of F_b pass through a minimum
and, when the temperature is reduced further, they begin to increase again right up to the melting point. The minimum is particularly evident in the case of silicon and less so in the case of
germanium. Before discussing the deviations from the linear dependence of F_b on temperature, we shall discuss the general properties of silicon and germanium at higher temperatures.

It is known [60, 148] that

$$F_b = H_b - TS_b, \tag{68}$$

where H_b and S_b are the activation enthalpy and entropy of viscous
flow. Comparing the above expression with Eq. (67), which is valid
at high temperatures, we can see that

$$a = H_b; \ b = -S_b. \tag{69}$$

Consequently, using the curves in Fig. 19, we can determine the
activation entropy of viscous flow from the tangent of the slope;
this entropy is constant in that range of temperatures at which
Eq. (67) is valid. The results of such calculations are presented
in Table 11. Changes in the activation entropy of viscous flow outside this range of temperatures can be estimated from

$$\Delta S_b = -\left(\frac{\partial F_b}{\partial T}\right), \tag{70}$$

i.e., from the tangent of the slope at the corresponding point on the
$F_b = f(T)$ curve.

Bearing in mind that

$$H_b = E_b - RT, \tag{71}$$

Table 11. Values of E_b and S_b for Molten Silicon
and Germanium

Element	Δt^*, °C	$E_b \cdot 10^{-3}$, cal/mole	$-S_b$, cal $\cdot$ mole$^{-1} \cdot$ deg^{-1}
Si	120	8.63	~2.1
Ge	100	2.74	~2.85

*Δt is the range of temperatures above which the free activation energy of viscous flow increases linearly with temperature, i.e., the range of temperatures at which Eq. (67) is valid.

and substituting this expression into Eq. (68), we find that

$$F_b = E_b - RT - S_b T, \tag{72}$$

where E_b is the activation energy of viscous flow. Substituting this expression into Eq. (65), we find

$$\nu = \frac{Nh}{M} \exp{(E_b/RT)} \exp{(-1)} \exp{(-S_b/R)}. \tag{73}$$

In the temperature range within which Eq. (67) is valid, we can re-write Eq. (73) in the form:

$$\nu = A \exp{(E_b/RT)}. \tag{74}$$

Consequently, in this range of temperatures, the logarithm of the kinematic viscosity should be a linear function of the reciprocal of the absolute temperature and the tangent of the slope of the function can be used to determine the activation energy of viscous flow E_b. The curves used in such a calculation are shown in Fig. 20. It follows from this figure that at high temperatures the logarithm of the kinematic viscosity does indeed depend linearly on the reciprocal of the absolute temperature. At lower temperatures, there are considerable deviations from this linear dependence and these deviations are stronger for silicon than for germanium.

The values of the activation energy and entropy of viscous flow (E_b and S_b), found in the range of temperatures within which Eq. (67) is satisfied, are listed in Table 11.

Analyzing the data presented in Table 11, we must point out first of all that the absolute values of the quantities given there are typical of metals [74], which is in agreement with the conclusion, given in the preceding subsection, that silicon and germanium become metallic in the molten state.

The increase in the free activation energy of viscous flow with rising temperature at temperatures about 100° above the melting point can be easily explained by changes which take place during the heating of any liquid (increase in the intermolecular distances, weakening of the binding forces between individual molecules in the liquid, increase in the number of vacancies, etc.) [23]. Changes in such energy parameters as the activation energy and entropy of viscous flow, taking place in a temperature interval Δt, indicate considerable changes in the nature of the intermolecular interaction. The free activation energy of viscous flow passes through a minimum below this range of temperatures and then increases when the temperature is reduced further.

The increase in the free activation energy of viscous flow during cooling is obviously due to important changes which take place in the liquid during the precrystallization stage.

The nature of the changes in the free activation energy of viscous flow with temperature is described well by the temperature dependence of the kinematic viscosity. During the precrystallization stage, the viscosity increases fairly rapidly when the temperature is reduced very slightly. Therefore, the influence of the viscosity on the free energy predominates, in this range of temperatures, over the influence of temperature so that the free energy increases when the temperature is reduced. Physically, this means that the viscosity increases due to an increase in the binding forces and the formation of structure units with directional bonds.

To obtain information on these changes, we can use the dependence of the viscosity on the density, deduced by Bachinskii [62-63],

$$\frac{1}{\nu} = \frac{1}{c} + \frac{b}{c}\,d, \tag{75}$$

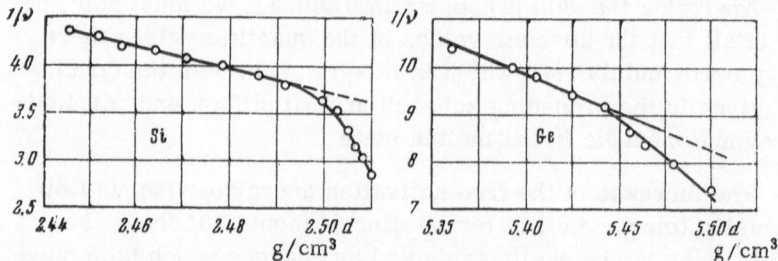

Fig. 21. Dependences of the reciprocal of the kinematic viscosity on the density of molten silicon and germanium.

where ν is the kinematic viscosity (in stokes), c and b are constants, and d is the density in g/cm^3.

Bachinskii's equation is based on the assumption that the viscous flow of a liquid (and, consequently, the viscosity) are governed by the nature of the interaction between molecules. Experiments show that a linear dependence of the reciprocal of the kinematic viscosity on the density is obeyed by a large number of liquids, including metals [74]. Figure 21 shows the dependence of the reciprocal of the viscosity on the density. It follows from this figure that, during the precrystallization stage, there are considerable deviations from Eq. (75), of the type which is observed for water. These deviations indicate a change in the nature of the intermolecular interaction at these temperatures. In the case of water, the deviations from Bachinskii's equation during the precrystallization stage are due to the transition from a quartz-type structure, typical of water above 4°C, to a tridymite ice structure (below 4°C) [149].

We can draw the following conclusions from these results: when the temperature is lowered during the precrystallization stage, considerable changes take place in the short-range order structure of silicon and germanium: the coordination number decreases and the liquid-state structure approaches the solid-state structure.

If this interpretation of the experimental data and conclusions is correct, then the phenomena described should be observed in many substances [20].

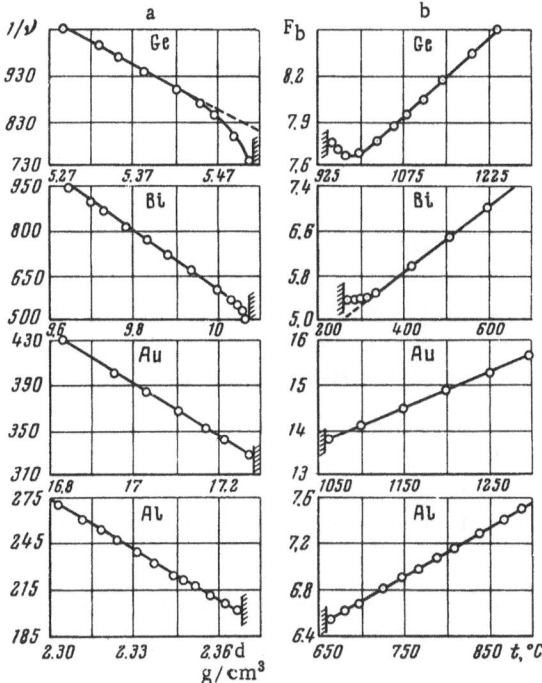

Fig. 22. Dependences of the reciprocal of the kinematic viscosity on the density (a) and of the free activation energy (kcal per mole) of viscous flow on temperature (b).

In such a case, if we take two substances such that in one of them the short-range order in the liquid state is very different from that in the solid state, while in the other substance the differences between the short-range order structures in the liquid and solid states are slight, it follows that during the precrystallization stage the first substance should exhibit anomalies in the temperature dependence of the structure-sensitive properties, while the second substance should not exhibit such anomalies.

For the sake of comparison, we shall consider aluminum, gold, bismuth, and germanium. The first two metals have the fcc structure in the solid state with a coordination number of 12. According to x-ray diffraction and electron-diffraction data, the short-range order in gold [136, 150] and aluminum [151-155] changes only very slightly after melting (K = 10.8-11).

Solid bismuth has the rhombohedral structure with a coordination number 3 + 3 (the three nearest atoms and the three atoms further away). In the liquid state, the coordination number of bismuth is 7 + 8 [136, 151-152, 154, 156].†

Figure 22 shows the dependence of the reciprocal of the viscosity on the density and the temperature dependence of the free activation energy of viscous flow for these four elements. It is evident from Fig. 22 that, in the case of gold and aluminum, these dependences obey well Eqs. (67) and (75) over the whole temperature range right down to the melting point. In the case of bismuth, small departures are observed during the precrystallization stage. These departures are much stronger in the case of germanium.

Thus, a comparison of substances with different degrees of compactness of the structure in the solid and liquid states shows that the precrystallization anomalies are observed in those substances whose short-range structures differ strongly in the liquid and solid states.

Danilov et al. [178] have shown that although the melting of bismuth is accompanied by changes in the short-range structure (in this respect, bismuth differs from gold and aluminum), the short-range structure immediately after melting remains very similar to the structure in the solid state. Danilov concludes that in the case of elements with loose packing in the solid state, we

---

† The viscosity and density of gold has been investigated by Gebhardt et al. [157-161]. Analysis of the investigation method used in [157], as well as the good agreement between the results of Gebhardt et al. [157, 158, 161-163] on the viscosity of tin, silver, and copper and the values reported in [164-166], suggest that the data on the viscosity of gold reported by Gebhardt et al. are very reliable. The viscosity of aluminum has been investigated by many workers [74, 167, 173], but the results obtained are contradictory. In view of this, we investigated the viscosity of aluminum again [174]. The viscosity was determined in 10^{-4} mm Hg vacuum. Aluminum cylinders were placed in graphite cans of 12 mm internal diameter, with screw-on lids made of spectroscopically pure graphite. The ratio of the height of the cylinder to its radius was 10. These measurements were in good agreement with the results reported in [74, 170, 173]. The density of aluminum was taken from the paper of Gebhardt et al. [171]. The viscosity of bismuth was taken from [74, 175-176] (the results of different workers were in good agreement) and the density of bismuth was taken from [74].

can expect considerable local differences in the distribution of atoms in the liquid state near the melting point. One distribution is under the dominant influence of those forces which are responsible for the short-range order in the solid state, while another distribution is governed primarily by the thermal motion which tends to produce a closer packing when the temperature is increased. The former distribution is gradually destroyed and replaced by the latter. Consequently, any similarity between the structures of the solid and liquid phases, for example, an additional maximum in the x-ray diffraction intensity, should be observed most clearly at temperatures close to the melting point.

Later, it was shown experimentally that the coordination numbers of bismuth [151, 179] and of its analog antimony [151] change in the liquid state, increasing with temperature up to t_{mp} + 30° and decreasing at still higher temperatures.

We may assume that the characteristic features of the structure of liquid bismuth, pointed out by Danilov, should be observed also in liquid silicon and germanium, which exhibit departures from the linear dependence of structure-sensitive properties during the precrystallization (or post-melting) stage, which are of the same type as in the case of bismuth (these departures are stronger in silicon and germanium than in bismuth).

For this reason, A. S. Lashko (Metal Physics Institute, Academy of Sciences of the USSR, Kiev) carried out in 1960 (at the suggestion of the present authors) an investigation of the structure of liquid germanium at various temperatures.

Figure 23 shows the diffraction intensity curves obtained by Lashko for molten germanium at 1000°C and 1300°C; the corresponding atomic radial distribution curves are also included in Fig. 23 (A. S. Lashko, private communication, 1965).

It is evident from Fig. 23 that the intensity and radial distribution curves obtained for molten germanium are not identical at these two temperatures.

The intensity and the radial distribution curves obtained at 1000°C have a definite additional maximum, which is absent in the curves recorded at 1300°C.

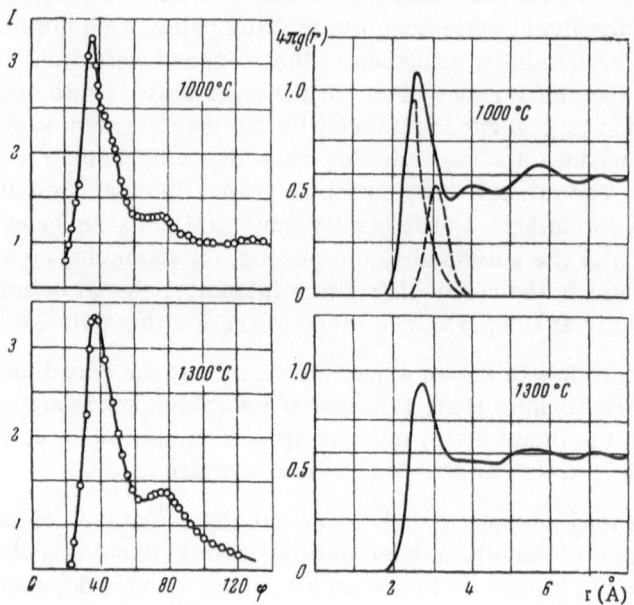

Fig. 23. X-ray diffraction intensity and atomic radial distribution curves of molten germanium at 1000 and 1300°C (results of A. S. Lashko).

The first peak and the additional maximum in the radial distribution curve can be interpreted as being due to the superposition of two maxima (shown by dashed curves), of which one (the stronger) represents the short-range structure dominant in molten germanium at 1000°C and the other represents the residue of the structure typical of lower temperatures; the latter disappears completely at 1300°C.

It follows that molten germanium still contains traces of the solid germanium structure at 1000°C and that these traces disappear finally at much higher temperatures.

Similar conclusions have been obtained by Krebs and V. B. Lazarev (Lazarev's lecture presented in October 1966 at the Institute for General and Inorganic Chemistry of the Academy of Sciences of the USSR), who investigated the structure of molten germanium at 960 and 1270°C.

Thus, the conclusions on the nature of molten silicon and germanium during the precrystallization stage, deduced from investigations of structure-sensitive properties, are confirmed by the results of x-ray diffraction investigations of molten germanium. Investigations of the viscosity of silicon and germanium during their rapid cooling from high temperatures have indicated departures from the normal temperature dependence of the viscosity (i.e., from the dependence observed during heating): the viscosity during cooling has been found to have lower values. This has happened in those cases where the melt has been subjected to strong supercooling.

However, in the absence of supercooling or in the case of slight supercooling, these anomalies in the temperature dependence of the viscosity are not observed. Thus, the branching of the viscosity curve is due to supercooling of the kind observed earlier in the case of tin [180]. The viscosity is a structure-sensitive property and therefore the branching of the viscosity curve due to supercooling must be related to changes in the structure of the liquid.

In view of this, we carried out a special thermographic investigation of the supercooling of germanium and of the influence on supercooling of the initial temperature, cooling rate, and soluble impurities.

5. Influence of Cooling Rate and Initial Temperature on the Degree of Supercooling of Molten Germanium

There have been very few investigations of the supercooling of germanium. Turnbull et al. [182, 183] have investigated the supercooling of several substances, including germanium, in small drops of 10-100 μ diameter. Experiments on small drops have established a formal rule, according to which the minimum temperature to which any substance not forming glasses can be supercooled is $0.82T_{mp}$. According to this rule, a germanium drop of 15-μ diameter can be supercooled by about 235°. The larger the size of the drop, the weaker is the supercooling. For a drop of 400-μ diameter, the supercooling is 219°. Both these results apply to 99.99% pure germanium. The rate of cooling and the initial temperature are not given by Turnbull et al.

The observation that the smaller the drop the stronger the maximum supercooling, which may be observed only in one of the many investigated drops, can be attributed to the inhomogeneity of the distribution of suspended impurities in these drops so that at least one of these drops becomes completely free of impurities and exhibits maximum supercooling. The phase transitions in small particles, whose size is comparable with the dimensions of critical nuclei, cannot be attributed only to the absence of catalytic occlusions in some of the drops. Other factors, which do not apply to bulk samples, are also important. Petrov [184, 185] has investigated the melting and crystallization of aerosol particles $(2.5 \cdot 10^{-6}$ cm) of many metals and semimetals and has concluded that a uniform pressure of several thousands of atmospheres, due to the surface tension, plays an important role in phase transitions in such particles. During the cooling of molten particles, this uniform pressure increases the "orientational viscosity" of ordered regions and prevents their crystallization.

The influence of the initial temperature on the degree of supercooling of "bulk" samples (2-20 g in weight) of molten 99.99% pure germanium has been investigated by Fehlig and Scheil [186]. However, the considerable scatter of the experimental points prevented them from establishing exactly the upper temperature limit of this influence. The cooling rate is not given. Fehlig and Scheil [186] attribute this influence to impurities which are not adsorbed sufficiently by flux at temperatures below 1100°C. It follows that the influence of the initial temperature and of the cooling rate on the degree of supercooling of molten germanium is best investigated on "bulk" samples free of suspended impurities.

We investigated a germanium single crystal containing less than $10^{-6}\%$ of impurities. To carry out a thermal analysis, germanium was crushed and placed in a carefully washed and dried Stepanov-type quartz ampoule evacuated to 10^{-3}-10^{-4} mm Hg. In each experiment, we used new samples, since after one run the ampoule usually cracked and the sample oxidized. The melting was carried out in a TG-3 resistance furnace. Various cooling rates were obtained by reducing the current in the furnace windings. A gradual adjustment of an autotransformer made the cooling relatively uniform. For rapid cooling, a sample was pulled out of an open furnace. We recorded the results in "temperature–

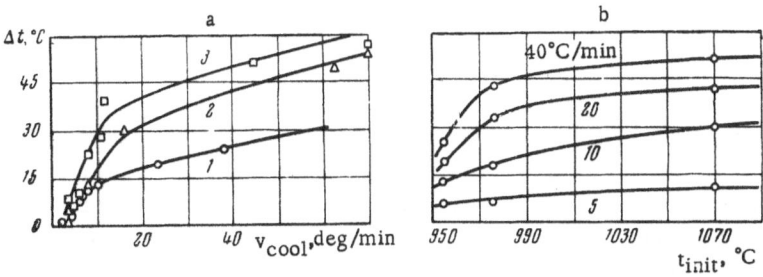

Fig. 24. Dependences of the degree of supercooling of molten germanium on the rate of cooling from a given initial temperature (a) and on the initial temperature at a constant cooling rate (b).

time" coordinates. The degree of supercooling was defined as the difference between the melting point and the minimum temperature to which a melt could be cooled, and after which the temperature began to rise due to the evolution of the latent heat of crystallization. The cooling curves were used to plot the dependence of the degree of supercooling on the cooling rate from a given temperature. Figure 24a shows such dependences for temperatures of 18 (1), 38 (2), and 133° (3) above the melting point. The temperature to which a sample was initially heated was maintained to within ±5°. At low cooling rates, the curves rose steeply from the abscissa: a small (several deg/sec) increase in the cooling rate increased considerably the degree of supercooling. At high cooling rates, the degree of supercooling depended little on a cooling rate.

Some scatter of the experimental points was probably due to fluctuations of the initial temperature from which the cooling was started (±5°), and to a possible nonuniformity of the cooling process.

The data presented in Fig. 24a were used to plot the dependence of the degree of supercooling on the initial temperature from which the cooling was started (Fig. 24b). A rise in the initial temperature increased the supercooling (for a constant cooling rate), and this dependence was strong at high cooling rates but very weak in the case of slow cooling.

We also investigated the influence of the same factors on the supercooling of germanium doped with soluble impurities (very pure gallium, indium, and antimony) and we found that the nature

of the dependences was still the same but the degree of supercooling was somewhat less.

Our investigations were carried out on very pure germanium single crystals in vacuum, while Fehlig and Scheil [186] used
a flux. Even in our case, the influence of suspended impurities
was not suppressed in spite of the use of single crystals from
which most of the impurities were expelled during the crystallization. The walls of the container could also act as a catalyst in the
crystallization, as demonstrated in [187]. Nevertheless, the dependences obtained may be attributed to those structural changes
which take place in liquid germanium when the temperature is increased above the melting point and which have been described in
preceding sections. The observation that an increase in the initial
temperature increases the degree of supercooling (at a fixed cooling rate) may be attributed to a gradual destruction of the residual
short-range order when the melt is heated. The higher the temperature of the melt, the greater is the difference between its
structure and the solid-state structure, but the main changes take
place immediately above the melting point.

Consequently, at a given cooling rate, the probability of
formation of a nucleus which has the same atomic distribution as
in the solid phase decreases when the temperature of the melt is
increased. This decrease is greatest immediately above the melting point.

The dependence of the degree of supercooling on the initial
temperature may be associated, to some extent, with the kinetics
of structural transformations during cooling.

At high cooling rates, the high-temperature structure may
still be observed at lower temperatures, since the time may be insufficient for all the changes to be completed. At low cooling
rates, the structural changes are completed in the available time
and the structure is of the equilibrium type.

Insoluble impurities (whose influence was minimized in our
investigations) can, as shown in [188], retard those structure transformations which are caused by temperature changes. If the short-
range order of the liquid is not affected by the variation of the temperature (with the exception of changes which are due to the thermal motion of the particles), the degree of supercooling does not

depend on the initial temperature, as found in the experiments on mercury carried out by Danilov and Neimark [178].

Thus, all the available experimental data on the structure and physical properties of molten germanium and silicon allow us to conclude that the change from the solid-state structure and chemical binding to the liquid-state properties takes place in two stages. The first stage consists in melting, which destroys the three-dimensional system of rigid covalent bonds and forms a short-range order characterized by a mixed covalent-metallic binding. The second stage takes place when the melt is heated further: during this stage, the residual covalent bonds are finally destroyed, free electrons are transferred to the collective state, and a structure with a higher coordination number is formed. The characteristic features of the crystal structure of silicon and germanium indicate that the lattice is fractured along {111} planes at the melting point. Partial conservation of microregions with parts of such planes is possible in the liquid. Further heating destroys these regions.

We shall now briefly discuss the papers of Kontorova [189, 190], who has suggested that the melting of silicon and germanium is not accompanied by any change in the type of binding between atoms and does not produce any new electron configuration. In her opinion, the melting results in the loss of the spatial rigidity by the "valence bridges" between atoms (the sp^3 hybrids) so that these bridges can now rotate. By analogy with the orientational melting of polar crystals, predicted and investigated by Frenkel' [191-192], such a change in the initial orientations of covalent bonds in germanium and silicon crystals has been called by Kontorova the orientational melting. As a result of an increase in the coordination number, each bond is only partly saturated and this is responsible for the metallic nature of the electrical conductivity of the melt. However, the hysteresis observed in the case of rapid cooling from high temperatures and the dependence of the degree of supercooling on the cooling rate (described in the present section) indicate that above the melting point changes take place not only in the structure but in the nature of the chemical binding and the retention of the initial structure is difficult. It is possible that the covalent resonant bonds described by Kontorova do exist close to the melting point but there is no doubt that they become metallic at higher temperatures.

§2. Selenium and Tellurium

1. Characteristic Features of Crystal Structure, Nature of Chemical Binding, and the Principal Physicochemical Properties of Solid Selenium and Tellurium

Selenium and tellurium belong to Group VI in the periodic table.

Selenium has several allotropic modifications. These modifications are obtained by the crystallization of selenium from different solutions. Liquid selenium is easily supercooled to form amorphous or glassy selenium [99, 112, 193]. Different crystalline and amorphous modifications of selenium can coexist but below the melting point at atmospheric pressure the thermodynamically stable crystalline form is gray selenium, which has the hexagonal structure [99, 112, 193]. However, due to the characteristic features of the structure of this modification, other crystalline modifications are transformed to the hexagonal form very slowly and the rate of formation of the hexagonal structure during crystallization from the melt is also very slow. This makes it difficult to prepare homogeneous samples, particularly single crystals. Tellurium has one modification (hexagonal), similar to gray selenium.

The structure of tellurium and gray selenium consists of zigzag-like helical chains parallel to the z axis (Fig. 25). Each helical chain has three atoms in each complete turn. Each atom lies above a similar atom in the chain so that the projection on a plane perpendicular to the axis of the chain is an equilateral triangle [99, 194]. The main crystal lattice parameters of gray selenium and tellurium are listed in Table 12.

Table 12. Principal Crystal Lattice Parameters
of Gray Selenium and Tellurium [99]

Element	a, Å	c, Å	Shortest interatomic distances		Valence angle, deg
			in chains	between chains	
Se (gray)	4.36	4.95	2.32	3.46	105
Te	4.45	5.91	2.86	3.74	102

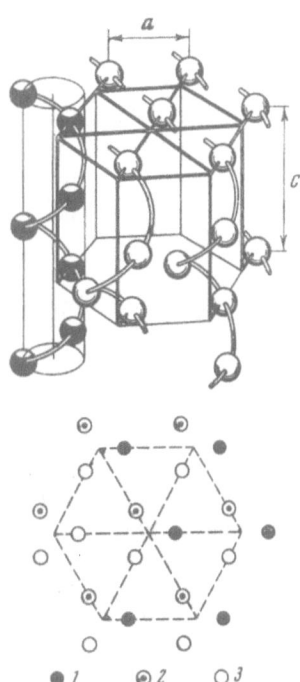

Fig. 25. Structure of gray selenium and tellurium: 1, 2, and 3
represent the first, second, and
third atoms in a chain.

The characteristic features of the structure of selenium and tellurium can be understood by examination of the structure of the electron shells of these elements. The outer electron states of selenium and tellurium are, respectively, the $(4s)^2$ $(4p)^4$ and $(5s)^2$ $(5p)^4$ configurations. Out of six outer electrons, two p electrons are paired and therefore they do not take part in the formation of chemical bonds. Atoms are bound by two unpaired p electrons. The electron clouds of the unpaired p electrons of each atom are overlapped by the p shells of neighboring atoms so that filled p orbitals are formed and each such orbital contains two electrons with opposite spins. Rows of atoms join to form a helical chain [100]. According to Mooser and Pearson, the binding between chains is complex and d electrons participate in it [195]. However, in the opinion of the majority of other in-

Table 13. Principal Properties of Gray Selenium and Tellurium

Property	Se (hex.)	Te	Reference
d, g/cm^3	4.81	6.23	[199—202]
t$_{mp}$, °C	220	452—453	[203]
Q$_f$, kcal/g-atom . . .	1.5	4.17	[113]
t$_{boil}$, °C	685	990	[193,204]
ΔE_{opt}, eV (300°K)	1.79—1.8	0.33—0.37	[110—112]
u$_n$, cm$^2\cdot$V$^{-1}\cdot$sec^{-1}(300°K)	—	1700	[110]
u$_p$, cm$^2\cdot$V$^{-1}\cdot$sec^{-1}(300°K)	0.1—0.3	1200	[197,198,110,205]

vestigators [110, 112, 196], this binding is due to weak van der
Waals forces, which explains the easy cleavage along planes
parallel to the chain axes, the low melting points of these elements,
etc. The structure of gray selenium and tellurium is extremely
anisotropic, which is to be expected in the presence of two types
of binding. For example, the resistivity of selenium and tellurium
single crystals along the chain axes $(\rho_{\parallel})$ is 3-10 times less than at
right angles to these axes $(\rho_{\perp})$ [112, 197, 198].

The main physical properties of gray selenium and tellurium
are listed in Table 13.

It is evident from Table 13 that the properties of these two
elements are not as similar as those of silicon and germanium.
They are isomorphous in respect of the structure but their other
properties differ considerably. The electrical resistivity of pure
selenium is close to the value typical of insulators. Eckart [206]
was able to obtain oxygen-free selenium with a resistivity of 10^9-
10^{10} $\Omega \cdot$ cm at room temperature, and Ugai [110] reports that the
resistivity of purified selenium can be increased to 10^{12} $\Omega \cdot$ cm.
Measurements of the Hall coefficient [197] and of the thermoelectric power [197, 207, 208] of hexagonal selenium have shown that
it is always p-type. The nature of the p-type conduction of selenium has not yet been finally resolved, but it has been suggested that
it is due to structure defects [110, 112, 209], namely: the finite
lengths of the chains whose unsatisfied valence bonds give rise to
acceptor levels.

The hole density decreases slowly when the temperature is
increased but the mobility rises. The majority of investigators
agree that the cause of such (for semiconductors) anomalous properties is the existence of some potential barriers which holes have
to overcome by thermal activation. In the opinion of Henkels [207]
and Kozyrev [208], amorphous layers between grains in hexagonal
selenium act as such barriers. According to Stubb [198] and Stuke
[210], the crystal structure of selenium itself is responsible for
these properties because of an inhomogeneous distribution of defects which are always present in it. The latter explanation is
more likely to be correct because these anomalous temperature
dependences are observed also in single crystals. Moreover, the
latter explanation agrees with the proposed mechanism of hole
generation [110, 112, 209]. It is suggested that the structure be-

comes less imperfect when the temperature is increased and this is the cause of the fall in the carrier density and one of the causes of the increase in the carrier mobility. It is interesting to note that microwave measurements have shown that the temperature coefficient of the conductivity is negative because of the usual decrease in the carrier mobility with temperature [211]. The principal carrier scattering mechanism at high temperatures must be the lattice scattering, while the scattering by defects becomes much weaker.

In contrast to selenium, tellurium is a typical semiconductor. Even at room temperature, pure tellurium exhibits intrinsic conduction. The resistivity of tellurium at room temperature is 0.29 $\Omega \cdot$ cm along the z axis and 0.59 $\Omega \cdot$ cm at right angles to this axis [111, 112, 212].

The Hall effect and the thermoelectric power of tellurium are anomalous because they twice change their sign when heated from low temperatures [111, 112, 212]. At low temperatures, in the impurity conduction region, tellurium is a p-type semiconductor. Beyond the intrinsic conduction region, tellurium becomes n-type, which indicates that the electron mobility becomes higher than the hole mobility. Near 500°C, the signs of the Hall coefficient and of the thermoelectric power again become positive. Several hypotheses have been suggested to explain these observations. According to these hypotheses, the observed behavior is due to the characteristic features of the band structure of tellurium, the thermal generation of lattice defects (which act like acceptors), etc. [111, 112].

Thus, hexagonal selenium and tellurium have similar structures but very different electrical properties.

The differences between these two elements are even greater in the liquid state. Since their behavior at the solid-to-liquid transition is quite different, we shall consider each of them separately.

2. Changes in Short-Range Order and Chemical Binding of Selenium Due to Melting

The temperature dependence of the electrical conductivity of selenium near its melting point and in the liquid state has been investigated on many occasions [3, 5, 6, 8, 11, 15, 16, 213-220].

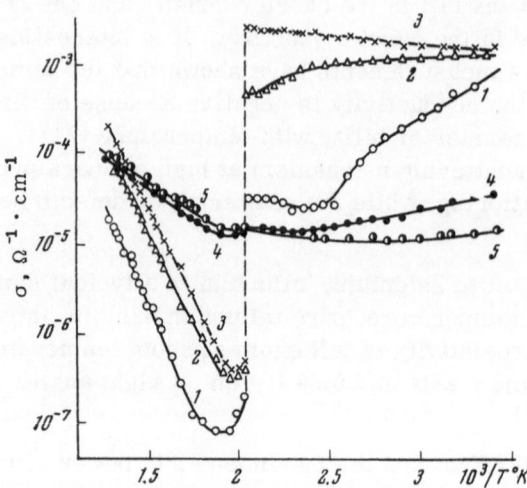

Fig. 26. Temperature dependence of the electrical conductivity
of hexagonal selenium in the solid and liquid state [214].

The electrical conductivity of those selenium samples which had not been subjected to oxygen-removing treatments decreases by several orders of magnitude at the melting point but in the liquid state it increases with rising temperature (curves 1-3 in Fig. 26).

The electrical conductivity discontinuity of oxygen-free selenium is vanishingly small because the removal of oxygen reduces the electrical conductivity of the solid and increases the electrical conductivity of the liquid [213, 214] (curves 4 and 5 in Fig. 26). Similar temperature dependences of the electrical conductivity are obtained for ordinary selenium at high frequencies. It is not clear why the electrical conductivity discontinuity disappears at high frequencies.

Measurements of the thermoelectric power of ordinary selenium indicate that p-type conduction is retained in the liquid state [214, 218, 219, 221]. The thermoelectric power of oxygen-free selenium has a negative sign in the molten state [214]. The same is true of the thermoelectric power of ordinary selenium at high frequencies [211].

Thus, selenium belongs to those substances which retain their semiconducting properties in the liquid state.

The fall of the electrical conductivity of selenium at the melting point may be due to a reduction in the density of carriers and a reduction in their mobility.

Busch and Vogt [222] have found that the diamagnetic susceptibility of selenium increases in the molten state. The increase in the total susceptibility may be due to a reduction in the paramagnetic component; it has been shown in [223] that the paramagnetic contribution is due to the ends of chains in selenium at which holes are formed.

It follows that the melting of selenium is accompanied by a decrease in the hole density. This may be due to the closure of some chains to form ringlike molecules. X-ray diffraction analysis of molten selenium has demonstrated that it contains, in addition to hexagonal selenium chains, Se_6 rings [224, 225].

Se_8 rings have been observed in glassy selenium, which is obtained by the supercooling of liquid selenium [226]. Since the transition from the glassy to liquid state is not accompanied by any appreciable change in the electrical conductivity [8], it seems that these rings must exist also in the liquid state. A theory of equilibrium between chains and rings for liquid selenium, similar to that for liquid sulfur, has already been developed [227].

The coordination number is not affected by the formation of ringlike molecules but the form of the structural units is and, consequently, the nature of packing becomes different. In fact, the melting is accompanied by a considerable loosening of the structure, which is indicated by a strong (~15.2%) decrease in the density of selenium in the molten state. The temperature dependence of the density (constructed from the results of investigations by one of the present authors as well as from other published results) is shown in Fig. 27.

Thus, the experimental results indicate that the short-range order and the nature of the chemical binding in hexagonal selenium are basically retained in the liquid state. Only the weak van der Waals forces, acting between chains, are destroyed by melting. The covalent bonds, which act between atoms, are retained and even strengthened.

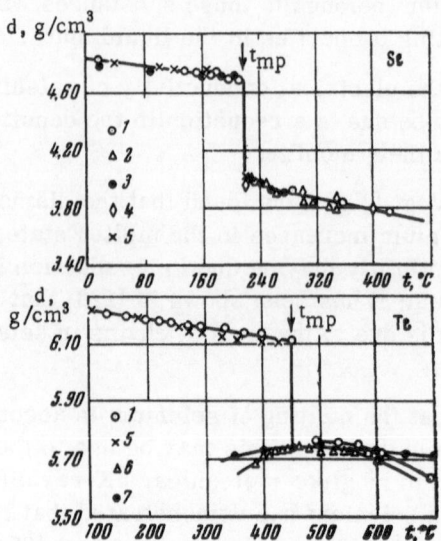

Fig. 27. Temperature dependences of the density of hexagonal selenium and tellurium in the solid and liquid states. 1) Results of the present authors; 2) Dobinski and Wesolowski [228]; 3) Campbell and Epstein [200]; 4) Lucas and Urbain [229]; 5) Bonnier et al. [201, 230]; 6) Klemm et al. [246]; 7) Regel' et al. [16].

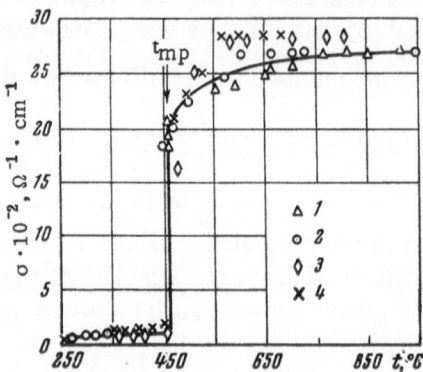

Fig. 28. Temperature dependence of the electrical conductivity of tellurium in the solid and liquid states. 1) Results of the present authors; 2) Perron [231]; 3) Regel' [15, 16]; 4) Epstein et al. [212, 235].

3. Changes in Short-Range Order and Chemical
Binding of Tellurium Due to Melting

Figure 28 shows the temperature dependence of the electrical conductivity of tellurium in the solid and liquid states, according to the results obtained by us and by other workers.

It is evident from Fig. 28 that the results are in fairly good
agreement, in spite of the fact that materials of different degrees
of purity were used by different workers and the measurements
were carried out using different methods. The electrical conductivity of solid tellurium increases when the temperature is increased, rises suddenly (up to ~2 · 10^3 Ω^{-1} · cm^{-1}) at the melting
point and continues to increase smoothly when the temperature is
increased further.

The nature of the temperature dependence of the electrical
conductivity and its absolute values above the melting point indicate that tellurium belongs to substances which undergo the
"semiconductor − semiconductor" type of transition (Regel's classification).

More detailed information on the nature of the conduction in
molten tellurium is provided by its thermoelectric and thermomagnetic properties.

The thermoelectric power of tellurium has been measured
at high temperatures [142, 212, 221, 236], and it has been found that
the power decreases suddenly by factors of 10-30 at the melting point.

The sign of the thermoelectric power is positive before and
after melting but at higher temperatures (560-580°C) it becomes
negative.

The Hall coefficient of solid and liquid tellurium has been
measured over a wide range of temperatures [142, 212, 236, 237].
Figure 29 shows the temperature dependence of the Hall coefficient
reported in [212].

It is evident from Fig. 29 that the Hall coefficient decreases
by about an order of magnitude at the melting point. The positive
sign of the Hall coefficient exhibited by crystalline tellurium is also retained at higher temperatures above the melting point. The
Hall coefficient decreases when the melt is heated and at 575°C its

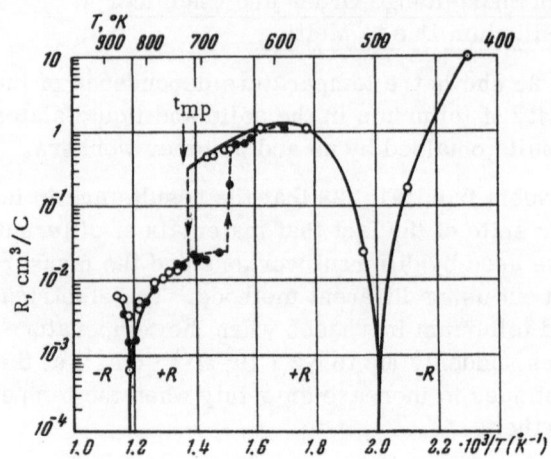

Fig. 29. Temperature dependence of the Hall coefficient of tellurium in the solid and liquid states [212]. O) During heating; ●) during cooling.

sign becomes negative. The retention of the same carrier sign below and above the melting point of tellurium has enabled Regel' [144] to calculate the mobility in molten tellurium using a simple formula without allowance for the existence of carriers of two signs. He has found that the carrier mobility above the melting point is somewhat higher than below the melting point ($u_S = +30$ cm$^2 \cdot$ V$^{-1} \cdot$ sec^{-1}, $u_l = +40$ cm$^2 \cdot$ V$^{-1} \cdot$ sec^{-1}). This means that the sudden increase in the electrical conductivity at the melting point and the corresponding fall of the Hall coefficient and thermoelectric power are all due to an increase in the carrier density.

This increase in the density of free electrons is responsible also for the fall of the (experimentally measured) total diamagnetic susceptibility of tellurium due to an increase in the spin paramagnetism.

A simple calculation shows that the carrier density of tellurium increases by about one order of magnitude at the melting point and that it then reaches ~$3 \cdot 10^{20}$ cm^{-3}.

Thus, the melting of tellurium is accompanied by the destruction not only of the weak van der Waals forces acting between

chains but also some of the covalent bonds; this liberates an additional number of carriers.

However, the melting does not destroy all the covalent bonds in tellurium. A considerable proportion of these bonds is retained above the melting point. Busch and Tieche [142] have compared the experimental values of the Hall coefficient of molten metals with the values obtained by calculation on the assumption that the number of carriers per atom is equal to the number of valence electrons.

For the majority of molten metals, they have obtained good agreement (within the limits of the experimental error) between the empirical and calculated values. The same has been reported by other investigators [238-242], who have concluded that the electrical properties of liquid metals can be described by the theory of free electrons.

Tellurium does not obey this theory. The discrepancy between the experimental and calculated values of the Hall coefficient for tellurium is far outside the limits of the possible experimental error and can be explained only by assuming that some electrons in molten tellurium are used in the formation of covalent bonds and therefore they do not take part in conduction.

These conclusions are also confirmed by investigations of the optical properties of liquid tellurium, carried out by Hodgson [243], who has shown that the number of effective electrons per atom is only ~ 0.17 at $500°C$ (and not ~ 2, which would be expected if all valence electrons were free).

We shall now consider the structural changes which take place in tellurium at the melting point.

Investigations of the structure of molten tellurium, reported in [244, 245], have shown that the coordination number is 2 at $465°C$ and the interatomic spacing at this temperature is 2.9 Å (in the solid state, $K_S = 2$, $a_S = 2.86$ Å). Consequently, the main features of the short-range order of solid tellurium, including the chainlike structure, are retained in the molten state, although the distance between atoms in a chain increases.

These structure investigations are in good agreement with the nature of the change in the density of tellurium at the melting point.

Figure 27 shows the temperature dependence of the density of solid and liquid tellurium, according to the results obtained by one of the present authors. The same figure includes the results of Bonnier, Hicter, and Aléonard [201] on the density of solid tellurium, as well as the data on the density of molten tellurium, obtained by Klemm et al. [246] and Regel' [16]. The same figure includes the value of the specific volume discontinuity obtained by Ball [247].

These results are in good agreement with those obtained by one of the present authors. The only difference is in the nature of the temperature dependence of the density of molten tellurium, which we shall consider later.

It is evident from Fig. 27 that the melting of tellurium is accompanied by a sudden decrease in the density by about 5.9% (cf. 15.2% for Se). The fall in the density confirms the results of x-ray diffraction investigations and indicates some loosening of the structure, which is evidently due to an increase not only in the interatomic distances in chains but also in the distances between chains, because the binding between chains is destroyed.

Thus, the results of investigations of various physical properties, combined with the results of x-ray structure determinations, show that the melting of tellurium destroys completely the van der Waals bonds between chains and destroys partly the covalent bonds between atoms in chains; this shortens the chains and increases the carrier density. We shall now consider the changes which take place in the structure and chemical binding when molten selenium and tellurium are heated.

4. Changes in Short-Range Order and Chemical Binding during the Heating of Molten Selenium and Tellurium

Structural changes which take place in melts can be deduced from temperature dependences of structure-sensitive properties, particularly from the temperature dependence of the viscosity.

Figure 30 shows the temperature dependences of the kinematic viscosity of selenium and tellurium, which are based on the data obtained by the present authors [232] and other workers; it also includes the temperature dependences of the dynamic viscosity calculated from the kinematic viscosity and density.

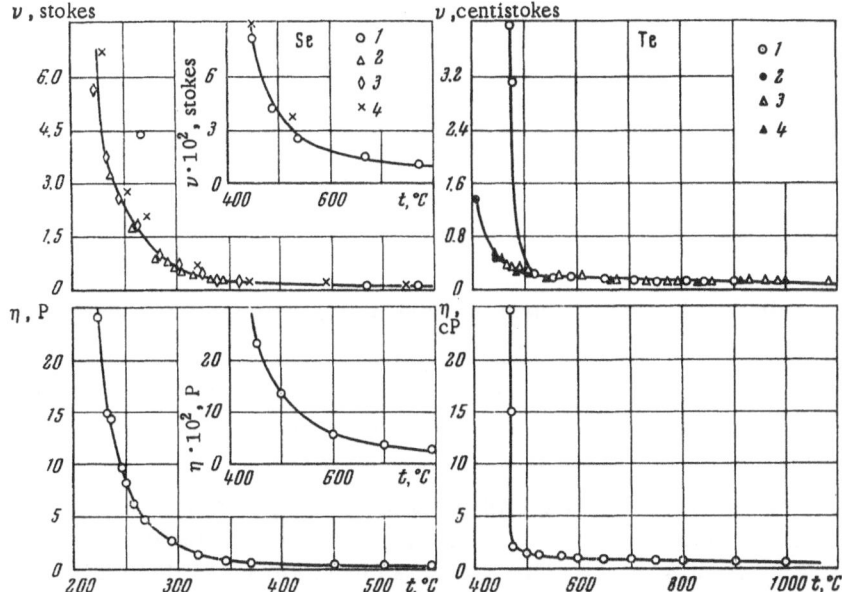

Fig. 30. Temperature dependences of the kinematic and dynamic viscosity of selenium and tellurium. 1) Results of the present authors; 2) Regel' [15]; 3) Dobinski and
Wesolowski [250]; 4) Khalilov [249].

It is evident from the curves in Fig. 30 that the viscosity of
tellurium and selenium at the melting point is very high: it is one
to three orders of magnitude higher than the viscosity of nonasso-
ciated liquids, for example, molten metals. Such high values of the
viscosity of molten selenium and tellurium are in full agreement
with the nature of the structure of these elements in the molten
state near the melting point.

The heating of molten selenium and tellurium above the melting point strongly reduces ·the viscosity. The rate of fall of the
viscosity is such that an increase in the temperature by 60-70°
above the melting point reduces the viscosity of selenium by an
order of magnitude. At much higher temperatures, the temperature dependences of the viscosity of molten selenium and tellurium
become much flatter and the viscosity of tellurium decreases to
values typical of liquid metals. However, the viscosity of selenium
is 2-3 times higher than that of molten metals even at several
hundreds of degrees above the melting point.

Table 14. Values of E_b and S_b for
Molten Selenium and Tellurium

Element	$E_b \cdot 10^{-3}$, cal/mole	$-S_b$, cal $\cdot$ mole^{-1} deg^{-1}	Δt, °C
Se	3.94	6.7	300—320
Te	1.18	6.7	200—220

We shall analyze these results using the theory of activated complexes.

We find that the dependences $F_b \propto f$ (t) have minima and at high temperatures [above a certain temperature $(t_{mp} + \Delta t)$] and they satisfy Eq. (67).

Consequently, as shown in the preceding section, we can determine the activation entropy of viscous flow of these elements at high temperatures from the tangents of the slopes of the curves $F_b \propto f$ (t). The results of such calculations are presented in Table 14.

Above a certain temperature (Δt), Eq. (74) is valid and the logarithm of the kinematic viscosity should be a linear function of the reciprocal of the absolute temperature and the tangent of the slope of this dependence should give the activation energy of viscous flow (E_b). The results of plotting $\log \nu \propto f(1/T)$ show that the linear dependence is indeed observed at high temperatures.

A considerable departure from this dependence is observed at lower temperatures. The values of the activation energies of viscous flow of selenium and tellurium at high temperatures are also listed in Table 14.

Analysis of the thermodynamic parameters of viscous flow of molten selenium and tellurium, listed in Table 14, shows that the activation energy of both elements is relatively low at high temperatures. This indicates weak binding between the structure units in these liquids. At lower temperatures, the values of E_b, found from the tangent of the slope of the $\log \nu \propto f(1/T)$ curve, are much greater, and the closer the temperature is to the melting point the higher the value of the activation energy.

At temperatures near the melting point, the activation energy is about an order of magnitude larger than at high temperatures. The decrease of the activation energy above the melting point indicates that the potential barriers, which have to be overcome by structure units in the liquid during viscous flow, become much lower when temperature is increased. This cannot be due solely to the weakening of the interatomic interaction because of an increase in temperature. The rapid changes in the thermodynamic parameters of viscous flow above the melting point indicate considerable structural changes. Changes in the nature of the intermolecular interaction in molten selenium and tellurium are indicated also by departures from the linear dependence of the reciprocal of the kinematic viscosity on the density (Bachinskii's equation [62, 63]) in this range of temperatures.

According to Stranski et al. [251], the structural changes taking place during the heating of molten selenium and tellurium are greater than those at the melting point and the heat evolved exceeds the heat of fusion.

Judging by the temperature dependences of the viscosity and of the thermodynamic parameters of viscous flow, these changes involve a reduction in the size of the structure units in the liquid and a weakening of the bonds between them.

The fall of the absolute value of the viscosity of liquid tellurium to values typical of liquid metals suggests that, in the limit, the structure of this element tends to become monatomic. The range of temperatures in which this process is completed is about 200-220° above the melting point.

The high absolute value of the viscosity of molten selenium, even at temperatures close to the boiling point, indicates that the final state of this element is not a monatomic liquid with a metal-like type of binding. At the same time, it is perfectly clear that considerable changes in the structure of liquid selenium do take place at 300-320° above the melting point. In this connection, it is interesting to consider the results of Urazovskii and Lyuft [252]. They subjected liquid selenium to quenching from various temperatures, and obtained glassy selenium, which has crystallized in the hexagonal modification under the action of special chemical reagents. They found that the ability to undergo such

crystallization decreases rapidly and the specific gravity of the
samples increases strongly when selenium is quenched from temperatures above 450°C. It is obvious that above 450°C the short-
range order of molten selenium differs considerably from the
solid-phase structure and this makes it difficult to crystallize the
quenched liquid. The tendency of liquid selenium to form rings,
consisting of several atoms (Se_6, Se_8) [224-225] suggests that
structural changes in selenium tend to produce a molecular liquid.
This would account for the difference between the temperature dependences of the density of liquid selenium and tellurium (Fig. 27).

Unfortunately, because of the high vapor pressure of selenium, these results on the density extend up to only 430°C. It is
evident from Fig. 27 that the density of molten selenium decreases
linearly when the temperature is increased. The density of tellurium increases slightly with temperature above the melting point,
passes through a maximum, and then begins to decrease smoothly.
These different temperature dependences of the density of molten
selenium and tellurium are due to the differences in their structure. Near the melting point liquid selenium consists of long
chains and rings. Tellurium chains are partly destroyed by melting and the melt consists of chain fragments and ions. Therefore,
the heating of molten tellurium somewhat increases the density by
producing a closer packing of short chains and ions although the
interatomic spacings may increase at the same time. In the case
of selenium, an increase in the interatomic spacings is the main
factor influencing the density and the long fairly immobile chains,
which are still retained in the molten state, cannot be packed more
closely.

The absolute values of the electrical properties of these two
elements, as well as their temperature dependences, also differ
considerably.

Molten ordinary selenium retains its p-type conduction at
temperatures well above the melting point and the low absolute
values of the conductivity are evidently due to a low carrier density.

All this confirms the conclusion that a molecular liquid is
the final state of selenium at high temperatures.

In contrast to that of selenium, the electrical conductivity of
molten tellurium has higher absolute values, which reach

$(2.7-2.8) \cdot 10^3 \ \Omega^{-1} \cdot cm^{-1}$ at 550-560°C; beyond this temperature
the conductivity is practically unaffected by further heating. At
approximately the same temperature (560-580°C), the thermoelectric power and Hall coefficient change from positive to negative,
indicating mixed conduction in liquid tellurium.

The processes which take place during the heating of molten
tellurium can be described as follows [212]. Near the melting
point liquid tellurium consists of chains of various lengths (atoms
in these chains are bound by covalent bonds) and the products of
the dissociation of these chains (these are ions, free electrons,
and holes, which are formed at points where covalent bonds are
broken and which migrate along the chains). At temperatures immediately above the melting point, the chain structure is dominant,
so that p-type conduction predominates. We must mention that the
mobility of ions is very low and their contribution to the electrical
conductivity can be ignored [221]. When the temperature is increased, the concentration of ions and electrons increases, because of the increasing degree of dissociation of the chains, and
the number of covalent bonds decreases. Therefore, at some temperature, the conduction in molten tellurium becomes n-type.

Physicochemical Properties of Molten Compounds with the Zinc-Blende Structure

In this chapter, we shall consider the results of investigations of compounds with the zinc-blende structure, whose properties in the molten state are hardly known; these compounds are:

Type	Compound
$A^{III}B^V$	AlSb, GaSb, InSb, GaAs, InAs
$A^{II}B^{VI}$	ZnTe, CdTe
$A^{I}B^{VII}$	CuI
$A_2^{III}B_3^{VI}$	Ga_2Te_3, In_2Te_3

Some of these compounds have not been investigated before because of the high vapor pressure of the components in the molten state.

§1. Crystal Structure, Chemical Binding, and Other Properties of ZnS-Type Compounds in the Solid State

1. General Nature of the Chemical Interaction in Systems with Compounds Crystallizing in the ZnS-Type Structure

The phase equilibria in $A^{III}-B^V$ systems are described by diagrams with one congruent-melting compound, which is formed

when the ratio of the components is $1:1$ (at.%) and which divides each diagram into two simple eutectic regions [203]. We have used the differential thermal analysis [22] to show that the maxima of the liquidus curves of the A^{III}–Sb systems are narrow, while the maxima in the phase diagrams of the A^{III}–As systems are broad [203]. These observations indicate that antimonides of elements in the third group melt without decomposition, while arsenides of the same group are thermally unstable.

At the melting points of antimonides of Group III elements the vapor pressure of the components is low, but the vapor pressures of gallium and indium arsenides are 0.9-1.0 and 0.25-3 atm, respectively [253-255].

The melting points of the $A^{III}B^V$ compounds are higher than the melting points of their components, with the exception of indium antimonide, whose melting point lies between the melting points of indium and antimony.

The problem of the existence of solid solutions based on the $A^{III}B^V$ compounds has been discussed on many occasions [256-259]. The usual investigation methods, including the precision x-ray structure analysis, carried out by Ozolin'sh et al. [258] with an accuracy up to 0.0001 Å, have indicated no changes in the lattice period in the presence of an excess of one of the components. Semiletov [259] has used x-ray diffraction data to show that if the $A^{III}B^V$ compounds have solid-solution regions, these regions are not wider than 0.01-0.02 at.%. Therefore, Semiletov concludes that departures from the stoichiometric composition are possible only in the presence of Frenkel' or Schottky defects.

Systems formed by elements of Groups II and VI have not been investigated much because of the high vapor pressures of the components [203]. Only the phase diagram of the Cd–Te system has been investigated in full [260, 261]. However, the published data indicate that only one compound is formed in each of the A^{II}–B^{VI} systems and the eutectics of the components are degenerate [203, 260, 261].

Among the tellurides of the zinc subgroup, zinc and cadmium tellurides are thermally stable. This is supported by the following observation. De Nobel [260] and Lorenz [261] have established that the maximum of the Cd–Te liquidus curve is narrow. Analysis of

the products of sublimation of zinc and cadmium tellurides, carried out at various temperatures up to 650-700°C, has shown that their compositions are stoichiometric [262]. The solubility of the components in chalcogenides of the zinc subgroup is less than 0.01-0.02 at.% [259].

The Cu−I phase diagram has not yet been investigated; all that we know is that copper iodide has extensive homogeneity regions on the copper and iodine side and that a defect solid solution (similar to that in the case of CuO) is formed in the presence of excess iodine. The mechanism of the formation of this solid solution is as follows [263-266]: a copper ion and an electron migrate to the surface of the sample and they combine with adsorbed iodine atoms. This gives rise to a copper vacancy and a hole in place of an electron. The concentration of excess iodine at 132°C and 100 mm Hg is $5 \cdot 10^{20}$ cm^{-3} [265].

Iodine is lost by evaporation from cuprous iodide films when they are heated in vacuum. Kurdyumova and Semiletov [267] have used electron diffraction results to show that the excess copper atoms are then located in octahedral interstices, forming an interstitial solid solution. The solubility of excess copper in cuprous iodide reaches several percent.

The phase diagrams of the A^{III}−Te systems are complex. In addition to the compound $A_2^{III}Te_3$, which melts congruently, one further congruent-melting compound $A^{III}Te$ and several other compounds which do not melt congruently are formed in these systems [203, 268-270]. According to recent data [268-269], a maximum in the liquidus curve is displaced from the ordinate corresponding to $A_2^{III}Te_3$ in the direction of the Group III element, and it corresponds to the composition $A_{27}Te_{40}$. The solubility of the components in In_2Te_3 is, according to the approximate results of Holmes, Jennings, and Parrott [271], 0.3-0.4 at.% at low temperatures and 1.6-1.8 at.% at high temperatures. There is no published information on the homogeneity (solid-solution) region of gallium telluride.

This discussion shows that the phase diagrams of the A^{III}−B^V and A^{II}−B^{VI} systems are identical, but that they differ strongly from the diagrams of the A^{III}−B^{VI} systems.

Within each group of systems, the phase diagrams vary in a regular manner with the positions of the components in the periodic

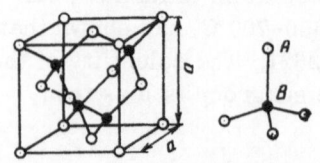

Fig. 31. Zinc-blende crystal structure.

table: when the atomic numbers of the components increase, the melting point of the compound decreases and, in the case of the $A^{III} - B^V$ systems, the eutectic composition on the "B" side is displaced toward the ordinate of the compound.

2. Structure and Chemical Binding of Compounds with the ZnS-Type Structure

1. Structure. The zinc-blende structure is similar to that of diamond, but its symmetry is lower, because it consists of atoms of two kinds (Fig. 31).

The unit cell is a face-centered cube, whose vertices and centers of faces are occupied by A (B) atoms, and four out of the eight tetrahedral interstices are occupied by B (A) atoms. Thus, the unit cell consists of four A atoms and four B atoms, i.e., it contains four formula units. Each A atom is surrounded tetrahedrally by B atoms and, conversely, each B atom is surrounded by four A atoms. Thus, the coordination number is 4 both for A atoms and for B atoms [99].

Some $A^{III}B^V$ compounds crystallize from the gaseous phase in the sphalerite as well as in the wurtzite form. This has been observed by Semiletov et al. for indium and gallium antimonides and for indium arsenide [272-273]; other investigators have observed the same effect for zinc and cadmium tellurides [274-279].

Mercury selenide and telluride have only one modification [280]. Hexagonal zinc telluride films are observed only in the case of departure from the stoichiometric composition in the direction of excess zinc [279]. It is reported in [273] that gallium antimonide and indium arsenide films always contain an excess of one of the components. This excess may be responsible for the appearance of the wurtzite-type crystals. The hexagonal modification of this class of compounds is metastable, and it transforms spontaneously (though slowly) into the cubic modification [281].

Cuprous iodide has three modifications (α, β, γ). The phase transition temperatures [282-285] are:

$$\gamma \to \beta \quad \text{at} \quad 369° \text{ C,}$$
$$\beta \to \alpha \quad \text{at} \quad 407° \text{ C.}$$

The low-temperature modification γ has the zinc-blende structure while the β modification has the wurtzite structure [282-286]. Japanese investigators have shown [282] that the x-ray diffraction patterns of the high-temperature α-phase are very similar to the patterns obtained for the γ-phase both in respect of the line distribution and their intensities. This suggests that the crystal structures of the α- and γ-phases are similar. However, an anomalous difference between the entropies of these phases as well as the nature of the temperature dependence of the intensity of the Debye lines indicate some structural difference between these two modifications. The Japanese workers [282] are of the opinion that the high-temperature modification of cuprous iodide has a special structure, similar to the zinc-blende structure but not exactly the same. Krug and Sieg [286] have compared the interference pattern obtained for α-CuI with several theoretical variants. The experimental and calculated results agree best on the assumption of a defect ZnS-type structure in which some copper

Table 15. Crystal Structure Parameters and Densities of $A^{III}B^{V}$, $A^{II}B^{VI}$, $A^{I}B^{VII}$, $A_2^{III}B_3^{VI}$ Compounds (ZnS-type structure, K = 4)

Compound	Lattice period, Å	d, g/cm³	Reference
AlSb	6.1355±0.0001	4.24	[294,297]
GaSb	6.0954±0.0001	5.65	[294,109]
InSb	6.47877±0.00005	5.71	[294,297]
GaAs	5.6534±0.0002	5.24	[294,297]
InAs	6.0584±0.0001	5.58	[294,297]
ZnTe	6.087	5.54	[109,99,298]
CdTe	6.46	5.8	[109,203,299]
CuI	6.16 (α-phase) 6.052 (γ-phase)	6.63 (γ-phase)	[282,109]
Ga₂Te₃*	5.886	5.57	[99,295,296,109]
In₂Te₃*	6.158	5.73 (β-phase) 5.79 (α-phase)	[99,295,296,290]

*K = 4 and 2.66.

ions are in octahedral interstices of the iodine sublattice; this may have been due to the nonstoichiometric nature of the investigated samples [267].

Some chalcogenides of elements in the third group, in particular gallium and indium tellurides, also have the zinc-blende structure. The structure of these compounds differs from the ZnS structure by the presence of 1.34 vacancies at "cation" sublattice sites in each unit cell. At high temperatures, these vacancies are distributed at random. At low temperatures, the distribution is ordered and the x-ray diffraction pattern shows superstructure lines [287, 290]. Possible variants of the ordering of vacancies in the "cation" sublattices of the $A_2^{III}Te_3$ compounds are considered in [288, 291]. Harbeke and Lautz [292, 293] have observed a phase transition in gallium telluride under the influence of small amounts of copper (10^{-5} at.%). The nature of this transition has not been determined. The main crystal structure data for the compounds considered in the present chapter are listed in Table 15.

2. Chemical Binding. The discovery (about 15 years ago) of semiconducting properties in compounds with the ZnS-type structure has been followed by very extensive investigations of their properties and applications but the nature of chemical binding in these compounds is still not clear. Hilsum and Rose-Innes [300], Goryunova [109], Ugai [110], Suchet [301] in their books, as well as other workers in their papers, have all offered different explanations. The vigorous discussions between Mooser and Pearson on the one hand and Goodman on the other [302-305], as well as between Syrkin and Batsanov [306-308] have not resolved this problem.

Let us consider first the available experimental data. Back in the 1920's, Gol'dshmidt [309] drew attention to the similarity of the lattice periods and the densities of compounds with the ZnS-type structure formed from elements of one period and the corresponding elements in Group IV. This similarity was attributed by Gol'dshmidt to a basically nonionic interaction between the components (in the ionic interaction case, the interatomic spacings should decrease as the difference between the valences of the atoms increased because the Coulomb attraction would get stronger).

The discovery of semiconducting properties in these compounds (they have a low carrier density, high carrier mobility, and the carrier density increases with temperature, etc.), combined with investigations of the crystallochemical properties, also support the covalent nature of the binding between atoms in these materials. A direct confirmation of the covalent binding has been provided by x-ray diffraction investigations of the electron density, carried out by Sirota et al. [310-315]. The electron "bridges," found between the nearest A and B atoms in the $A^{III}B^V$ compounds, can be regarded as regions of electron-cloud overlap, similar to those observed in germanium and silicon. However, because the atoms forming these compounds are chemically different, the binding between them cannot be purely covalent. Such quantities as the ionization potential and the electronegativity,† which characterize the binding energy of the valence electrons with the core, are larger for B atoms than for A atoms. Therefore, the electron clouds which bind the atoms should be asymmetrical. Many observations confirm the asymmetry of the electron clouds, which results in the appearance of effective charges at the component atoms. It has been found that the main cleavage plane of ZnS-type compounds is a {110} plane [321-323], while in crystals with the diamond structure the plane of the highest reticular density is a pair of {111} planes. This can be explained only by assuming that the {111} planes in the zinc-blende structure are packed either with A atoms or with B atoms, i.e., with "ions" of opposite signs which are mutually attracted by electric forces and which make it difficult to split these crystals along the {111} planes. Conversely, the Coulomb repulsion of neighboring ions with like charges aids the cleavage along the {110} planes.

---

† The concept of electronegativity, which provides a measure of the ability of atoms in a molecule to attract an electron, was introduced by Pauling [316]. This concept has been criticized by Syrkin [306], Goryunova [109], and Goodman [302, 317]. The critics point out that the value of the electronegativity is not a constant of the element but depends on the degree of the element's ionization in the formation of a bond, on the particular orbital which participates in binding, on the presence of orbital hybridization, etc. The values of the electronegativity obtained by Pauling [316], Gordy and Thomas [318], Allred and Hensley [319], Batsanov [320], and Suchet [301] are different. Bearing in mind that the electronegativity is not a rigorous chemical concept, we shall nevertheless use it as a rough qualitative characteristic which can be employed to predict changes in the "ionic component" of binding in series of analog compounds.

Table 16. Effective Charges of Atoms
in $A^{III}B^V$ Compounds

Compound	Reference			
	[324]	[325]	[326]	[331]
AlSb	—	0.48	0.53	0.67
GaSb	—	0.30	0.33	0.58
InSb	0.34	0.34	0.42	0.38
GaAs	—	0.43	0.51	0.80
InAs	—	0.56	0.56	0.49

Information on the effective charges of atoms in $A^{III}B^V$ crystals has been obtained from the infrared lattice-vibration absorption spectra. The relevant theory is given in detail in Hilsum and Rose-Innes's book [300]. The effective charges of atoms are listed in Table 16 [324-326]. However, the signs of these charges have not been determined. In this connection, investigations of the x-ray absorption spectra, the theory of which has been developed by Barinskii and Nadzhakov [327-328], are of great interest. Ugai and Tomashevskaya [329] have carried out an x-ray spectroscopic analysis of the $A^{III}Sb$ compounds. The shift of the L_β band maximum of antimony in compounds in the direction of longer wavelengths, compared with pure antimony, is attributed by Ugai and Tomashevskaya to the contraction of the electron clouds around the antimony lattice sites. Published almost simultaneously with the work of Ugai and Tomashevskaya, the papers of Sirota [330, 331] give the effective charges of atoms for nine compounds of the $A^{III}B^V$ type; the changes calculated from electron density maps obtained by x-ray structure analysis (Table 16).

Ugai and Sirota have shown convincingly that A atoms in $A^{III}B^V$ compounds are positively charged, while B are negatively charged. Their investigations should put an end to the controversy as to which of the components in compounds with the ZnS-type structure, particularly in the $A^{III}B^V$ compounds, is positively charged.

To understand the cause of this controversy, we shall consider first theoretical aspects of the formation of bonds in the ZnS-type structure, specifically in the $A^{III}B^V$ compounds.

Each A^{III} atom is valence-bound to four B^V atoms, and conversely. This means that A^{III} and B^V atoms should be quadrivalent, in accordance with their coordination numbers. In its normal state, an atom of Group III has three electrons in its outer shell (configuration s^2p^1) while an atom of Group V has five electrons (configuration s^2p^3). Thus, an A^{III} atom has one unpaired electron and a B^V atom has three unpaired electrons. According to Mooser and Pearson [332], bonds are formed when one electron is transferred from a B^V atom to a vacant p state in an A^{III} atom.

This results in an overlap of the s and p states and formation of mixed sp^3 hybrid orbitals as well as a redistribution of electrons between these orbitals so that all the electrons become unpaired. Consequently, both atoms, A^{III} and B^V, become quadrivalent and can form four covalent bonds. All these bonds, irrespective of the origin of electrons forming them, are equivalent.

We have considered only two atoms, but our conclusions apply also when we consider all the atoms in the lattice. According to Mooser and Pearson, the formation of covalent bonds in the $A^{II}B^{VI}$ and $A^{I}B^{VII}$ compounds proceeds similarly except that the number of electrons transferred from a B atom to an A atom is now different.†

Mooser and Pearson do not assume that the complex process of the chemical bond formation in these semiconductors can be reduced simply to the scheme just described, but they allow also for the ionic component of binding [303, 305, 334]. However, they assume that the polarization of an electron cloud in a valence bond

---

† Zhuze, Sergeeva, and Shelykh [333] have used Mooser and Pearson's theory to suggest a scheme for the formation of the covalent bonds in In_2Te_3. In the crystal lattice, each B atom is surrounded, on the average, by 2.66 A atoms, but each A atom is surrounded by four B atoms. When In_2Te_3 is formed, two tellurium atoms contribute one electron each to two indium atoms so that each indium atom now has four electrons, which can be in the sp^3 hybrid state and which can form bonds with unpaired tellurium electrons. All the tellurium electrons, both paired and unpaired, are also in the sp^3 hybrid state. Since the indium sublattice has three vacancies, tellurium atoms surrounding a vacancy and having four free electron pairs form bonds with this vacancy so that each atom in In_2Te_3 has four directed (tetragonal) bonds. In such a crystal, all atoms should be equivalent and there should be resonance between the Te^I and Te^{II} states.

bridge is not sufficiently strong to alter the sign of the charge (i.e., to make "anions" positive and "cations" negative). Mooser and Pearson's theory of the formation of tetrahedral bonds by the transfer of an electron from a B atom to an A atom predicts that B atoms become positive and A atoms negative, which is not in agreement with experiment. Recently, a different scheme has been proposed for the formation of tetrahedral covalent bonds, which does not rely on the questionable transfer of an electron from a B atom to an A atom [110, 306, etc.]: it is suggested that of the four covalent bonds formed in $A^{III}B^V$ crystals, three are formed by unpaired valence electrons of A and B atoms and the fourth is formed by an s-electron pair of a B atom.

Such binding is known as the donor–acceptor type. The formation of donor–acceptor bonds is favored by energy considerations because of the transfer of a pair of electrons from a "donor" to a level which is common to "donors" and "acceptors" [335]. The donor–acceptor binding differs from the ordinary covalent binding simply by the origin of the fourth electron pair. Again, the sp^3 hybridization should be observed and the four electron pairs should tend, irrespective of their origin, to orient themselves along angles in a regular tetrahedron. The formation of bonds in the lattices of the $A^{II}B^{VI}$ and $A^I B^{VII}$ compounds should be similar. The only difference should be in the number of bonds formed by unpaired electrons of A and B atoms and by s and p electron pairs of B atoms.

Thus, we may conclude that the chemical binding in the zinc-blende type structure can be regarded as intermediate between the covalent binding, when valence electrons belong simultaneously to two bound atoms, and the ionic binding, in which electrons are bound to one of the atoms. In real compounds, these bonds are electron bridges between atoms, in which the electron-density maximum is shifted in the direction of the element with the higher electronegativity. The magnitude of this shift or the ionic component of the binding, is different for different compounds.

We shall now consider how the nature of the chemical binding and the properties of a compound depend on the positions of the components in the periodic table.

We note two relationships governing the chemical binding. One is observed in series of analog compounds, i.e., compounds whose components belong to the same group. Examples of such series are $A^{III}B^V$, $A^{II}B^{VI}$, $A^{I}B^{VII}$, and $A_2^{III}B_3^{VI}$ compounds (cf. Table 15).

Another relationship is observed in the isoelectronic series, i.e., in series of compounds whose components have the same total charge and, consequently, the same total number of electrons. The isoelectronic series may be horizontal, if the components belong to the same period, or diagonal if the components belong to different periods.

The compounds listed in Table 15 form the following series:

$$Ge-GaAs;$$
$$InSb - (In_2Te_3) - CdTe;$$
$$GaSb - (Ga_2Te_3) - ZnTe - CuI.$$

We shall consider first the relationships governing the chemical binding in series of analog compounds, considering, by way of example, the $A^{III}B^V$ compounds whose binding has been investigated most thoroughly.

When the atomic weight of the components in a series of analog compounds is increased, the binding of the valence electrons with the nucleus weakens because of an increase in the screening of the nucleus by a large number of electron shells (the metallic component of the binding increases). The ionic component also varies along the series. The dependence of the ionic component on the positions of the atoms in the periodic table can be understood by considering the effective charge in the $A^{III}B^V$ compounds. The effective charge deduced from electron-density maps [331] is given in Table 17. It is evident from this table that the effective charge decreases when atoms in a compound are replaced by heavier elements. If we arrange these compounds in decreasing order of the ionic component of the binding, we obtain

$$GaAs - AlSb - GaSb - InAs - InSb.$$

The order obtained from the infrared absorption data is somewhat different:

InAs — AlSb —GaAs — InSb — GaSb.

An approximate correlation between the Knoop microhardness and the forbidden band width has been used in [336] to arrange these compounds in the decreasing degree of ionicity, as follows:

AlSb — GaAs — GaSb — InAs — InSb.

Microcleavage data [321-323] indicate that gallium and indium antimonide and indium arsenide exhibit, in addition to cleavage along the $\{110\}$ planes, appreciable cleavage along the $\{111\}$ planes, whereas in aluminum antimonide and gallium arsenide cleavage along the $\{111\}$ planes is much weaker. This means that the last two compounds exhibit a stronger ionicity than the first three substances, which is in agreement with the results obtained from electron density maps.

A regular variation of the chemical binding along a series of analog compounds is responsible for a regular variation of the physical properties. Some of these properties are listed in Table 17. Weakening of the binding of the valence electrons with the nuclei causes a decrease in the activation energy or in the forbidden band width. However, the forbidden band width is affected not only by the metallic but also by the ionic component of the binding.

Ormont [337, 338] has deduced a rule, according to which the largest forbidden band in a series of analogs with the same total atomic number is exhibited by that analog which has the smallest "cation" and the largest "anion," i.e., by the analog with a maximum polarization of the electron shell of the "anion." For example, in the AlSb – GaAs – InP series, AlSb has the largest forbidden band.

An increase in the atomic weight of the components reduces the amplitude of the thermal vibrations of the atoms and this reduces the scattering of carriers by these vibrations. Since a reduction in the ionicity also reduces the intensity of the carrier scattering, the carrier mobility increases rapidly (Table 17).

If we consider a vertical series of analogs, we find that such quantities as the melting point and heat of fusion and the standard enthalpy of formation depend in a regular manner on the positions of the components of these compounds in the periodic system. All

these quantities decrease when the metallic component of the binding increases, since the metallization of bonds reduces the change
in the energy of a system at the melting point (the melting point
and the heat of fusion both decrease) and consequently the heat of
formation also decreases. In series of analog compounds (and only
in such series) these quantities represent qualitatively the strength
of the binding when its nature changes.

Ormont [339, 340] has suggested that the atomization energy†
can be used as a quantitative characteristic of the strength of binding in compounds with mainly nonionic bonds. We can see from
Table 17 that the atomization energy as well as the Debye temperature, which also represents the strength of bonds in the lattice, decrease when the "cation" or the "anion" are replaced with heavier
elements.

We shall now consider the isoelectronic series. There have
been no direct investigations of the nature of changes in the chemical binding in these series as a function of the positions of the components in the periodic table. However, indirect investigations show
convincingly that the ionicity of the binding increases when the components move away from Group IV in the periodic table. By way of
example, we shall consider the isoelectronic series $GaSb - ZnTe -$
CuI. Gallium antimonide is externally very similar to germanium:
both are dark, opaque, and have a metallic luster. Zinc telluride
is red and semitransparent while copper iodide forms white crystals; thus, the latter two compounds resemble salts. Matyas [341]
has investigated the magnetic susceptibility of compounds with the
sphalerite structure. Analyzing the results obtained, Matyas has
concluded that the increase in the magnetic susceptibility along
isoelectronic series of $A^{III}B^{V}$ and $A^{I}B^{VII}$ compounds is associ-

---

† The atomization energy is the energy required for the formation of a monatomic pair
in the standard state, i.e., at 298.16°K, from a crystal lattice which is also in the
standard state. The atomization energy (Ω) can be calculated from

$$\Omega = \Delta H_{298} + S_m + \frac{1}{2} D_x,$$

where ΔH_{298} is the heat of formation of a compound in the standard state, S_m is the
heat of sublimation of the metal, and D_x is the heat of dissociation of the nonmetal
in the gaseous state.

Table 17. Physicochemical Properties

Property	AlSb	GaSb	InSb	GaAs	InAs
ΔE, eV (room temp.)	1.52	0.7	0.17	1.35	0.36
u_n, $cm^2 \cdot V^{-1} \cdot sec^{-1}$	200	4000	100 000	8500	33 000
u_p, $cm^2 \cdot V^{-1} \cdot sec^{-1}$	420	1400	1 000	420	460
t_{mp}, °C	1080	712	536	1238	942
Q_f, kcal/g-atom*	7.1	6.0	5.8	11.6	6.3
ΔH_{298}, kcal/g-atom*	11.5	4.97	3.67	11.25	6.9
Debye temperature, °K	350	270	228	~310	~250
Ω, kcal/mole	160	134	121	146	120

* Somewhat different results have been reported for Q_f of InSb in [378, 379, 383], for ΔH_{298} of InSb [373-377], and for ΔH_{298} of GaSb [376, 380, 381].

ated with an increase in the ionicity of the binding, which contributes to the total susceptibility. Wolff and Broder [323] have investigated the microcleavage and have also found the ionicity of $A^{II}B^{VI}$ compounds (ZnTe, CdTe) is greater than that of $A^{III}B^V$ compounds (GaAs, InSb). If we analyze carefully the results on the interatomic spacings, we can see that these spacings are not as constant as assumed by Gol'dshmidt. When we consider the third place after the decimal point, we find that the lattice period decreases when the components move away from Group IV, which is obviously due to an increase in the Coulomb interaction between "ions."

The variation of the properties in an isoelectronic series is more complex than in a chemical analog series.

Thus, for example, the melting points of the $A^{III}B^V$ compounds are lower than those of the corresponding isoelectronic analogs, the Group IV elements, while the melting points of the $A^{II}B^{VI}$ compounds are higher and those of the A^IB^{VII} compounds are lower. Such a complex dependence of the melting point on the positions of the components in the periodic table is evidently due to the fact that the nature of changes in the chemical binding at the melting point and in the heat of formation is different for different

of Compounds with ZnS-Type Structure

ZnTe	CdTe	CuI	Ga$_2$Te$_3$	In$_2$Te$_3$	References
2.15	1.51	2.8	1.16	1.02	[109—112, 300, 301, 359—362]
300	600	—	50	70	[109—112, 333, 364—366]
30—50	50	—	—	17	
1239	1092	602	790	667	[22, 109, 110, 203, 260, 261, 266, 300, 382]
—	—	1.3	—	—	[113, 368—371]
14.4	12.25	8.1	13.0	9.5	[115, 367, 372, 384]
250	200	—	—	—	[118, 385]
109	99	—	—	—	[337, 386]

substances in an isoelectronic series. Both these quantities depend on many factors and do not reflect changes in the strength of the binding in an isoelectronic series. The strength of the interatomic bonds in an isoelectronic series decreases from elements of Group IV to compounds of the $A^I B^{VII}$ type. This is indicated by the fall in the heat of atomization and of the Debye temperature (Table 17), as well as by the fall of other properties not listed in this table (microhardness, surface energy, microstrength, etc.) [109]. Among the electrical properties listed in Table 17, only the forbidden band width increases regularly in the isoelectronic series from the elements of Group IV to the $A^I B^{VII}$ compounds. Welker [342] explains this by an increase in the difference between the potential maxima and minima in the periodic field in a crystal (the forbidden band width is proportional to this difference).† The carrier mobility varies in different ways in the isoelectronic series. The mobility in the $A^{III} B^V$ compounds is higher than in elements which are their isoelectronic analogs. There is as yet no satisfactory explanation

---

† The potential variations in the periodic field of a covalent crystal are less than the potential variations in an ionic crystal, since in the latter the electron density between the atoms decreases almost to zero, while in a covalent crystal the presence of valence bridges reduces the difference between the potential maxima and minima.

of this behavior. The fall in the carrier mobility in the $A^{II}B^{VI}$ compounds is evidently due to an increase in the scattering of carriers on the optical lattice vibrations because of the stronger polarization of the bonds.†

It follows from our discussion that the physicochemical and electrical properties of compounds with the ZnS-type structure are related to the positions of the components of these compounds in the periodic system.

There have been many attempts to establish a correlation between the values of two or more properties of these compounds [121, 343-358]. For example, attempts have been made to relate the forbidden band width to the ionicity of the binding [344], the bond energies [353], the atomization energy [351, 355], and the thermodynamic potential [357, 358]. Ormont [121, 337] has proposed a formula according to which the forbidden band width can be found from quantities representing the strength of bonds, the nature of the binding, and the electronic structure of the components of the compounds. Attempts have been made to relate the electron mobility to the heat of formation and to the ionicity of the bonds [344, 349]. Suchet [301] has also proposed methods for estimating theoretically the forbidden band width and the carrier mobility.

However, all these attempts to relate the energy, crystallochemical, and other parameters with the electrical properties of ZnS-type semiconductors have not yet been successful, although some relationships allows us to predict, in the first approximation, the properties of substances not yet investigated [121, 301, 357, 358].

---

† In discussing the properties of the isoelectronic series, we have not considered the $A_2^{III}B_3^{VI}$ compounds. These compounds fit well an isoelectronic series joining the $A^{III}B^V$ and $A^{II}B^{VI}$ compounds both in respect of the positions of their components in the periodic table and some of their properties (for example, ΔE). However, other properties of these compounds do not fit the isoelectronic series sequence because their structure is of the defect type. Thus, the lattice periods of these compounds are much smaller (the difference can be seen even in the first decimal place) than the periods of the corresponding compounds in an isoelectronic series. Disturbance of the field periodicity because of a large number of intrinsic defects distorts the potential field and reduces strongly the carrier mobility.

Information on the properties of molten semiconducting compounds with the ZnS-type structure is of relatively recent origin. Beginning from 1951, Regel' and his colleagues have published a number of papers on the density and electrical conductivity of various semiconductors near the melting point and in the liquid state. Among the ZnS-type compounds, they have investigated indium and gallium antimonide, mercury selenide and telluride, as well as cadmium telluride [5-12]. These investigations have provided a stimulus for systematic studies of semiconducting compounds belonging to different structure groups, which we began in 1954 for the purpose of establishing general relationships governing changes in the properties at the melting point and during subsequent heating when the chemical composition of the compounds is varied in a regular manner in accordance with the positions of the components of these compounds in the periodic table.

§2. Changes in the Short-Range Order and the Chemical Binding at the Melting Points and during the Heating of Molten A^{III}B^{V}-Type Compounds

1. Changes in Short-Range Order and Chemical Binding at the Melting Points of A^{III}B^{V}-Type Compounds

We investigated samples of the A^{III}B^{V} compounds in which the carrier density had the following values (cm^{-3}):

Single crystals		Coarse-grained crystals	
AlSb	. . . . 10^{17}	GaAs	. . . . 10^{16}
GaSb	. . . . $2 \cdot 10^{17}$	InAs	. . . . 10^{16}
InSb	. . . . $5 \cdot 10^{14}$		

The methods used in the investigation of the electrical conductivity, magnetic susceptibility, and thermoelectric power have already been described. The electrical conductivity and magnetic susceptibility were measured using cylindrical samples (8 mm in diameter and 8 mm high, or 10 mm in diameter and 10 mm high). Samples of all these compounds, with the exception of aluminum antimonide, were placed in evacuated (10^{-3} mm Hg) and sealed

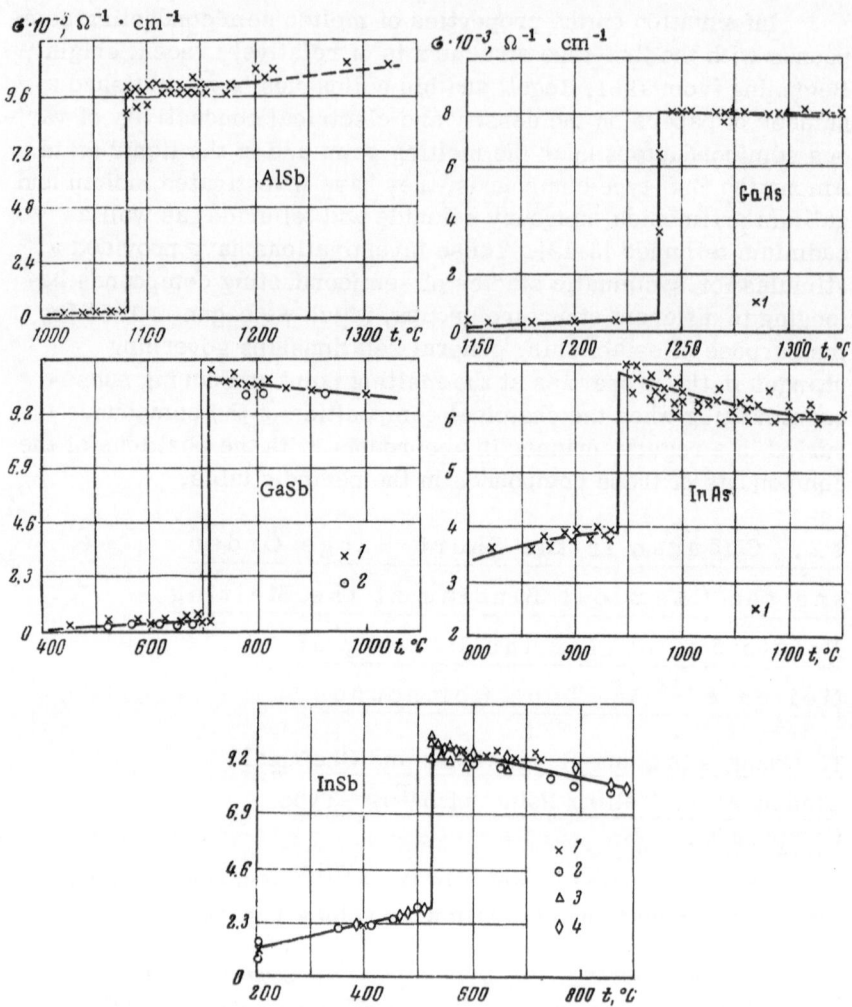

Fig. 32. Temperature dependences of the electrical conductivity of aluminum, gallium, and indium antimonides and of gallium and indium arsenides in the solid and liquid states. 1) Results of the present authors; 2) Regel' et al. [5, 6, 8, 12, 15]; 3) Hamilton and Seidensticker [126]; 4) Busch and Vogt [145].

quartz ampoules. Aluminum antimonide reacts chemically with quartz and therefore it was placed in corundum cans which were sealed into evacuated quartz ampoules. As already mentioned at the beginning of Chapter 4, gallium and indium arsenides have high arsenic vapor pressures at their melting points. Therefore, to

Table 18. Electrical Conductivity of A^{III}B^V
Compounds at Their Melting Points in Solid
and Liquid States

Compound	σ_s, $\Omega^{-1} \cdot cm^{-1}$	σ_l, $\Omega^{-1} \cdot cm^{-1}$	σ_s/σ_l
AlSb	~160	~ 9900	~60
GaSb	~280	~10600	~40
InSb	~2900	~10000	~3.5
GaAs	~300	~ 7900	~26
InAs	~3600	~ 6800	~1.9

maintain the stoichiometric relationship between the components
in these compounds in the molten state, a certain amount of arsenic was placed in each ampoule to ensure an equilibrium pressure
of its vapor. In measurements of the various properties, each
sample was heated to 15-20° above its melting point and held at
this temperature for 4-5 h before the measurements were begun;
this guaranteed that the sample melted completely. This long exposure was necessary because each of the investigated substances
had a high heat of fusion. Measurements below the melting point
were started after 15-20 min at a given temperature.

The results of measurements of the electrical conductivity
of the A^{III}B^V-type compounds are presented in Fig. 32 [387]. The
same figure includes the results of other authors, with which our
data agreed satisfactorily. Aluminum antimonide and gallium and
indium arsenide had not been investigated before.

It is evident from Fig. 32 that the electrical conductivity of
all these five compounds increases in the solid state when the temperature is increased and it rises suddenly at the melting point to
reach values of the order of 10^4 $\Omega^{-1} \cdot cm^{-1}$. In the liquid state the
electrical conductivity of all these compounds (with the exception
of aluminum antimonide) decreases slowly when the temperature
is increased. The electrical conductivity of aluminum antimonide
rises slightly above the melting point. Table 18 lists the values of
the electrical conductivity of these five A^{III}B^V compounds at their
melting points in the solid and liquid states.

The temperature dependences of the electrical conductivity
of these A^{III}B^V compounds are very similar, indicating that similar

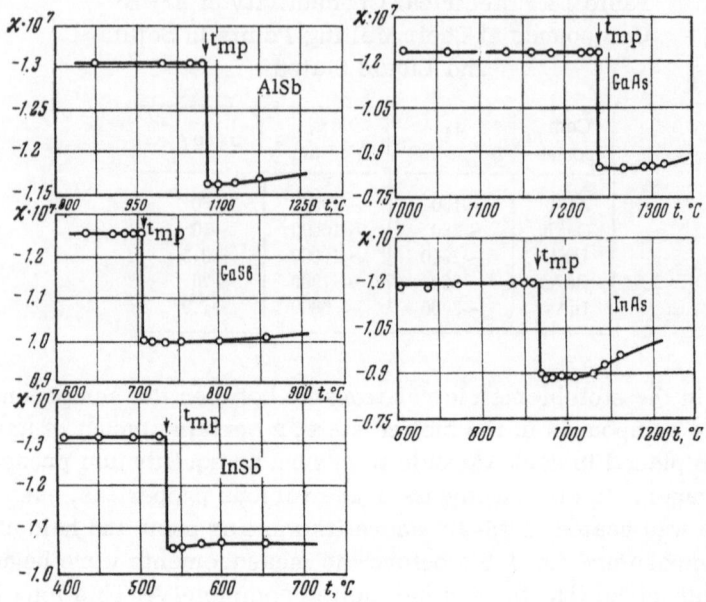

Fig. 33. Temperature dependences of the magnetic susceptibility of aluminum, gallium, and indium antimonides and of gallium and indium arsenides
in the solid and liquid states.

processes take place during the melting and further heating of the
melts. However, there are some quantitative differences in the
magnitude of the sudden rise of the electrical conductivity at the
melting point, which are related to the positions of the components
of these compounds in the periodic table.

The magnitude of the rise of the electrical conductivity at
the melting point decreases along a series of chemical analogs in
which the "cation" is replaced with a heavier element (AlSb –
GaSb – InSb). This is due to an increase in the electrical conductivity of the solid phase at the melting point. The observed change
in the rise of the electrical conductivity at the melting point is
evidently related to a decrease in the difference between the degrees of metallization of the chemical binding in the solid and liquid phases. This suggests that the electron density in the solid
phase before melting is important and it shows that the difference
between the short-range order in the solid and liquid phases is

least for InSb and that it increases gradually due to the "cation" substitution. In the two series of chemical analogs in which the "anion" is replaced (GaSb – GaAs, InSb – InAs), the situation is somewhat different: the rise of the electrical conductivity at the melting point decreases when the "anion" is replaced with a lighter element. This is obviously due to the considerable influence of the ionicity of the binding, which is considerably higher in arsenides than in antimonides of Group III elements (cf. Table 16). The absolute values of the electrical conductivity of the $A^{III}B^V$ compounds in the molten state as well as the negative temperature coefficients of the electrical conductivity indicate that these compounds become metallic in the molten state.

The results of the measurements of the magnetic susceptibility are presented in Fig. 33 [127]. It is evident from this figure that all the compounds are diamagnetic in the solid state and that their magnetic susceptibilities are practically unaffected by an increase in temperature in the solid state. At the melting point, the susceptibility falls suddenly and in the liquid state there is some tendency for the susceptibility to rise. The fall of the susceptibility at the melting point has been reported also by Busch et al. [145, 388].

Figure 34 gives the temperature dependences of the thermoelectric power for aluminum, gallium, and indium antimonides below and above the melting point. The graphs for gallium and indium antimonides include also the results of Blum and Ryabtsova [84]. The thermoelectric power of aluminum antimonide had not been measured before.

The temperature dependences of the thermoelectric power of gallium and indium antimonides are fairly complex. At high temperatures, the thermoelectric power of the solid begins to fall probably because of the appearance of intrinsic conduction; at the melting point the thermoelectric power decreases suddenly to low values of the order of 30-60 μV/deg and continues to decrease at still higher temperatures. The sign of the thermoelectric power is negative at high temperatures in the solid state and it remains negative after melting.

When the melts are heated, the thermoelectric power tends to become positive.

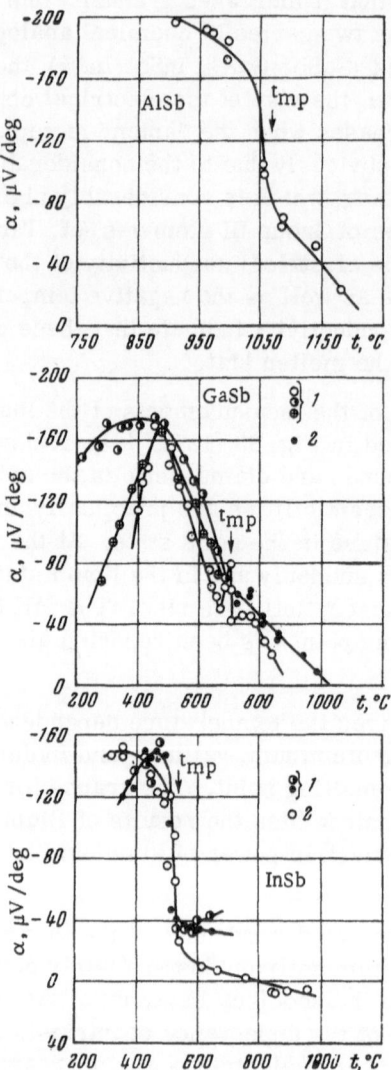

Fig. 34. Temperature dependences of the thermoelectric power of aluminum, gallium, and indium
antimonides in the solid and liquid states. 1) Blum
and Ryabtsova [84]; 2) results of the present authors.

Comparing the temperature dependences of the electrical
and magnetic properties of the $A^{III}B^V$ compounds with the corresponding dependences for germanium and silicon, we note that the
nature of changes in the properties is similar for all these substances. As mentioned in Chapter 3, a strong increase in the electrical conductivity at the melting point is, according to Regel', due
to a sudden increase in the carrier density. Unfortunately, the
carrier density in the molten state can be estimated only for indium antimonide, for which experimental Hall effect data are available [145]. Using the simple formulas σ = enu and R = 1/ne, we
find that the carrier density in indium antimonide increases at the
melting point by a factor of 400-450, reaching a value of the order
of 10^{21} cm^{-3}. However, the carrier mobility, calculated by Regel'
[144], falls by two orders of magnitude and therefore the electrical
conductivity increases only severalfold. The strong rise of the
carrier density is supported also by the sudden drop of the magnetic susceptibility, which is mainly due to an increase in the spin
paramagnetism of free electrons because of an increase in their
density. The strong fall in the thermoelectric power at the melting
points of the $A^{III}B^V$ semiconductors is also largely due to an increase in the carrier density.

Thus, all the electrical and magnetic properties indicate that
the melting of semiconducting $A^{III}B^V$ compounds is accompanied
by a very strong increase in the carrier density as a result of
which the electrical conductivity reaches values typical of molten
metals, such as mercury. The sudden increase in the carrier density can only be due to the destruction of a system of rigid covalent
bonds, typical of this group of compounds of the solid state, and the
transition of free electrons to the collective state. In accordance
with Regel's classification, the $A^{III}B^V$ compounds, like germanium
and silicon, undergo a "semiconductor–metal" transition at the
melting point.

The change in the nature of the chemical binding in the
$A^{III}B^V$ compounds from mainly covalent in the solid state to mainly metallic in the liquid state should be accompanied by a change
in the short-range order. Structure changes during the melting of
the $A^{III}B^V$ compounds can be deduced from changes in the density,
which is an integral structure-sensitive characteristic reflecting
changes in the coordination number and interatomic spacing. The
density was measured by the method described in Chapter 2.

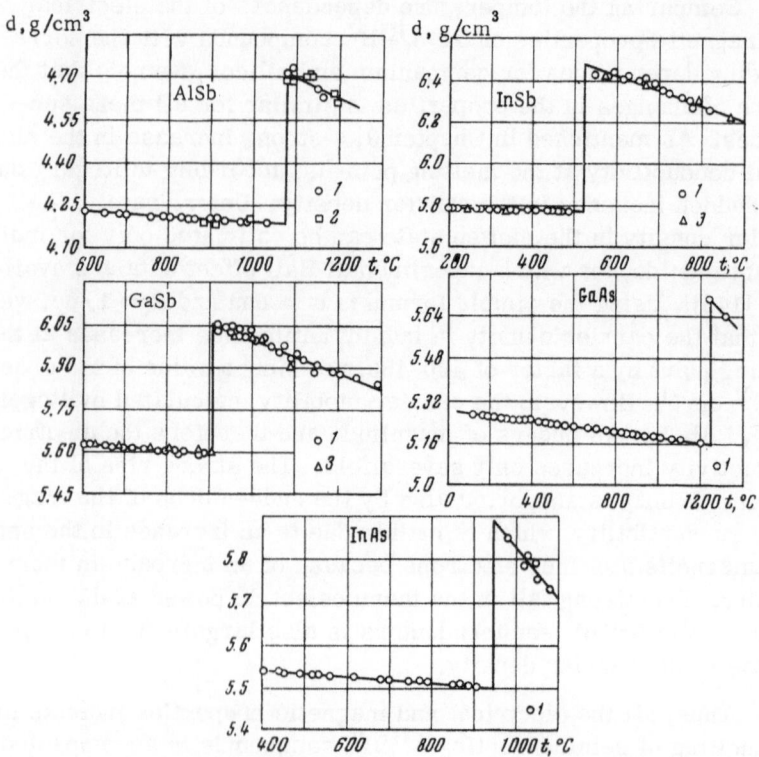

Fig. 35. Temperature dependences of the density of aluminum, gallium, and indium antimonides and of gallium and indium arsenides in the solid and liquid states. 1) Results of the present authors; 2) Sauerwald [635]; 3) Regel' et al. [5, 6, 8, 12].

Table 19. Density of $A^{III}B^{V}$ Compounds
At Their Melting Points in Solid and
Liquid States

Com- pound	d_s, g/cm³	d_l, g/cm³	$\dfrac{d_l - d_s}{d_s}$, %
AlSb	4.18	4.72	~12.9
GaSb	5.60	6.06	~ 8.2
InSb	5.76	6.48	~12.5
GaAs	5.16	5.71	~10.7
InAs	5.5	5.89	~ 7.1

The results of our investigations of Group III antimonides, together with the results of other workers, are presented in Fig. 35. The same figure includes also the results for indium and gallium arsenides, which had not been investigated before.

It is evident from Fig. 35 that, for the solid state, the density decreases linearly when temperature is increased, it rises suddenly at the melting point, and continues to decrease above the melting point. Table 19 lists the values of the density of these compounds at their melting points in the solid and liquid states.

It is evident from Table 19 that the jump in the density of the A^{III}B^V compounds at their melting points is not very large (7-13%).

The increase in the density of the A^{III}B^V compounds at their melting points may be due to a change in the short-range order resulting in a closer packing. This conclusion is in agreement with the results of x-ray diffraction investigations of indium antimonide, carried out by Krebs, Weyand, and Haucke [389], who have analyzed the atomic radial distribution curves (found from the experimental values of the x-ray diffraction intensity) to show that the short-range order in liquid indium antimonide near the melting point is similar to that of rocksalt with a coordination number of 6 and an interatomic spacing of 3.15 Å (cf. K = 4, r = 2.8 Å in the solid state). Because of this considerable increase in the interatomic spacing, the change in the density at the melting point of indium antimonide is relatively small, in spite of a considerable change in the coordination number. We may assume that identical or similar changes in the short-range order take place also in the other four A^{III}B^V compounds.

Thus, the melting of the A^{III}B^V semiconductors, like the melting of germanium and silicon, is accompanied by the destruction of the three-dimensional system of rigid covalent bonds and the corresponding destruction of the short-range order resulting in the formation of metallic bonds and a structure with a closer packing.

It seemed interesting to find how the short-range order of this group of compounds would be affected by heating above the melting point. To obtain this information, we investigated the temperature dependence of the viscosity of these compounds over a wide range of temperatures.

2. Changes in Short-Range Order during the Heating of $A^{III}B^V$-Type Melts

The viscosity of the $A^{III}B^V$ compounds was investigated by the method of torsional oscillations, described in Chapter 2.

Samples of gallium and indium antimonides and arsenides were placed in cylindrical quartz ampoules of 12 mm diameter. The ampoules with gallium arsenide were then placed in cylindrical graphite containers with screw-on lids in order to prevent the deformation of the ampoule and of the sample at high temperatures. Aluminum antimonide was placed in cylindrical corundum cans, which were sealed into quartz ampoules. In the investigation of gallium and indium arsenides, we added arsenic to the ampoules in amounts necessary to maintain the stoichiometry of the melts.

The results of measurements of the kinematic viscosity are presented in Figs. 36-40. The values of the kinematic viscosity and density were used to calculate the dynamic viscosity, whose temperature dependences are also presented in Figs. 36-40 [22, 387]. The dynamic viscosity was not calculated for gallium arsenide, because its density in the liquid state was not known sufficiently reliably.

It is evident from these figures that the viscosity immediately above the melting point decreases rapidly as the temperature rises in the range from t_{mp} to $t_{mp} + 30-60°$; further heating results in slower changes in the viscosity; and at still higher temperatures the viscosity of antimonides of Group III elements decreases rapidly again.

These results allow us to draw some conclusions about changes in the short-range order by comparing the experimental results with known theoretical relationships, in exactly the same way as we have done for silicon and germanium.

The results for the kinematic viscosity were used to calculate the free activation energy of viscous flow [23]. Analysis showed that the dependences $F_b \propto f(t)$ of all these compounds had definite minima at temperatures $t_{mp} + 15-75°$; these minima were similar to those found for germanium and silicon. The high-temperature parts of the $F_b \propto f(t)$ dependences of all these compounds obeyed satisfactorily Eq. (67). The tangents of the slopes

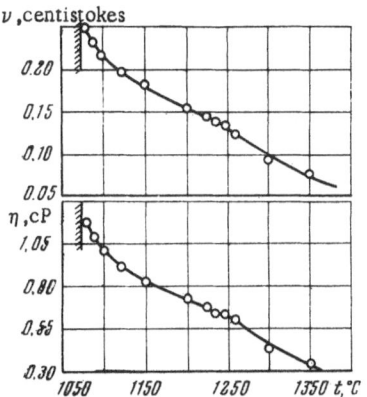

Fig. 36. Temperature dependences of the kinematic and dynamic viscosity of molten aluminum antimonide.

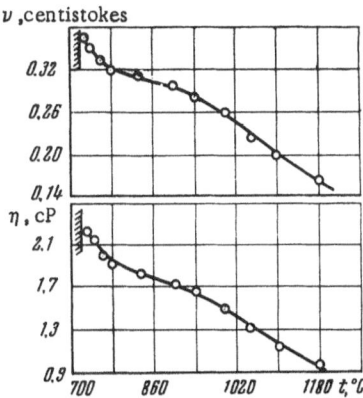

Fig. 37. Temperature dependences of the kinematic and dynamic viscosity of molten gallium antimonide.

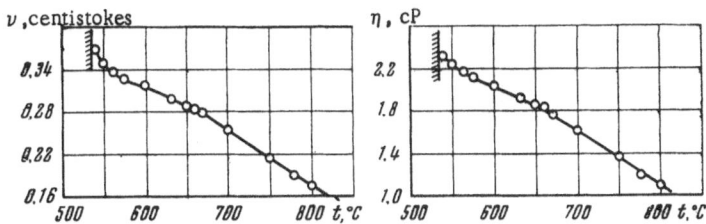

Fig. 38. Temperature dependences of the kinematic and dynamic viscosity of molten indium antimonide.

of the rectilinear parts of the $F_b \propto f(t)$ curves were used to calculate the values of the activation entropy of viscous flow of all five compounds (cf. Table 20).

Since the activation entropy of viscous flow at high temperatures was constant, the temperature dependence of the kinematic viscosity should obey Eq. (74).

In fact, we found that not too far from the melting point the logarithm of the kinematic viscosity depended linearly on the reciprocal of the absolute temperature. However, at temperatures close to the melting point, as well as at much higher temperatures, there were deviations from the linear dependence. The slopes of the rectilinear parts of the $\log \nu \propto f(1/T)$ curves were used to find

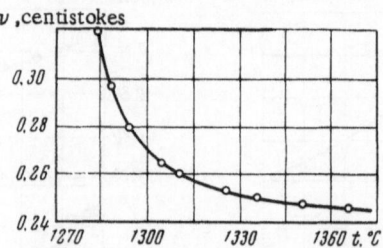

Fig. 39. Temperature dependence of the kinematic viscosity of molten gallium arsenide.

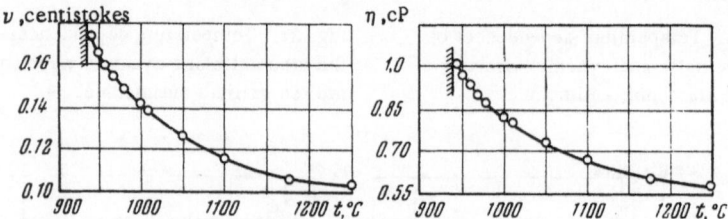

Fig. 40. Temperature dependences of the kinematic and dynamic viscosity of molten indium arsenide.

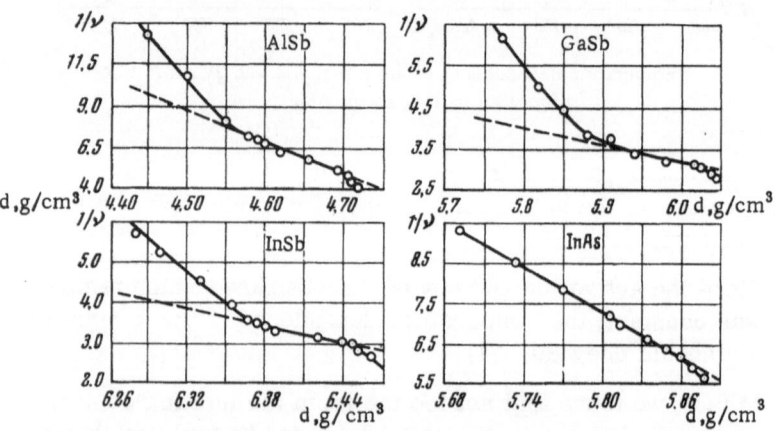

Fig. 41. Dependences of the reciprocal of the kinematic viscosity on the
density of molten aluminum, gallium, and indium antimonides and of indium arsenide.

Table 20. Values of E_b and $-S_b$ for $A^{III}B^V$ Melts

Com- pound	$E_b \cdot 10^{-3}$, cal/mole	$-S_b$, cal $\cdot$ mole$^{-1}\cdot$deg^{-1}	Δt, °C
AlSb	~10	~2.2	~60-70
GaSb	~2.7	~5.0	~40
InSb	~2,0	~9.4	~15
GaAs	~6.5	~7.8	~70
InAs	~6.2	~3	~50

the activation energy of viscous flow, whose values are listed in Table 20.

Analyzing the data given in Table 20, we note that the absolute values of the energy and entropy are close to those obtained for silicon and germanium and are of the same order as the corresponding values for molten metals [cf. Eq. (74)]. This is in agreement with the conclusion, given in the preceding section, that the $A^{III}B^V$ melts have the metallic binding.

The results presented in Table 20 show that the absolute values of the activation energy and entropy vary in a regular manner with the position of the components of the compound in the periodic table. The activation energy of viscous flow decreases and the entropy increases when the "cation" or the "anion" is replaced with a heavier element. This is probably due to the weakening of the binding forces between unlike atoms and the weakening of the potential barriers which have to be overcome by structural units in a liquid in the process of viscous flow. All the thermodynamic parameters of viscous flow begin to behave anomalously near the crystallization temperature: the activation energy and the entropy begin to vary with temperature and the entropy also changes its sign. This indicates the possibility of changes in the structure of activated complexes due to a change of temperature near the melting point.

Figure 41 shows the dependence of the reciprocal of the kinematic viscosity on the density for four compounds of the $A^{III}B^V$ type. It follows from this figure that near the crystallization temperature there are deviations from the linear dependence $1/\nu \propto f$ (d) and these deviations are of the same type as those observed for silicon and germanium.

In Chapter 3, we have compared the density dependence of the reciprocal of the kinematic viscosity with the temperature dependence of the free activation energy of viscous flow for substances with structures of different degree of compactness in the solid state, i.e., substances which suffer different structural changes at the melting point and above it. We have shown that deviations from a regular variation of the structure-sensitive properties of silicon and germanium near the crystallization temperature are associated with the rearrangement of the short-range order structures of their melts so that they become similar to the structures in the solid phase. The basically similar dependences of the structure-sensitive properties of the $A^{III}B^{V}$ compounds and of silicon and germanium suggest that in the case of these compounds the processes tending to loosen the structure of the liquid, i.e., to lower the coordination number and to produce covalent bonds, take place during cooling just before the crystallization point; thus, the structure of a liquid is prepared for crystallization. These processes do not affect the density, since they may be accompanied by a reduction in the interatomic spacing. The proportion of covalent bonds formed just above the melting point is obviously small and the temperature dependence of the electrical conductivity in the precrystallization stage is not affected because the carrier density is very high. Consequently, the observed anomalies in the temperature dependence of the electrical conductivity must be attributed to the melting kinetics [387].

We shall now consider the phenomena which take place in the $A^{III}B^{V}$ compounds at temperatures considerably above their melting points. Analysis of the temperature dependence of the viscosity of the $A^{III}Sb$ compounds show that above a certain temperature there are deviations from regular dependences. This can be seen even more clearly in the dependence of the logarithm of the viscosity on the reciprocal of the absolute temperature and in the dependence of the reciprocal of the kinematic viscosity on the density. If we compare the results obtained for the $A^{III}Sb$ compounds with the results reported for germanium and silicon (Chapter 3), we find that they are related to changes in the interatomic interaction which take place only in compounds. We may assume that they are due to the onset of an appreciable dissociation of $A^{III}Sb$ compounds. The same factor must be responsible for the change of the sign of the Seebeck coefficient (cf. Fig. 34) observed when

the A^{III}Sb compounds are heated well above their melting points. The thermal stability of these compounds decreases in a regular manner along the series AlSb−GaSb−InSb. The A^{III}As compounds are (as already mentioned at the beginning of the present chapter) thermally unstable even at the melting point, although the dissociation is partly suppressed by the addition of arsenic to the melt. The thermal stability of the A^{III}Sb compounds will be discussed in greater detail later.

§3. Changes in the Short-Range Order and the Chemical Binding at the Melting Points and during the Heating of Molten A^{II}B^{VI}-Type Compounds

1. Changes in the Short-Range Order and the Chemical Binding at the Melting Points of A^{II}Te-Type Compounds

The only A^{II}B^{VI} compounds investigated by us were zinc and cadmium tellurides. These were prepared from tellurium of grade I (powder of 99.998% purity), zinc of 99.999% purity, and cadmium of 99.95% purity. The tellurium was remelted under a layer of carbon in order to remove the oxides and to obtain the tellurium in a more compact form. All three elements had fairly high vapor pressures in the molten state and could be easily purified by sublimation in vacuum. After two or three sublimation runs, we obtain spectroscopically pure materials (the concentration of selenium in tellurium was not checked) containing not more than 10^{-4}% of the purities. The synthesis of zinc and cadmium telluride meets with considerable difficulties because of the high vapor pressure of components at the melting points of these compounds (the vapor pressure of zinc at the melting point of ZnTe is 10 atm; the vapor pressure of cadmium at the melting point of CdTe is 13 atm). However, we found that even at 550-570°C (in the case of zinc telluride) and 350-360°C (in the case of cadmium telluride) the components reacted violently to form compounds in the form of a loosely packed porous mass with local unreacted regions. This allowed us to synthesize these compounds in the following manner. The components were taken in their stoichiometric ratios and sealed

into quartz ampoules evacuated to 10^{-3} mm Hg and coated internally with graphite because pure cadmium and cadmium telluride react strongly with quartz (zinc telluride also reacts with quartz but less severely). In some cases the synthesis was carried out in corundum crucibles, sealed into evacuated quartz containers. The charge was subjected to heating in steps: 2-3 h at 600-800-1000°C (ZnTe) and at 500-700-900°C (CdTe). The compounds obtained in this way were crushed, sealed again into evacuated containers, and heated for 3 h at temperatures 50-60° higher than their melting points. During this stage they were stirred strongly from time to time. They were cooled very slowly (5°/h) by the gradual reduction of the current in the furnace.

This method of synthesis produced large coarse-grained ingots consisting of one phase throughout. Samples used in the investigations of the physical properties were cut from these ingots. The extremely narrow range of the homogeneity (solid-solution region) allowed us to assume that the compositions of the compounds so obtained were close to stoichiometric.

Zinc and cadmium telluride were found to have a relatively low electrical conductivity both in the solid state before melting and in the molten state. In order to increase the sensitivity of the electrical conductivity measurement method, we used larger samples than in the case of germanium, silicon, or the $A^{III}B^V$ compounds (the diameters of these samples were 15 mm and their heights were also 15 mm); moreover, we used also a thinner suspension filament (30 μ in diameter). During the measurements of the electrical conductivity and magnetic susceptibility, the samples of zinc and cadmium telluride were placed in cylindrical corundum crucibles, which were first baked in vacuum, and then sealed into quartz containers evacuated to 10^{-3} mm Hg. The electrical conductivity was measured using several samples of each compound. Before the measurements at the melting point, the samples were kept at the requisite temperature for 3-4 h. The results of the measurements are presented in Fig. 42 [26].

It is evident from Fig. 42 that the electrical conductivity of solid zinc and cadmium telluride before melting is very low: of the order of several $\Omega^{-1} \cdot cm^{-1}$. At the melting point, the conductivity increases approximately by an order of magnitude. Further heating produces a further increase in the electrical conductivity.

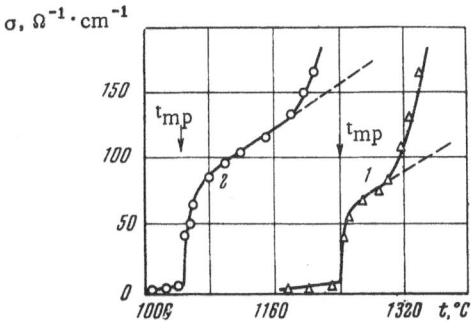

Fig. 42. Temperature dependences of the electrical conductivity of zinc telluride (1) and cadmium telluride (2)
in the solid and liquid states.

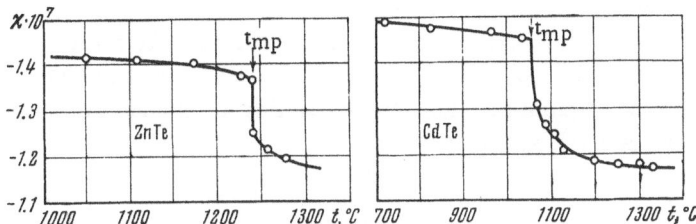

Fig. 43. Temperature dependences of the magnetic susceptibility of zinc and
cadmium tellurides in the solid and liquid states.

This increase is fairly rapid in a narrow range of temperatures
above the melting point but it is followed by a slower rise. At still
higher temperatures (~120° above the melting point for cadmium
telluride and ~60° above t_{mp} for zinc telluride), the electrical conductivity begins to rise again more rapidly.

Figure 43 shows the temperature dependences of the magnetic susceptibility of zinc and cadmium telluride. The results obtained indicate that both compounds are diamagnetic in the solid
and liquid state. The magnetic susceptibility of both tellurides
drops sharply at the melting point and continues to decrease at still
higher temperatures.

The absolute values of the electrical conductivity of molten
zinc and cadmium tellurides, as well as the positive temperature
coefficients of the conductivity, indicate that both compounds melt

in accordance with the "semiconductor-semiconductor" scheme.†
The rise of the electrical conductivity at the melting points of
these compounds may be due to an increase in the carrier density,
because we can hardly expect the mobility to increase in the transition to a liquid structure more defective than the crystal structure.
Moreover, the available information on the behavior of semiconductors which retain their semiconducting properties in the liquid
state (Te, Bi_2Te_3) indicates that the carrier mobility is not greatly
affected by melting [9, 144]. If we assume that the carrier mobility in zinc and cadmium tellurides remains practically constant
below and above the melting point, it follows that the carrier density rises only by one order of magnitude (this should be compared
with $n_l/n_S \simeq 10^4$ for silicon and germanium). Consequently, a considerable proportion of valence electrons is used to form covalent
bonds. This is supported also by the nature of the temperature dependence of the magnetic susceptibility. The fall in the susceptibility of these $A^{II}Te$ compounds is 2-3 times smaller than in the
case of the $A^{III}B^V$ compounds, which indicates that the changes in
the chemical binding are less far-reaching in our $A^{II}Te$ compounds
than in the $A^{III}B^V$ semiconductors.

We did not investigate the change in the density of zinc and
cadmium telluride at their melting points, but qualitative observations showed that (in contrast to silicon, germanium, and the
$A^{III}B^V$ compounds) no expansion was observed at the crystallization point. The absence of any marked increase in the degree of
packing in the short-range order of molten ZnTe and CdTe is in
agreement with the retention of covalent bonds in these compounds
in the liquid phase. It is known that their covalent bonds have definite directions in space and are rigid. However, a three-dimensional system of covalent bonds cannot exist in liquids. It is more
likely that a chainlike or a molecular structure is formed in the
liquid state.

---

† In principle, the same type of transition and the same temperature dependence of the
electrical conductivity in the liquid state is also possible in ionic liquids. However,
the conductivity of molten ionic liquids is usually two orders of magnitude lower than
the conductivity of zinc and cadmium tellurides [390]. Moreover, it is known that the
interatomic spacing of all substances increases after their melting. Batsanov and
Pakhomov [391] have shown that this reduces the polarity of bonds in predominantly
covalent compounds (ionicity less than 50%). As shown earlier (cf. §1), the chemical
binding between atoms in the $A^{II}Te$ compounds is basically covalent and, therefore,
we may assume that these compounds do not exhibit any tendency to form ionic liquids.

The possibility of the formation of a molecular structure in molten mercury selenide has been pointed out by Regel' [5, 8, 11, 12]. According to Regel', the electrical conductivity and density of mercury selenide decrease after melting. These observations are attributed to the loosening of the initial short-range structure and the formation of a molecular structure with a lower coordination number. Electrons are then localized at molecules, and their wave functions do not overlap, which explains the low conductivity of the melt.

The rise in the carrier density at the melting points of zinc and cadmium tellurides contradicts the hypothesis of a stable molecular structure. The assumption of a strongly dissociated molecular structure also seems incorrect, because these compounds are not decomposed by melting (cf. §1). It is more likely that a chainlike (A−B−A−B) structure, with carrier emission at the ends of the chains, is formed in these compounds in the liquid state.

The structural changes in the liquid state of these compounds can be deduced from the temperature dependences of the viscosity and from analysis of these temperature dependences by comparison with known theoretical relationships.

2. Changes in the Short-Range Order during the Heating of Molten Zinc and Cadmium Tellurides

To investigate the viscosity, the synthesized ZnTe and CdTe compounds were sealed into quartz ampoules with graphite-coated internal walls; these ampoules had an internal diameter of 12 mm and were evacuated to 10^{-3} mm Hg. The ampoules were placed in cylindrical graphite cans with screw-on lids, in order to avoid the deformation of the ampoules and samples at high temperatures. The temperature dependences of the kinematic viscosity of zinc and cadmium tellurides are presented in Fig. 44. It is evident from this figure that the temperature dependences are smooth curves.

Using the kinematic viscosity data, we calculated the free activation energy of viscous flow. Analysis of the $F_b \propto f(t)$ curves showed that at high temperatures this energy depended linearly on temperature. However, near the melting point (up to $\Delta t = 110°$

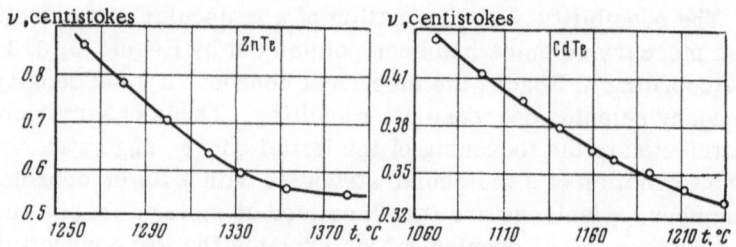

Fig. 44. Temperature dependences of the kinematic viscosity of molten zinc
and cadmium tellurides.

Table 21. Values of E_b and $-S_b$ for Molten Zinc
and Cadmium Tellurides

Compound	$\Delta t,°C$	$E_b \cdot 10^{-3}$, cal/mole	$-S_b$, cal $\cdot$ mole$^{-1}\cdot$deg^{-1}
ZnTe	~110	~9.0	~6.5
CdTe	~110	~5.75	~7.7

above this point), we observed considerable departures from the
linearity which were greater for zinc telluride than for cadmium
telluride. For zinc telluride these departures were similar to
those observed for selenium and tellurium. The values of the activation entropy of viscous flow (Table 21) were calculated from
the slopes of the high-temperature branches of the $F_b \propto f(t)$ curves.

At high temperatures, the logarithm of the kinematic viscosity is a linear function of the reciprocal of the absolute temperature. However, in the temperature range between the melting
point and about 110° above it, there are departures from the linearity of the $\log \nu \propto f(1/T)$ dependence. The activation energy of
viscous flow (Table 21) can be found from the slopes of the rectilinear parts of the $\log \nu \propto f(1/T)$ dependences. The results in
Table 21 show that the value of this energy (E_b) is less and the
value of the entropy (S_b) is greater for the compound with a heavier
component (CdTe) than for its analog ZnTe.

We have already seen that the same applies to other groups
of chemical analogs. The results presented in Table 21 show that
the activation energy of viscous flow of the molten $A^{II}Te$ com-

pounds is approximately of the same order of magnitude as the corresponding energy of the $A^{III}B^V$ compounds. However, if we compare the values of E_b for compounds in the same isoelectronic series, we see that they are 2-4 times higher for the $A^{II}Te$ group than for the $A^{III}B^V$ group. This may be due to the fact that larger structural units take part in the process of viscous flow in zinc and cadmium tellurides and they have to overcome large potential barriers. This assumption is in agreement with the hypothesis of a chainlike structure of molten ZnTe and CdTe.

We shall now consider the deviations of $F_b \propto f(t)$ and $\log \nu \propto f(1/T)$ from linearity. Immediately above the melting point (in a range of temperatures Δt) the entropy not only varies but even changes its sign. For ZnTe, all this indicates changes in the structure of activated complexes during the precrystallization stage. In this range of temperatures, the cooling or heating of the melt gives rise to complex processes which – in the case of heating – are due to the rearrangement of the structure from a three-dimensional system of covalent bonds to a linear structure, as well as due to dissociation processes. The dissociation immediately above the melting point is slight, but it begins to increase rapidly some tens of degrees above the melting point, as indicated by the rapid increase in the temperature coefficients of the electrical conductivity of molten ZnTe and CdTe. The problem of the thermal stability of these compounds will be considered in more detail later.

The hypotheses put forward about the structure of the molten $A^{II}Te$ compounds require a further experimental check by diffraction methods.

§4. Changes in the Short-Range Order and the Chemical Binding at the Melting Point and during the Heating of Molten Cuprous Iodide

1. Changes in the Short-Range Structure and the Chemical Binding at the Melting Point of Cuprous Iodide

Cuprous iodide was prepared from copper sulfate and potassium iodide in the presence of an excess of sulfuric acid [392]:

$$CuSO_4 + 2KI = K_2SO_4 + CuI_2. \qquad (76)$$

CuI_2 decomposes in accordance with the reversible reaction

$$2CuI_2 \rightleftarrows 2CuI \downarrow + I_2. \qquad (77)$$

To make sure that this reaction was completed, free iodine was removed by means of hyposulfite:

$$I_2 + 2Na_2S_2O_3 = 2NaI + Na_2S_4O_6. \qquad (78)$$

All the substances used were of the "chemically pure" grade. Unwanted reaction products were dissolved in distilled water. Cuprous iodide was strongly hygroscopic and, in the presence of traces of moisture, it was decomposed by heating and illumination. For this reason, special attention was paid to careful drying of this compound. The precipitate was dried first in a vacuum furnace (slow heating to 100°C and exposure to this temperature for five days). The dried substance was purified by vacuum distillation; this also removed the excess iodine. Cuprous iodide was deposited on the cold walls of a bulb in the form of semitransparent white crystals. The compound prepared in this way was stable and melted in vacuum without any trace of decomposition.

To measure the electrical conductivity, which was low, we prepared samples of 15-mm diameter and 15 mm high. These samples were sealed into quartz ampoules which were evacuated first to 10^{-3} mm Hg. To avoid oxidation in air during the sealing, the lower end of the ampoule containing the sample was cooled with liquid nitrogen. The measurements of the electrical conductivity were carried out using a suspension filament 30 μ thick.

The results of the measurements are presented in Fig. 45 in a form of the temperature dependence of the electrical conductivity [25].

It is evident from Fig. 45 that the electrical conductivity of cuprous iodide in the solid state increases when the temperature is increased, rises strongly at the melting point, and continues to increase during the subsequent heating. The nature of the rise at the melting point and the temperature dependence in the liquid state suggest the existence of covalent or ionic bonds in the melt.

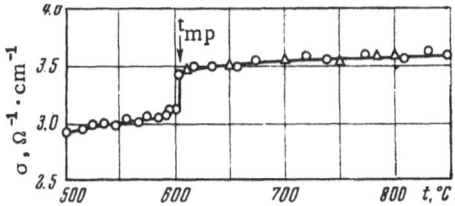

Fig. 45. Temperature dependence of the electrical conductivity of cuprous iodide in the solid and liquid states. O) Measurements during heating; Δ) measurements during cooling.

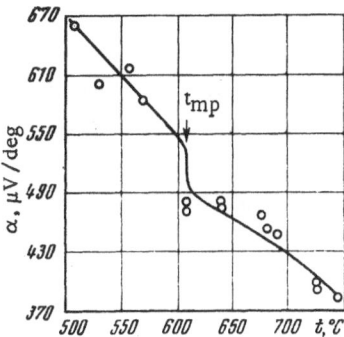

Fig. 46. Temperature dependence of the thermoelectric power of cuprous iodide in the solid and liquid states.

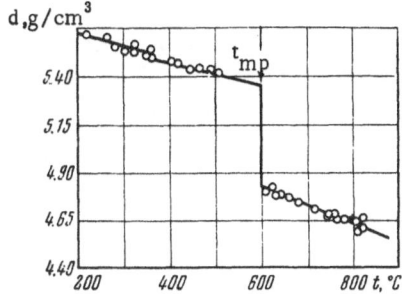

Fig. 47. Temperature dependence of the density of cuprous iodide in the solid and liquid states.

The absolute value of the electrical conductivity in the molten state is low, being 2-4 orders of magnitude lower than that of the compounds undergoing the "semiconductor – semiconductor" transition, but it does agree well with the values of the electrical conductivity of ionic liquids [390]. Moreover, as demonstrated in [266, 393, 394], the conductivity of the high-temperature α-modification of cuprous iodide has an appreciable ionic component. The rate of self-diffusion of cations (Cu^+) is high but that of anions (I^-) is very low, so that iodine ions play a negligible role in the conductivity [395]. With a high degree of certainty we may assume that

the ionic conductivity is also retained in cuprous iodide above the melting point. The relatively small rise in the conductivity at the melting point indicates that no radical changes take place in the nature of conduction.

This conclusion is supported also by investigations of the temperature dependence of the thermoelectric power (cf. Fig. 46). It is evident from Fig. 46 that the sign of the thermoelectric power is positive in the solid and liquid states. Melting reduces the thermoelectric power but the actual change at the melting point, like the change in the electrical conductivity, is small. The positive sign of the thermoelectric power indicates that the mobility of positive carriers is higher than that of negative carriers and the relatively small change at the melting point suggests that changes in the nature of charge transport are not very great. In the solid state, the charge is carried by positive ions and therefore we may assume that in the melt the conduction is still due to Cu^+ ions. The relatively high values of the thermoelectric power are evidently due to a low mobility, typical of ions [32, 397].

Figure 47 shows the temperature dependence of the density of cuprous iodide. It is evident from this figure that the density falls by about 9% at the melting point. This is probably due to the usual increase in the number of vacancies and in the interatomic spacing at the melting point.

The results obtained indicate that the melting of cuprous iodide is not accompanied by radical changes in the short-range order or in the chemical binding. The basically ionic nature of the chemical binding of the high-temperature modification is retained in the melt, which is an ionic liquid with structural units in the form of Cu^+ and I^- ions.

2. Changes in the Short-Range Order and the Chemical Binding during the Heating of Molten Cuprous Iodide

We investigated the temperature dependence of the viscosity and calculated the thermodynamic parameters of viscous flow (cf. Chapter 3).

We tried to find whether any changes took place in the short-range order and chemical binding when the melt was heated. The

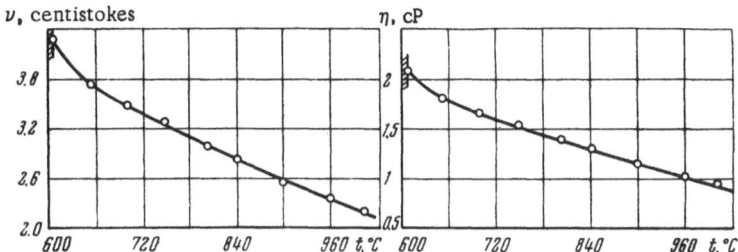

Fig. 48. Temperature dependences of the kinematic and dynamic viscosity
of molten cuprous iodide.

temperature dependence of the kinematic viscosity is presented in
Fig. 48. The same figure includes the temperature dependence of
the dynamic viscosity, calculated from the data on the kinematic
viscosity and density. It is evident that the dependences $\nu \propto f(t)$
and $\eta \propto f(t)$ are smooth curves. Using the kinematic viscosity data
and Eq. (67), we calculated the free activation energy of viscous
flow. We found that at all the investigated temperatures, beginning
from the melting point, a linear dependence $F_b \propto f(t)$ was observed.
Consequently, at all these temperatures the activation entropy of
viscous flow, calculated from the tangent of the slope of $F_b \propto f(t)$
was constant:

$$-S_v \approx 6.5 \ \text{cal} \cdot \text{mole}^{-1} \cdot \text{deg}^{-1}.$$

In view of the constancy of the entropy, the temperature dependence of the kinematic viscosity should obey Eq. (74), i.e., the
logarithm of the kinematic viscosity should be a linear function of
the reciprocal of the absolute temperature, which was indeed observed. From the tangent of the slope of $\log \nu \propto f(1/T)$, we calculated the activation energy of viscous flow:

$$E_b \approx 3.5 \cdot 10^3 \ \text{cal/mole}.$$

This activation energy is relatively small; it is approximate-
ly $\frac{2}{5}$ of the energy for zinc telluride and only slightly larger than
that for gallium antimonide (both these compounds are isoelectronic analogs of cuprous iodide). The small activation energy
may be the consequence of the small size of the structural units
in the melt. The constancy of the thermodynamic parameters of

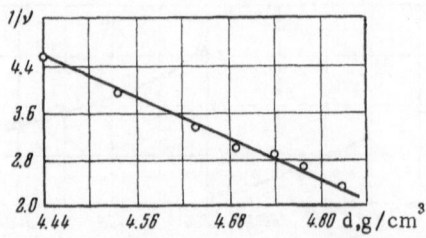

Fig. 49. Dependence of the reciprocal of the kinematic viscosity on the density of molten cuprous
iodide.

the process of viscous flow, E_b and S_b, at all the investigated temperatures indicates the absence of any marked changes in the
short-range order during heating of the melt, beginning from the
melting point. The linear dependence of the reciprocal of the kinematic viscosity on the density, given in Fig. 49, as well as the absence of a hysteresis in the viscosity during cooling (even from
temperatures 350°C above the melting point), also support the conclusion of the absence of significant structural changes.

From all these properties, we may conclude that the short-
range order and the chemical binding of cuprous iodide are not
radically altered by the melting or heating of the melt.

§5. Changes in the Short-Range Order and the Chemical Binding at the Melting Points and during the Heating of Molten $A_2^{III}Te_3$-Type Compounds

1. Changes in the Short-Range Order and the Chemical Binding at the Melting Points of $A_2^{III}Te_3$-Type Compounds

We investigated only two compounds of the $A_2^{III}Te_3$ type: gallium and indium telluride. These compounds were prepared from
gallium and indium containing not more than $10^{-4}\%$ impurities and
from tellurium in which the impurity concentration was reduced
to below $10^{-4}\%$ by triple sublimation in vacuum. These compounds
were synthesized by fusing stoichiometric proportions of the com-

ponents in quartz ampoules evacuated to 10^{-3} mm Hg. During synthesis, the ampoules were kept at temperatures 50-80° higher than the melting points of the compounds and then they were slowly cooled with the furnace. The synthesized compounds were purified by zone recrystallization, which yielded dense coarse-grained single-phase ingots, which had single-crystal parts. The samples were cut from these ingots. Since the absolute values of the electrical conductivity were low, large samples (15 mm in diameter and 15 mm high) were prepared. To increase the sensitivity of the instrument used in the measurement of the electrical conductivity, a thin (30 μ) suspension filament was used.

The electrical conductivity of gallium and indium telluride is known to rise by several orders of magnitude in the presence of the merest traces of surface oxidation. Such oxidation takes place even when the substances are slightly heated in air. Therefore, the quartz ampoules containing the samples were cooled continuously during the sealing-off stage.

The temperature dependences of the electrical conductivity of Ga_2Te_3 and In_2Te_3 are given in Fig. 50 [28, 29]. It is evident from this figure that the electrical conductivity of these compounds increases exponentially with temperature in the solid state, it rises suddenly at the melting point, and continues to increase rapidly immediately above the melting point. Further heating of these compounds in the molten state still increases the conductivity but the rate of increase is now slower and from about 1200°C (in the case of gallium telluride) and 1000°C (in the case of indium telluride) the conductivity ceases to depend on temperature. Cooling from any temperature gives exactly the same results as heating.

The temperature dependences of the magnetic susceptibility of Ga_2Te_3 and In_2Te_3 are given in Fig. 51 [127]. Both compounds are diamagnetic in the solid and liquid state. At the melting point, the magnetic susceptibility drops suddenly and continues to decrease quite rapidly above the melting point. At much higher temperatures, the susceptibility ceases to depend on temperature.

Figure 52 shows the temperature dependences of the thermoelectric power of gallium and indium tellurides; in this figure, our results (for two samples) are compared with those of Zhuze and Shelykh [400]. Melting reduces the thermoelectric power of both

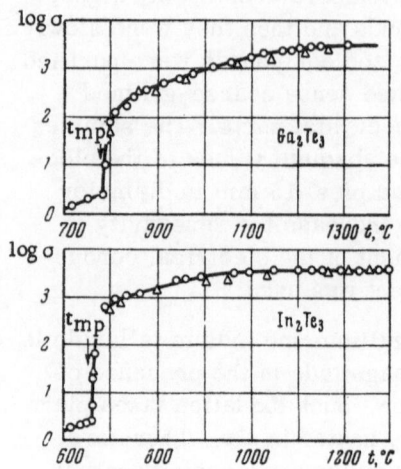

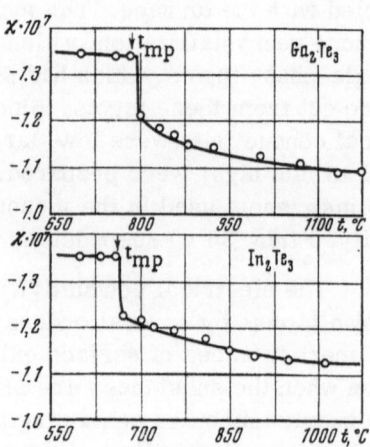

Fig. 50. Temperature dependences of
the electrical conductivity of Ga_2Te_3
and In_2Te_3 in the solid and liquid
states. O) Measurements during heating; $\triangle$) measurements during cooling.

Fig. 51. Temperature dependences of
the magnetic susceptibility of Ga_2Te_3
and In_2Te_3 in solid and liquid states.

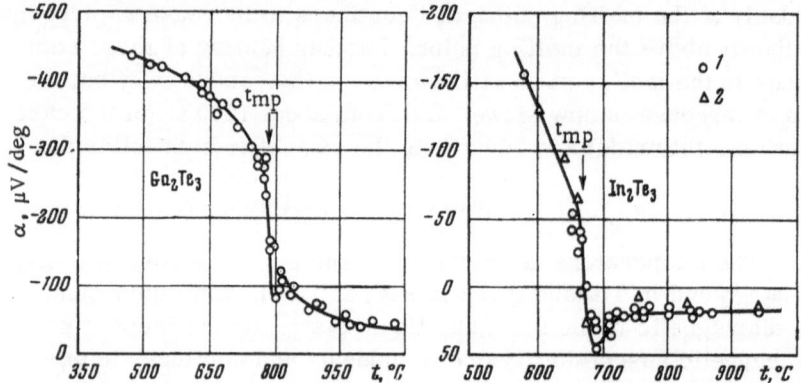

Fig. 52. Temperature dependences of the thermoelectric power of Ga_2Te_3 and In_2Te_3
in the solid and liquid states. 1) Results of the present authors; 2) Zhuze and Shelykh
[400].

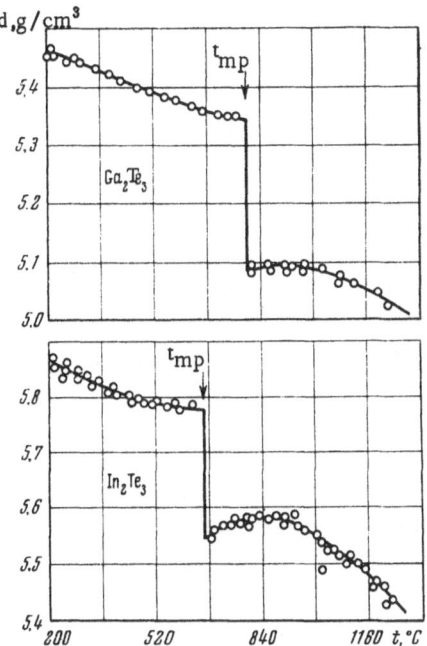

Fig. 53. Temperature dependences of the density of Ga$_2$Te$_3$ and In$_2$Te$_3$ in the solid and liquid states (results of five runs).

compounds by about an order of magnitude. The thermoelectric power of Ga$_2$Te$_3$ and In$_2$Te$_3$ is negative in the solid state and remains negative after melting in the case of Ga$_2$Te$_3$. However, in the case of indium telluride, melting reverses the sign of the thermoelectric power. The heating of gallium telluride above its melting point produces first a small rise in the thermoelectric power, which is followed by a gradual fall. In the case of indium telluride, heating above the melting point produces first a small fall of the thermoelectric power, beyond which this property remains constant.

These results allow us to draw some conclusions about changes in the short-range order and the chemical binding caused by the melting of gallium and indium tellurides. The nature of the temperature dependence of the electrical conductivity of both compounds indicates that they, like zinc and cadmium tellurides, under-

go the "semiconductor – semiconductor" transition at the melting
point. This means that gallium and indium tellurides retain covalent bonds in the molten state and these bonds are gradually destroyed by further heating. The large jump in the electrical conductivity at the melting point and the strong rise immediately
above it are due to an increase in the carrier density and a probable increase in the mobility. The absolute electrical conductivity
of solid gallium and indium telluride is very low because of a low
carrier mobility, due to the presence of vacancies in the "cation"
sublattice (cf. §1). The carrier mobility in these compounds may
increase at the melting point because of the filling of these vacancies, and this tends to increase the electrical conductivity. A
similar increase in the carrier mobility at the melting point has
been observed for tellurium [144].

The relatively small changes in the nature of chemical binding caused by the melting are supported also by the nature of the
temperature dependence of the magnetic susceptibility. The fall
of the magnetic susceptibility of the $A_2^{III}Te_3$ compounds at the melting point is, like that in the case of the $A^{II}Te$ compounds, 2-3 times
smaller than for germanium, silicon, and the $A^{III}B^V$ compounds.

It is worth stressing that the melting of gallium and indium
telluride is accompanied by relatively small (4-5%) volume
changes (Fig. 53). A decrease in the density at the melting point
indicates some loosening of the structure which is evidently due
to an increase in the interatomic spacings. However, the small
value of this change in the density shows that the short-range order
does not change radically.

Let us now consider the influence of melting on the thermoelectric power of these two compounds. The reversal of the sign
of the thermoelectric power of indium telluride indicates that the
hole mobility begins to predominate over the electron mobility.
The mixed nature of conduction is also responsible for the fairly
low (for semiconductors) values of the thermoelectric power and
the large change at the melting point. The hole mobility increases
also in the case of Ga_2Te_3, as indicated by the tendency toward a
sign reversal of the thermoelectric power when the melt is heated.

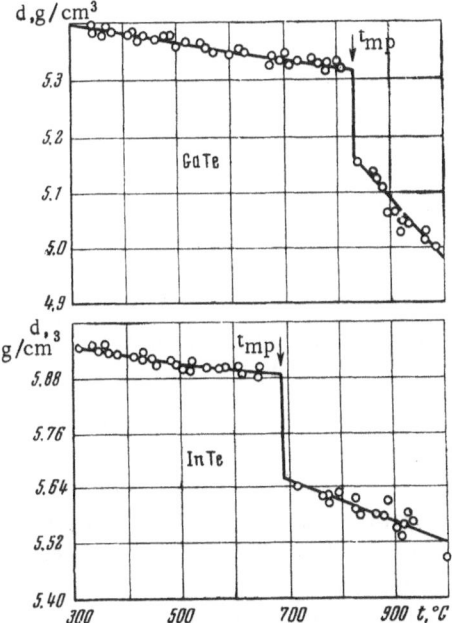

Fig. 54. Temperature dependences of the density
of GaTe and InTe in the solid and liquid states.

2. Changes in Short-Range Order and Chemical
Binding during Heating of Molten Gallium and
Indium Tellurides

Analysis of the curves presented in Figs. 50 and 51 shows
that, after an initial sudden change in the electrical conductivity
and magnetic susceptibility at the melting point, the electrical conductivity rises fairly rapidly immediately above the melting point
(in a temperature range Δt) and the susceptibility falls in approximately the same range. The anomalous changes in the thermoelectric power are also observed in the same temperature range. The
density of Ga_2Te_3 and In_2Te_3 (Fig. 53) rises immediately above the
melting point, passes through a maximum, and then falls with increasing temperature.

In order to determine whether the anomalies in the temperature dependence of the density are observed only in Ga_2Te_3 and
In_2Te_3 compounds, we also investigated gallium and indium mono-

telluride (Fig. 54), whose solid-state structure was laminar [402]. According to our data, these two compounds also retain their semiconducting properties in the liquid state [28-29]. Their density decreases by 3-4% at the melting point and continues to decrease monotonically at higher temperatures. Thus, gallium and indium monotelluride exhibit no anomalies in the density above the melting point. Such anomalies are not observed either in compounds belonging to other structure groups, which retain their semiconducting properties in the liquid state (cf. Chapters 6 and 7). It follows that the anomalies in the temperature dependence of the density of Ga_2Te_3 and In_2Te_3 are due to structural changes in the molten state.

Ga$_2$Te$_3$ and In$_2$Te$_3$ have a defect structure, with the lowest packing among the compounds with the ZnS-type lattice. Moreover, the ZnS-compounds as a whole have a low coordination number. We may therefore assume that the short-range order of Ga_2Te_3 and In_2Te_3 changes above the melting point and the change results in some increase in the packing density. Judging by the temperature dependence of the density, two competing processes take place in the molten state: the short-range order changes, producing a closer packing of the structural units in the melt; and the usual increase in the interatomic spacings, the appearance of local vacancies, etc., which make the structure looser. Immediately above the melting point (in the temperature range Δt), the changes in the short-range order play the dominant role. They alter the carrier density and mobility, and they are responsible for those features in the temperature dependence of the electrical conductivity, magnetic susceptibility, and thermoelectric power, which we have discussed above.

Table 22. Values of E_b and $-S_b$ for Molten $A_2^{III}Te_3$ Compounds

Compound	Δt, °C	$E_b \cdot 10^{-3}$, cal/mole	$-S_b$, cal. mole$^{-1} \cdot$ deg^{-1}
Ga$_2$Te$_3$	~100	11	7
In$_2$Te$_3$	~ 80	13	7.5

In view of this, it seemed interesting to study the temperature dependence of the viscosity of Ga_2Te_3 and In_2Te_3. The temperature dependences of the kinematic and dynamic viscosities of these two compounds are given in Fig. 55. For the sake of comparison, Fig. 56 gives the temperature dependences of the kinematic and dynamic viscosity of GaTe and InTe. Comparison of Figs. 55 and 56 shows that the viscosity curves of the $A_2^{III}Te_3$ compounds have maxima and the absolute values of the viscosity are fairly high. Above the temperature interval already referred to (Δt), the viscosity begins to fall. The temperature dependences of the viscosity of gallium and indium monotelluride are in the form of the usual smooth curves.

The results obtained were used to calculate the free activation energies of viscous flow of Ga_2Te_3 and In_2Te_3. It was found that F_b increased rapidly with rising temperature immediately above the melting point but this increase slowed down at a ~80-100° above the melting point. In the latter region, $F_b \propto f(t)$ was linear. The activation entropy of viscous flow, calculated for this range of temperatures, is given in Table 22.

A linear dependence of the logarithm of the kinematic viscosity of Ga_2Te_3 and In_2Te_3 on the reciprocal of the absolute temperature [$\log \nu \propto f(1/T)$] is observed only at temperatures 80-100° above the melting point. Considerable deviations from the linear law are observed close to the melting point. Consequently, the activation energy of viscous flow, calculated from the tangent of the slope of $\log \nu \propto f(1/T)$, also varies during heating of the melt. The rapid variation in the thermodynamic parameters of viscous flow in the temperature range Δt immediately above the melting point indicates considerable changes in the structure of activated complexes. At higher temperatures, the activation energy and entropy of viscous flow remain constant and the electrical conductivity, magnetic susceptibility, and thermoelectric power vary slowly with temperature. Consequently, we can conclude that at temperatures well above the melting point the short-range order of Ga_2Te_3 and In_2Te_3 tends to a stable state, which does not vary greatly with temperature.

Summarizing our discussion, we can draw the following conclusions. The melting of Ga_2Te_3 and In_2Te_3 does not greatly alter the short-range order and the chemical binding. However, con-

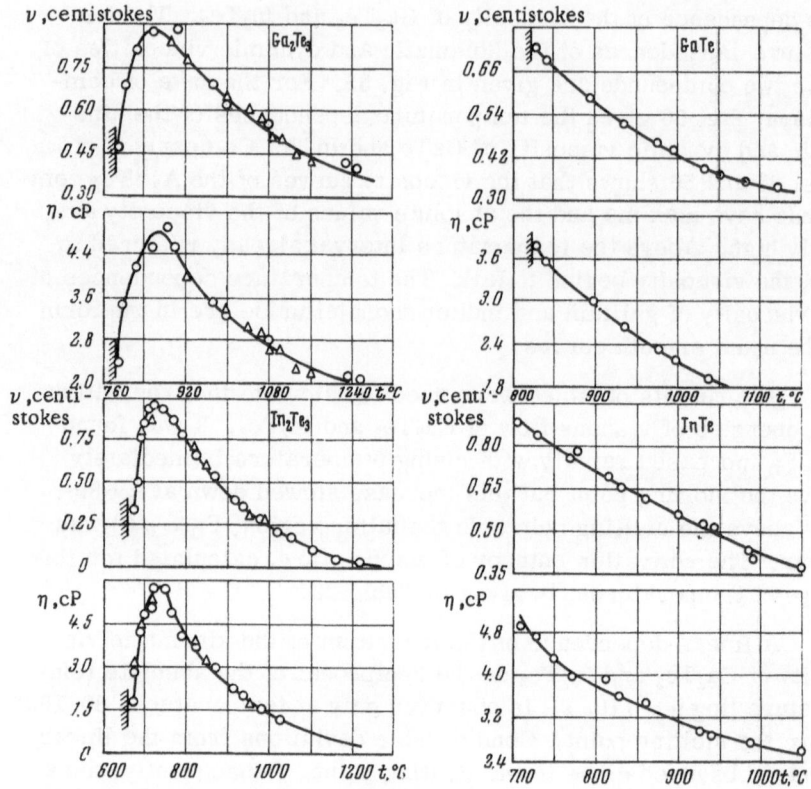

Fig. 55. Temperature dependences of the kinematic and dynamic viscosity of molten Ga_2Te_3 and In_2Te_3. ○) Measurements during heating; △) measurements during cooling.

Fig. 56. Temperature dependences of the kinematic and dynamic viscosity of molten GaTe and InTe.

siderable structural changes take place in the melt at temperatures between the melting point and 80-100° above it. These changes result in a structure with covalent bonds of the linear type, which are more stable when the mobility of particles in the melt is high.

Chapter 5

Physicochemical Properties of Molten Mg_2B^{IV} Compounds with the Antifluorite Structure

§1. Crystal Structure and Chemical Binding of Mg_2B^{IV} Compounds in the Solid State

The general nature of the chemical interaction in the $Mg-B^{IV}$ systems (where B is Si, Ge, Sn, Pb) is described by phase diagrams with a single congruent-melting chemical compound of the Mg_2B^{IV} type (Fig. 57) [203]. It follows from these diagrams that in all four systems the nature of the chemical interaction between components is similar. The Mg_2B^{IV} compounds divide the phase diagram of each system into two simple or partial eutectic diagrams. However, a more detailed examination shows changes in the general nature of the chemical interaction when the "anion" is replaced by a heavier element.

The Mg−Si and Mg−Ge systems exhibit practically no solubility (either in Mg or in B^{IV}). The eutectic points in the partial systems $Mg-Mg_2Si$ and $Mg-Mg_2Ge$ are close in composition to pure magnesium and the temperatures of the eutectic points are slightly lower than the melting point of magnesium. On the other hand, in the partial systems Mg_2Si-Si and Mg_2Ge-Ge, the eutectic points are located more or less symmetrically in the middle and the eutectic temperatures are much lower than the melting points of the components (in this case, one of the components is the

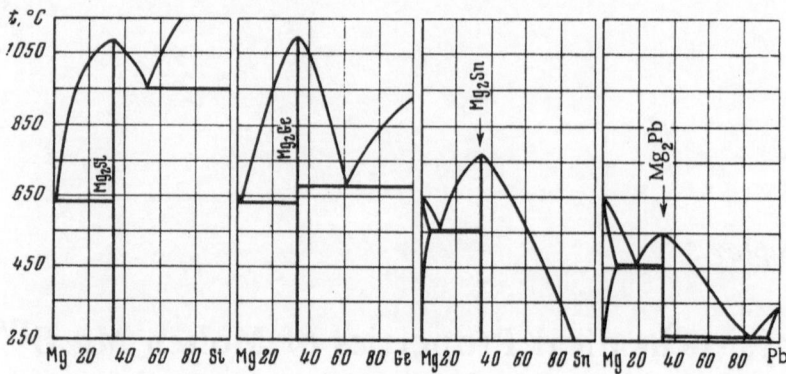

Fig. 57.   Phase diagrams of the Mg−B^{IV} systems (according to Hansen and A'nderko [203]).

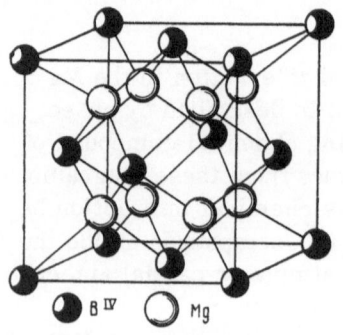

Fig. 58.   Crystal structure of the Mg$_2$B^{IV} compounds.

chemical compound Mg$_2$B^{IV}). The situation is somewhat different in the Mg−Sn system. A fairly extensive region of magnesium-base solid solution is observed. The eutectic point in the partial system Mg−Mg$_2$Sn is shifted slightly from the pure magnesium composition in the direction of the compound and the eutectic transition temperature is considerably below the melting points of the components. In the partial Mg$_2$Sn−Sn system, the situation is, in general, similar to the first two systems but the eutectic point is shifted considerably in the direction of the tin component. The Mg−Pb system exhibits a considerable solubility both in magnesium and in lead. The eutectic points lie more symmetrically in the corresponding partial systems and the temperatures of the eutectic transitions are much lower than the melting points of the components. It is worth pointing out the almost complete absence of solubility of Mg$_2$B^{IV} compounds in all four systems considered here.

Thus, even a simple description and comparison of the phase diagrams of the $Mg_2 - B^{IV}$ systems shows that the replacement of "anions" with heavier elements alters considerably the nature of the chemical interaction in these systems. This is obviously due to the increasingly metallic nature of the "anion" and due to changes in the nature of the chemical binding of the Mg_2B^{IV} compounds (the latter is a consequence of the former). Consequently, the nature of the chemical binding of the components plays an important role and exerts a considerable influence on the final results of the chemical interaction. These problems have been considered in detail by many workers — in particular, by Ugai [110, 403, 404].

We shall now consider in detail the chemical binding and structure of the Mg_2B^{IV} compounds formed in these systems. The Mg_2B^{IV} compounds crystallize with the antifluorite structure. The crystal lattice of these compounds is shown in Fig. 58. It is a face-centered cubic lattice in which the B^{IV} atoms form a cubic close packing and the magnesium atoms occupy all the tetrahedral vacancies. Each magnesium atom is surrounded by four B^{IV} atoms, forming a tetrahedron. Each B^{IV} atom is surrounded by eight magnesium atoms located at the corners of a cube. The unit cell contains 12 atoms or 4 Mg_2B^{IV} formula units. The ratio of the number of electrons to the number of atoms in the cell is $^8/_3$, i.e., it is exactly equal to the number of states in the Brillouin zones of the antifluorite structure [405].

The lattice parameters (average values taken from [406–414]) of the Mg_2B^{IV} compounds (CaF_2 structure, K = 4 + 8) are as follows (in Å):

$$Mg_2Si \ldots 6.338$$
$$Mg_2Ge \ldots 6.380$$
$$Mg_2Sn \ldots 6.765$$
$$Mg_2Pb \ldots 6.836$$

It is worth pointing out the considerable increase in the lattice parameter resulting from the replacement of "anions" with heavier elements, i.e., in the transition from Mg_2Si to Mg_2Pb.

The views of different workers on the nature of the chemical binding in the crystal lattices of Mg_2B^{IV} compounds are contradictory. The Mg_2B^{IV} compounds are Zintl phases; Zintl regards

them as valence compounds with a mainly heteropolar binding [415]. Kurnakov [4] is of the same opinion: he assumes that the binding between atoms is due to Coulomb interaction forces between positively charged magnesium ions and negatively charged B^{IV} ions. A. F. Ioffe suggested that these compounds have mixed chemical binding: partly ionic and partly metallic [416]. However, many observations do not agree with the conclusion of a predominantly ionic binding of compounds with the CaF_2 structure. It is known [417, 418] that substances with fluorite-type structure exhibit unusual heterotypic solid solutions with high-valence "cations" (for example, $CaF_2 - YF_3$). In such solid solutions, fluorine ions occupy vacancies between other fluorine ions, which is in strong disagreement with Pauling's electrostatic valence rule [316]. According to Neuhaus [419], the easy cleavage of fluorite along octahedral planes is difficult to explain on the assumption of purely ionic binding in this compound; is likely to be due to a covalent component. From these observations, Krebs [420] has concluded that if the covalent component of the binding is appreciable in such typical compounds as CaF_2, it is undoubtedly present in less saltlike compounds such as our Mg_2B^{IV} compounds. To account for the nature of the chemical binding in the Mg_2B^{IV} compounds, Krebs assumes a covalent component due to the partial filling of quantum states of four neighboring "cations" with electron pairs of the sp^3 hybrid of an anion. Examination of the crystal structure of the Mg_2B^{IV} compounds (Fig. 58) shows [420] that the "anion" atoms (B^{IV}) together with half the magnesium atoms, form a zinc-blende (ZnS) structure. We may also assume that the other half of the magnesium atoms also forms a ZnS-type structure with the B^{IV} atoms. Since neither of these zinc-blende sublattices (known as mesoforms) has any advantage over the other, it follows that they can exchange their roles or resonate. Welker [421] has suggested degeneracy along the direction of the sp^+ hybrids of the B^{IV} atoms in order to explain some aspects of this structure and some properties of the Mg_2B^{IV} compounds. He has shown that the distance between different atoms (Mg–Si, etc.) in the series Mg_2Si, Mg_2Ge, Mg_2Sn, and Mg_2Pb is approximately equal to the sum of the Pauling tetrahedral radii. Krebs [420] remarks that degeneracy along the direction of the sp^3 hybrid of the B^{IV} atoms may partly delocalize the valence pairs of electrons but they still remain localized in the space around the B^{IV} atoms. On the other hand, the exchange

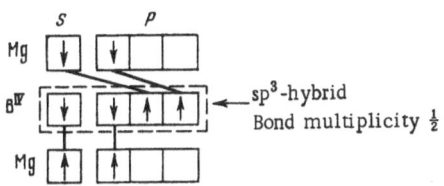

Fig. 59. Electron-valence scheme of bonds in the
Mg_2B^{IV} compounds [332].

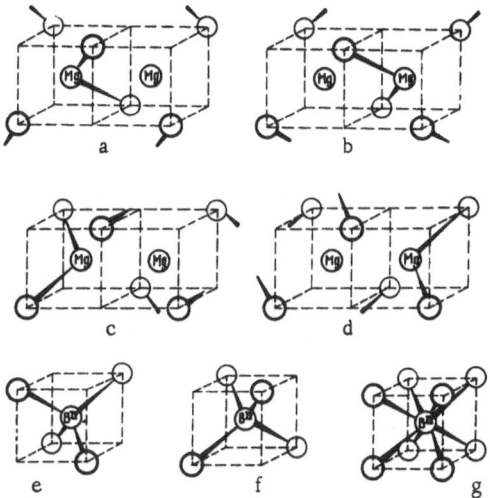

Fig. 60. Resonance of bonds with magnesium atoms (a-d) and with
B^{IV} atoms (e-g) in a Mg_2B^{IV} lattice [301].

interaction between the individual sp^3 hybrid eigenfunctions of the
magnesium atoms is very weak. Consequently, free electrons and
holes should form near one of the B^{IV} atoms in the immediate vicinity of such an atom.

According to Mooser and Pearson [332], the Mg_2B^{IV} compounds satisfy all the conditions for the formation of semiconducting phases. Figure 59 shows Mooser's and Pearson's electron-
valence system of bonds between atoms in the Mg_2B^{IV} compounds.

In this scheme, each B^{IV} atom forms four covalent sp^3 bonds
which — due to the presence of saturated orbitals of the magnesium

Table 23. Some Physicochemical Properties of the Mg$_2$B^{IV}
Compounds in the Solid State

Property	Mg$_2$Si	Mg$_2$Ge	Mg$_2$Sn	Mg$_2$Pb	Reference
ΔX	0.6	0.6	0.6	0.6	[476]
t_{mp}, °C	1102	1115	778	550	[203]
Q_f, kcal/g-atom	6.8	—	3.8	3.1	[424]
$-\Delta H_{298}$, kcal/g-atom	6.3	—	6.1	4.2	[424]
ΔE, eV (T=0)	0.77	0.74	0.36	0.1	[406, 425—428]
u_n, cm^2·V^{-1}·sec^{-1}	405	520	320	—	[406, 425, 426,
u_p, cm^2·V^{-1}·sec^{-1}	70	110	260	—	429]

atoms — may exhibit rotational resonance between 8 positions.
Thus, each B^{IV} atom forms eight semibonds with the surrounding
magnesium atoms. The s and p subshells of magnesium are only
half-saturated. At the same time each atom of magnesium uses its
two electrons to saturate bonds with four B^{IV} neighbors so that no
electrons are left to form Mg—Mg bonds. The bond scheme in a
Mg$_2$B^{IV} lattice at the moment of existence of one of the mesoforms
can be represented as shown in Fig. 60 [301], where thick circles
and lines represent the B^{IV} atoms forming bonds at a given moment.

The idea of the rotational resonance of bonds, developed by
Pauling and used by Krebs, Mooser, and Pearson, has been criticized [422] but in the present case the resonance concept gives a
clear picture which explains why semiconducting properties are
observed in the Mg$_2$B^{IV} compounds.

Several investigations, reviewed in Smith's book [111], have
demonstrated that the Mg$_2$B^{IV} compounds are semiconductors,
mainly due to the covalent chemical binding between atoms (cf.
also [423]).

Table 23 lists some of the physicochemical properties of
solid Mg$_2$B^{IV} compounds, which indicate the nature of the changes
in the chemical binding when the "anions" are replaced with heavier elements.

The results presented in Table 23 indicate that the "ionic" component of the binding between atoms is practically unaffected by the substitution of heavier "anions" because the difference between the electronegativities ΔX is the same for all four compounds. On the other hand, the melting point, the heat of fusion, the heat of formation, and the forbidden band width all decrease when the "anions" are replaced with heavier elements. This indicates gradual weakening of covalent bonds along the series $Mg_2Si \rightarrow$ $\rightarrow Mg_2Ge \rightarrow Mg_2Sn \rightarrow Mg_2Pb$. The semiconductor nature of the compound Mg_2Pb has not yet been considered since all authors have regarded this compound as metallic. In particular, it has been reported in [430] that the temperature dependences of the electrical resistivity and of the thermoelectric power of Mg_2Sn and Mg_2Pb indicate that the former is a semiconductor and the latter is a metal. Hence, it has been concluded that the crystal structure is not the factor which governs the presence or absence of semiconductivity. However, Busch and Moldovanova [431] have demonstrated that Mg_2Pb is a semiconductor with a very narrow forbidden band (0.1 eV). Busch and Moldovanova [431] are of the opinion that the reason why this compound has been regarded as a metal has been its contamination with gases (mainly with nitrogen).

Thus, these results indicate that the Mg_2B^{IV} compounds are distinguished by the covalent type of chemical binding together with a "metallic component" which increases in strength along the series $Mg_2Si \rightarrow Mg_2Ge \rightarrow Mg_2Sn \rightarrow Mg_2Pb$.

§2. Changes in the Short-Range Order and the Chemical Binding at the Melting Points of Mg_2B^{IV} Compounds

1. Changes in the Electrical and Magnetic Properties at the Melting Points of the Mg_2B^{IV} Compounds

Detailed investigations of the electrical conductivity and of the magnetic susceptibility of the Mg_2B^{IV} compounds were carried out primarily in order to determine changes in the nature of the chemical binding in the transition from the solid to liquid state. Investigations of changes in these properties at the melting point also yield information on the nature of the solid phase and on the chemical binding in the liquid state, i.e., they can be used to determine the validity of any particular model.

Mg_2Si, Mg_2Ge, Mg_2Sn, and Mg_2Pb were prepared from the following reagents: sublimated magnesium containing traces of iron, manganese, and copper (the overall purity was not less than 99.98% Mg); germanium and silicon with carrier densities of 10^{13} and 10^{14} cm^{-3}, respectively; tin of 99.9994% purity; lead of 99.997% Pb purity.

Since the magnesium reacted strongly with quartz and the Mg_2B^{IV} compounds were volatile, the compounds were synthesized using a modification (Fig. 61) of a method described in [432]. The components 6, taken in their stoichiometric ratio, were placed in a corundum crucible 5 which was closed with a corundum stopper. The stopper was in the form of a second crucible 4 (its outer walls were ground to produce a vacuum-tight joint). The main crucible and the stopper were bound with a Nichrome wire 3 located in special grooves. Next, the system was placed in a tall quartz test tube 2 with a connecting tube. A cover 1 with a quartz holder was soldered to the open end of the tube; the system was pumped down to 10^{-3} mm Hg and then

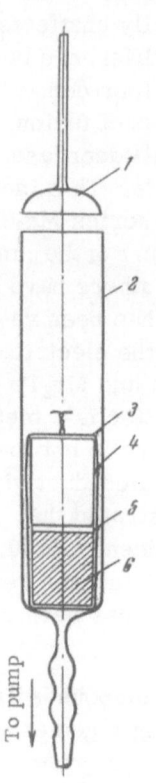

Fig. 61. Synthesis of the Mg_2B^{IV} compounds.

sealed. The whole system was placed in a furnace and heated slowly to a temperature about 50° higher than the melting point. The mixture was held at this temperature for 2-3 h. It was shaken periodically and finally cooled in the furnace at a rate of 20-30°/h. This method ensured no loss of the volatile magnesium and of the compounds themselves. The inner surface of the quartz ampoule, in which the crucible was placed, was perfectly clean. The quality of the materials obtained was checked by microscopic analysis and sections prepared for this analysis were polished with turpentine because water decomposed these compounds. A microscopic investigation of the structure indicated that the samples obtained consisted of one phase only. Because of the extremely narrow region of homogeneity of the Mg_2B^{IV} compounds, the substances obtained had a nearly stoichiometric ratio of the components.

To pump →

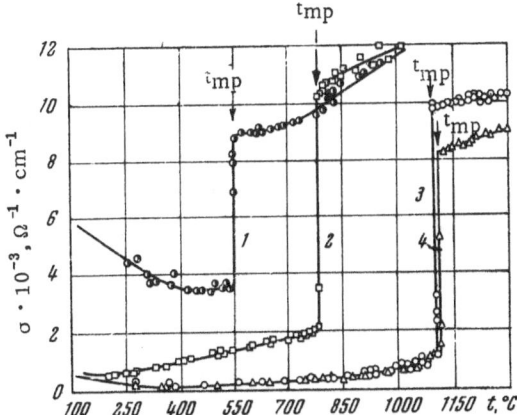

Fig. 62. Temperature dependences of the electrical conductivity
of the Mg_2B^{IV} compounds in the solid and liquid states: 1) Mg_2Pb;
2) Mg_2Sn; 3) Mg_2Si; 4) Mg_2Ge.

The preparation of the samples for the measurement of the
electrical conductivity and magnetic susceptibility was also difficult because Mg_2Ge and Mg_2Pb decompose fairly rapidly when ex-
posed to atmospheric moisture, and Mg_2Si and Mg_2Sn were decomposed if water was used as a lubrication in their cutting. Therefore, these compounds were crushed and placed in cylindrical
corundum cans of 10 mm diameter and 20 mm high; they were
closed with identical cans, whose outer walls were ground. The
closed cans were bound with Nichrome wire and sealed into quartz
ampoules, evacuated down to 10^{-3} mm Hg. The ampoules were
placed in a furnace heated to a temperature 20-30° above the melting point of a given compound and kept at this temperature for 20-
30 min; then they were cooled in the furnace. The wire binding the
cans prevented the cover from rising under the pressure of the
vaporized compound and, therefore, there was no loss during melting. However, even the smallest gaps between the two cans became filled with the substance so that after the removal of the binding wire the cans could not be separated. Next, these firmly joined
cans were sealed again in cylindrical ampoules evacuated down to
10^{-3} mm Hg, in which the measurements were carried out; when
this procedure was followed there was no loss of these volatile
compounds during the measurements.

Table 24. Electrical Conductivity of the
Mg_2B^{IV} Compounds at Their Melting Points
in the Solid and Liquid States

Compound	$\sigma_s \cdot 10^{-3}$, $\Omega^{-1} \cdot cm^{-1}$	$\sigma_l \cdot 10^{-3}$, $\Omega^{-1} \cdot cm^{-1}$	σ_s/σ_l
Mg_2Si	1.12	9.8	~0.1
Mg_2Ge	1.14	8.4	~0.1
Mg_2Sn	2.04	10.6	~0.2
Mg_2Pb	3.53	8.6	~0.4

The electrical conductivity and magnetic susceptibility [433] were measured using the method described in Chapter 2. The measurements were carried out on three separately prepared samples of each compound. We also measured the electrical conductivity of an Mg_2Sn single crystal (this crystal was prepared by pulling from the melt using the Czochralski method and was kindly supplied by Assistant Professor E. B. Sokolov).

The results of all these measurements are presented in summary graphs which give the temperature dependences of the electrical conductivity (Fig. 62). These graphs show that the electrical conductivity of solid Mg_2Si, Mg_2Ge, Mg_2Sn decreases when the temperature is increased from room temperature to about 200-300°C. Above this temperature, the electrical conductivity increases right up to the melting point. The electrical conductivity of Mg_2Pb decreases from room temperature to about 400°C. Further increase in temperature causes a very weak rise in the electrical conductivity of Mg_2Pb (two samples) or even a decrease right up to the melting point (one sample). At the melting point, the electrical conductivity of all four compounds increases suddenly up to about $10^4 \Omega^{-1} \cdot cm^{-1}$, which is of the order of the electrical conductivity of liquid mercury. Table 24 gives the electrical conductivity of the Mg_2B^{IV} compounds at their melting points.

The results for molten Mg_2Pb are in good agreement with those reported in [434-436], concerning the electrical conductivity of melts in $Mg-Pb$ systems. Table 24 shows that the greatest jumps in the electrical conductivity are observed for Mg_2Si and Mg_2Ge; in the case of Mg_2Sn and Mg_2Pb the jumps are, respectively,

a half and a quarter the magnitude. The decrease in the jump is due to an increase in the electrical conductivity of the solid phase at the melting point along the series $Mg_2Si \rightarrow Mg_2Ge \rightarrow Mg_2Sn \rightarrow \rightarrow Mg_2Pb$; in the liquid state, the electrical conductivities of all four compounds are similar. The absolute values of the electrical conductivity of the melts (of the order of $10^4 \ \Omega^{-1} \cdot cm^{-1}$) indicate that all four investigated compounds become metallike in the liquid state.

Thus, by analogy with germanium, silicon, and the $A^{III}B^V$ compounds, we may conclude that the Mg_2B^{IV} group of compounds melts, in accordance with the "semiconductor–metal" transition (the Regel' classification). This conclusion applies also to Mg_2Pb, since the data on the nature of the temperature dependence of the electrical conductivity of this compound in the solid state and the complete analogy with the other three compounds, except the jump of the electrical conductivity at the melting point, indicate that solid Mg_2Pb is a semiconductor.

The observation that all four compounds become metallike in the molten state indicates that a large number of free electrons is liberated by melting. By analogy with $A^{III}B^V$ compounds, we may conclude that the dominant chemical binding in the solid state is covalent, which confirms indirectly the correctness of our description of binding given in §1 of the present chapter. Once again we must stress the point, first put forward by A. R. Regel', that the nature of the jump in the electrical conductivity at the melting point, together with the absolute values of the conductivity in the solid and liquid states, can be used to draw fairly definite conclusions about the nature of binding in the solid state.

The heating of molten Mg_2B^{IV} compounds increases the electrical conductivity which, in general, is untypical of metals and – in this case – is evidently due to the thermal dissociation of the metallike compounds. This appears most clearly in the case of Mg_2Pb and Mg_2Sn and somewhat less clearly in the case of Mg_2Ge. Mg_2Si is evidently the most stable thermally. The conclusion about the thermal dissociation of the Mg_2B^{IV} compounds, drawn here from the nature of the temperature dependence of the electrical conductivity in the molten state, is in full agreement with the results of a physicochemical analysis of double Mg–Sn and Mg–Pb liquid systems, reported in [437-439], where the broad viscosity

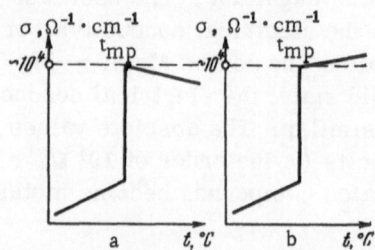

Fig. 63. Temperature dependences of the electrical conductivity in the solid and liquid phases for a "semiconductor–metal" transition in compounds with a metallic type of binding in the liquid phase. a) Thermally stable compounds; b) dissociating compounds.

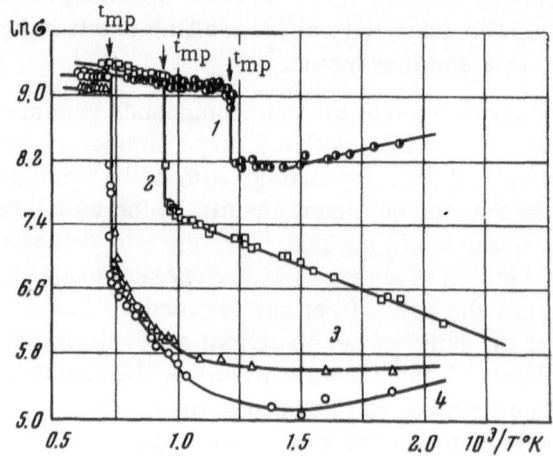

Fig. 64. Dependences $\ln \sigma = f(1/T)$ for the Mg_2B^{IV} compounds in the liquid and solid states. 1) Mg_2Pb; 2) Mg_2Sn; 3) Mg_2Ge; 4) Mg_2Si.

and electrical conductivity isotherms have been used to obtain information on the dissociation of the corresponding compounds in the liquid state. Thus, to supplement the Regel' classification, we may represent the temperature dependence of the electrical conductivity during a "semiconductor–metal" transition by a scheme shown in Fig. 63, which applies to dissociating compounds with the metallic binding in the liquid state.

Figure 64 shows the dependence of the logarithm of the electrical conductivity on the reciprocal of the absolute temperature for all four Mg_2B^{IV} compounds. It is evident from this figure that, beginning from certain temperatures, each of those compounds has a linear region which indicates intrinsic conduction. The dependences $\ln \sigma \propto f(1/T)$ for Mg_2Si and Mg_2Ge practically merge in the intrinsic conduction region. In general, this is to be expected because the values of the forbidden band widths of these two compounds are very similar. Similar band widths are responsible for similar values of the intrinsic carrier density at the same temperatures. Moreover, the nature of the scattering of intrinsic carriers is also similar in these two compounds and this results in similar values of the carrier mobility and the same temperatures.

The forbidden band widths of Mg_2Si and Mg_2Ge, calculated from the linear parts of the dependences $\ln \sigma \propto f(1/T)$ are ~0.7 and ~0.65 eV, respectively; the forbidden band width of Mg_2Sn, calculated in the same way, is ~0.23 eV, which is slightly less than the value given in Table 23. This again is expected, since the forbidden band width of a semiconductor decreases when the temperature is increased [32]. The forbidden band width of Mg_2Pb has not been determined because the slope of the dependence $\ln \sigma \propto f(1/T)$ is very slight and cannot be measured accurately. In any case, it is clear that the change in the slope of this dependence at high temperatures is due to the appearance of intrinsic conduction and the activation energy of intrinsic electrons in this temperature range is less than 0.1 eV.

Investigations of the magnetic susceptibility of the Mg_2B^{IV} compounds have shown that they are diamagnetic both in the solid and liquid states. A sudden change in the magnetic susceptibility is observed at the melting point and the magnitude of this change decreases along the series $Mg_2Si \rightarrow Mg_2Pb$. Thus, we have data similar to the results obtained for the $A^{III}B^V$ compounds [127]. Changes in the nature of the chemical binding during the melting of the $A^{III}B^V$ and Mg_2B^{IV} compounds are similar.

Thus, investigations of the electrical conductivity and magnetic susceptibility of the Mg_2B^{IV} compounds in the liquid and solid phases have shown that melting destroys the three-dimensional system of covalent bonds and produces a metallic state. More definite conclusions about the nature of changes in the short-range order

due to melting can be obtained by considering the results of an investigation of changes in the density at the melting point (next subsection).

2. Volume Changes at the Melting Points of the Mg_2B^{IV} Compounds and General Characteristics of the Short-Range Order in Melts

The density is an integral structure-sensitive property, reflecting changes in the coordination number and interatomic spacings. Consequently, since the interatomic spacings usually increase during heating, the direction of a sudden change in the density can be used as a measure of the change in the coordination number and in the nature of the chemical binding, particularly when combined with other physicochemical properties and when a comparison is made with other groups of compounds.

Compounds of the Mg_2B^{IV} group become metallike in the liquid state, as indicated by a sudden increase in the electrical conductivity at the melting point to values of the order of 10^4 $\Omega^{-1} \cdot cm^{-1}$, which are typical of liquid metals. The transition of these predominantly covalent compounds to the metallic state in the liquid phase is accompanied by an increase in the tightness of the packing of the structure, as represented by an increase in the coordination number. This has been observed in $A^{III}B^V$ compounds, as well as in germanium and silicon [3, 5-8]. We can expect the Mg_2B^{IV} compounds to exhibit an increase in their density, since the chemical binding in these compounds changes in the same way as in the $A^{III}B^V$ groups.

Some published data on the density of molten Mg_2Sn and Mg_2Pb are available; they indicate that these compounds become denser in the molten state [440, 441]. However, these data are very limited because of the narrow range of the investigated temperatures and they should be regarded as qualitative, since none of these investigations gives the overall graphs showing volume changes at the melting points of Mg_2Sn and Mg_2Pb. There are no published data on the density of molten Mg_2Si and Mg_2Ge.

To determine the density of the Mg_2B^{IV} compounds, we prepared them by a method already described. The density in the solid state at elevated temperatures was measured by the dilatometric method (Chapter 2, §2).

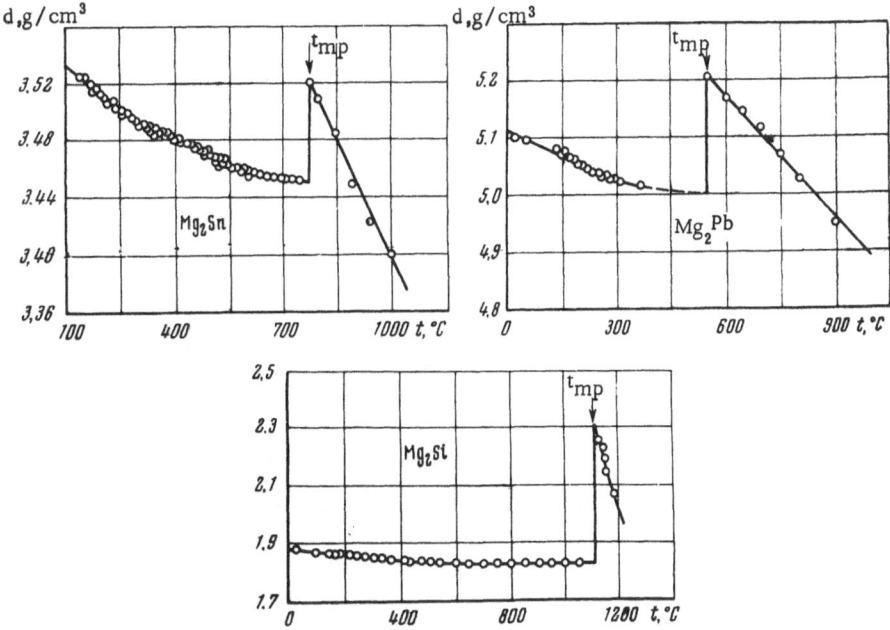

Fig. 65. Temperature dependences of the density of Mg_2Si, Mg_2Sn, and Mg_2Pb in the liquid and solid states.

The room-temperature values of the density of these compounds were taken from [406, 440]:

Compound	Mg_2Si	Mg_2Ge	Mg_2Sn	Mg_2Pb
Density, g/cm³	1.88	3.09	3.53	5.1

The density of the Mg_2B^{IV} compounds in the liquid state was measured by the hydrostatic weighing method using an analytic balance. Other methods for the measurement of the density of melts could not be applied to these compounds because of their high reactivity and volatility. The measurements of the density of molten Mg_2Ge did not yield reproducible results over an extended range of temperatures because of difficulty of finding a thermally stable flux of a suitable specific gravity. For each of the remaining three compounds (Mg_2Si, Mg_2Sn, and Mg_2Pb), we carried out three series of experiments (using new samples on each occasion) and we obtained reproducible results. In the solid state, we carried out three series of measurements on Mg_2Sn, and

Table 25. Density of the Mg_2B^{IV} Compounds
at Their Melting Points in the Solid and
Liquid States

Com-pound	d_s, g/cm³	d_l, g/cm³	Δd, g/cm³
Mg_2Si	1.84	2.27	0.43
Mg_2Ge	—	3.20	—
Mg_2Sn	3.45	3.52	0.07
Mg_2Pb	—	5.20	—

two series on Mg_2Pb and Mg_2Si (using samples obtained from two different melts). The results were reproducible.

The temperature dependences of the density in the solid and liquid states are given in Fig. 65 for Mg_2Si, Mg_2Sn, and Mg_2Pb. It follows from this figure that the density of the Mg_2B^{IV} compounds in the solid state decreases when the temperature is increased. Mg_2Pb becomes volatile well before the melting point and, therefore, its density just before the melting point cannot be determined. At the melting point, the density of each of these three compounds increases considerably but further heating causes a decrease.

Table 25 lists the values of the density of the investigated Mg_2B^{IV} compounds at their melting points in the liquid and solid states.

The values obtained were reliable because they are in good agreement with the data on Mg_2Sn and Mg_2Pb [440, 441]. The reliability of the results for the solid state and, consequently, of the magnitudes of the discontinuities at the melting points, were governed by the reliability of the initial values of the room-temperature density, taken from [406, 440].

From the results obtained (Fig. 65), it follows that the density of the Mg_2B^{IV} compounds increases strongly at the melting point, which is typical of the transition of a semiconductor with a predominantly covalent binding to a metallike state. It is likely that this is associated with the destruction of a three-dimensional system of covalent bonds, the destruction of the loose-packing units

typical of the solid state, and an increase in the coordination number. This conclusion is in agreement with the results of investigations of the electrical conductivity of these compounds in the solid and liquid states. We may assume that the increase in the density at the melting points of these compounds is directly due to the destruction of the loose ($K_1 = 4$) antifluorite structure.

Thus, the nature of the transition from the solid to the liquid state in the Mg_2B^{IV} compounds is completely similar to the corresponding transition in the $A^{III}Sb$ analogs, in which the substitution of a heavier element also alters the "metallic component" of the binding.

Thus, investigations of the electrical conductivity, magnetic susceptibility, and density give us a qualitative picture of the nature of changes in the chemical binding and short-range order due to melting.

We shall consider the influence of temperature on the nature of changes in the short-range order of the molten Mg_2B^{IV} compounds.

§3. Changes in the Short-Range Order during the Heating of Molten Mg_2B^{IV} Compounds

A large difference between the short-range structures in the solid and liquid states favors the development and appearance of precrystallization phenomena associated with definite changes in the melt structure as it approaches the solid phase structure during cooling. These changes appear most clearly in the viscosity and density of molten germanium and of molten aluminum, gallium, and indium antimonides [21-23]. In order to follow changes in the short-range order during the heating of molten Mg_2B^{IV} compounds, we used the viscosity method. The viscosity was measured with a high-temperature viscometer and calculated from the formula for weakly viscous liquids. The Mg_2B^{IV} compounds were prepared using the method described in §2 of the present chapter. Since the magnesium compounds were extremely reactive in the molten state, we were unable to use quartz containers and had to take the same measures as in the preparation of these compounds [442]. A

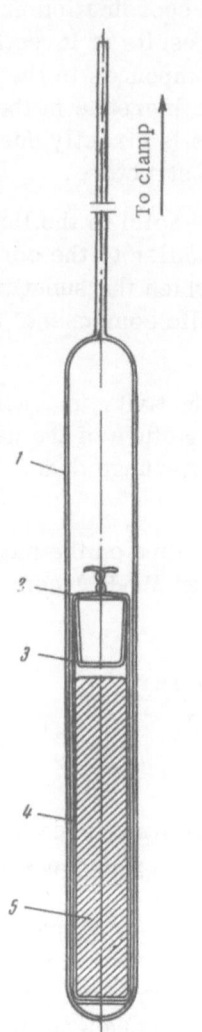

To clamp

Fig. 66. Suspension system used to measure the viscosity of molten Mg_2B^{IV} compounds.

sample 5 (Fig. 66) was placed in a special corundum test tube 4, 80 mm high and of 11-12 mm internal diameter; this tube was closed by a ground corundum stopper 3. The stopper and the test tube were tightly bound with a Nichrome wire 2 and placed in a quartz ampoule 1, which was then evacuated to 10^{-3} mm Hg and sealed. The moment of inertia of the suspension system (which was required in the calculations of the viscosity) was determined using the same test tube with the stopper and binding wire and in the same quartz container but without the sample. After the determination of the moment of inertia, the quartz ampoule was cut open, and the corundum test tube was extracted and filled with the substance being investigated. Then the ampoule was closed and the system pumped down to 10^{-3} mm Hg through a connecting tube and sealed off. Each sample was treated in this way before the measurements were carried out.

The viscosity of each compound was measured using two or three samples. We obtained reproducible results, which are given in Figs. 67-70. It is evident from these figures that the viscosity decreases smoothly when the temperature is increased and that initially the rate of decrease is somewhat more rapid than at higher temperatures.

The viscosity of Mg_2Sn has also been investigated by Gebhardt et al. [438], who have studied mainly the concentration dependence of the dynamic viscosity of the Mg−Sn system and have consequently paid little attention to the temperature dependence of the viscosity of the compound itself. They have obtained

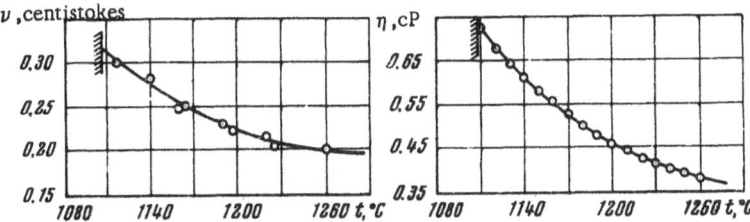

Fig. 67.   Temperature dependences of the kinematic and dynamic viscosity of molten Mg$_2$Si.

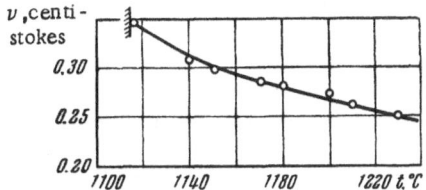

Fig. 68.   Temperature dependence of the kinematic viscosity of Mg$_2$Ge.

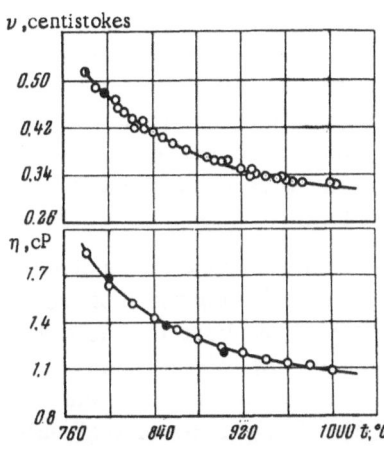

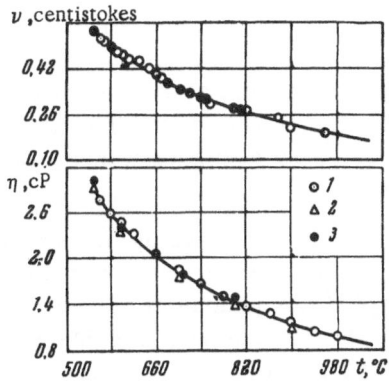

Fig. 69.   Temperature dependences of the kinematic and dynamic viscosities of molten Mg$_2$Sn.   ○) Results of the present authors;   ●) results of Gebhardt et al. [438].

Fig. 70.   Temperature dependences of the kinematic and dynamic viscosity of molten Mg$_2$Pb.   1) Results of the present authors; 2) Knappwost and Horz [439]; 3) Gebhardt et al. [437].

only three experimental points. Using our values of the kinematic viscosity and density, we calculated the dynamic viscosity, which agreed well with the results of Gebhardt et al.

The viscosity of Mg_2Pb has also been determined by Gebhardt et al. [437] in connection with an investigation of the $Mg-Pb$ system. Again, these results agreed well with our data. The viscosity of the $Mg-Pb$ system has also been investigated by Knappwost and Horz [439] and, although they have not investigated the compound Mg_2Pb itself, its viscosity can be deduced from the isotherms given by these authors. The results of Knappwost and Horz are similar to those obtained by us. There are no published data on the viscosity of Mg_2Si and Mg_2Ge.

From the results obtained, we can draw some conclusions about the changes in the short-range order during heating by comparison of the results with the known theoretical relationships.

The kinematic viscosity of the molten Mg_2B^{IV} compounds was used to calculate the free activation energy of viscous flow at various temperatures. From the graphs $F_b \propto f(t)$, it followed that above a certain temperature, whose value decreased regularly along the series $Mg_2Si \rightarrow Mg_2Ge \rightarrow Mg_2Sn \rightarrow Mg_2Pb$, the free activation energy of viscous flow increased practically linearly when the temperature was increased. Mg_2Pb exhibited some relatively slight deviations from the linear law, while Mg_2Sn exhibited a small minimum at a temperature which was only about 20° higher than the melting point; the same minimum was observed more clearly, but at higher temperatures, in the case of Mg_2Ge and Mg_2Si.

Before analyzing the deviations from the linear dependence of F_b, we shall give a general description of the behavior of the Mg_2B^{IV} compounds at higher temperatures (Δt above the melting point). The activation enthropy of viscous flow, which is constant in this range of temperatures, can be determined from the tangent of the slope of the dependence $F_b \propto f(t)$. The results of such calculations are presented in Table 26. In the range of temperatures with which we are concerned, the logarithm of the kinematic viscosity should be a linear function of the reciprocal of the absolute temperature and the tangent of the slope of this function can be used to determine the activation energy of viscous flow E_b. When

Table 26. Activation Energies and Entropies
of Viscous Flow of the Molten Mg_2B^{IV} Compounds

Compound	t_{mp}, °C	Δt, °C *	$E_b \cdot 10^{-3}$, cal/mole	$-S_b$, cal $\cdot$ mole$^{-1} \cdot$ deg^{-1}
Mg_2Si	1102	~90	13.9	2.3
Mg_2Ge	1115	~60	9.5	3.8
Mg_2Sn	778	~40	9.5	5.8
Mg_2Pb	550	~30	9.6	6.2

* Δt is the temperature range beyond which the free
activation energy of viscous flow increases linearly
when the temperature is increased.

the relevant functions are plotted, it is found that the logarithm of
the kinematic viscosity is indeed a linear function of the reciprocal
of the absolute temperature at high temperatures. Appreciable deviations are observed at lower temperatures and these deviations
are weaker for Mg_2Pb than for the other three compounds. Similar
behavior is observed also in the case of the dynamic viscosity. The
values of the activation energy of viscous flow E_b, found in the
range of temperatures where these linear laws apply, are listed in
Table 26.

Analyzing the data on E_b and S_b given in Table 26, we must
point out the fact that the absolute values of these thermodynamic
parameters are similar to those of metals (cf. [74]), which is in
agreement with the conclusion, given in the preceding section, according to which the Mg_2B^{IV} compounds are metallike in the molten state. The maximum value of the activation energy of viscous
flow of Mg_2Si is also in agreement with the conclusion that the
thermal stability of this compound is higher than the stability of the
other three compounds. The lower values of the activation energy
of viscous flow of Mg_2Ge, Mg_2Sn, and Mg_2Pb compared with Mg_2Si
are evidently due to an increase (due to dissociation) in the number
of migrating particles and a decrease in the heights of the potential
barriers which have to be overcome.

Thus, an analysis of the thermodynamic parameters describing the process of viscous flow of the molten Mg_2B^{IV} compounds indicates that these compounds are metallike in the molten state.
Naturally, the results can be regarded only as an indirect proof,

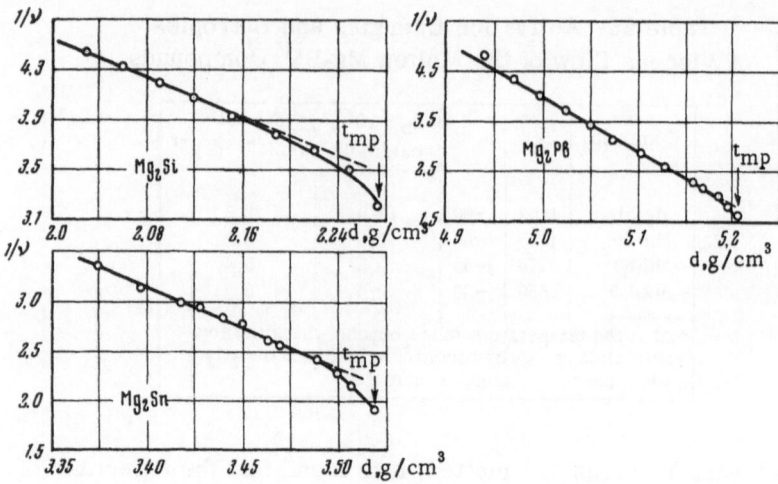

Fig. 71. Dependences of the reciprocal of the kinematic viscosity on the density of the molten Mg_2B^{IV} compounds at various temperatures (the dependences in accordance with Bachinskii's equation are shown dashed).

and the metallike nature of the molten Mg_2B^{IV} compounds can be deduced only by combining these results with those on the electrical conductivity and density. However, since the results of such an analysis of the kinematic viscosity lead to the same conclusion as the results deduced from other properties, the former results can be regarded as reliable.

The dependences $F_b \propto f(t)$ and $\log \nu \propto f(1/T)$ within the temperature interval ΔT above the melting point show that the activation entropy of viscous flow does not remain constant, and in the case of Mg_2Sn, Mg_2Ge, and Mg_2Si it even changes its sign. This indicates considerable changes in the short-range order and the possibility of changes in the structure of activated complexes immediately above the melting point. The nature of the intermolecular interaction obviously changes a little above the melting point. In our opinion, this may be associated with transcritical phenomena or, in the case of cooling, with precrystallization effects; on the other hand, we may be dealing with changes in the degree of dissociation of the Mg_2B^{IV} compounds. However, dissociation is likely to be less important because deviations of $F_b \propto f(t)$ and

log $\nu \propto f$ (1/T) from linearity decrease regularly along the series $Mg_2Si \rightarrow Mg_2Ge \rightarrow Mg_2Sn \rightarrow Mg_2Pb$, which supports the former explanation postulating precrystallization effects in the molten Mg_2B^{IV} compounds. In this respect, the magnesium compounds are analogous to aluminum, gallium, and indium antimonides, which behave similarly [19, 22].

The regular decrease in the deviations along the $Mg_2Si \rightarrow Mg_2Pb$ series may be due to a decrease in the difference between the chemical binding in the solid phase before melting and in the liquid phase, which is indicated by a gradual decrease in the forbidden band width (this width is very small for Mg_2Pb, cf. Table 23) and in the jump of the electrical conductivity at the melting point. This gradual lessening, along the $Mg_2Si \rightarrow Mg_2Pb$ series, of the changes in the thermodynamic parameters of viscous flow within the interval Δt and the decrease of the interval itself are thus correlated with a gradual decrease in the difference between the chemical binding in the solid and liquid phases or, more accurately, with a decrease in the difference between the degrees of metallization. This indicates that the electron density in solid Mg_2Pb before melting is more important than the difference between the short-range structure in the liquid and solid phases and that this difference becomes more important when lead is replaced, respectively, with tin, germanium, and silicon. The direction of changes in the short-range order during heating of the molten Mg_2B^{IV} compounds can be deduced by analyzing the results obtained on the basis of Bachinskii's equation [62, 63].

Figure 71 shows the dependence of the reciprocal of the viscosity on the density of molten Mg_2Si, Mg_2Sn, and Mg_2Pb (the data on the density of molten Mg_2Ge were not available). It is evident from this figure that all three compounds exhibit deviations from Bachinskii's equation and these deviations are of the type observed for water. The deviations observed for the Mg_2B^{IV} compounds may be attributed to the gradual rearrangement of the short-range structure in the melt so that it approaches the solid-phase structure. Since the nature of the volume changes during melting of the Mg_2B^{IV} compounds is the same as in the case of water, germanium, silicon, and $A^{III}B^V$ compounds, we may conclude that these changes amount to a loosening of the structure of the liquid (in the sense of coordination) and the formation of covalent bonds. The loosening

of the liquid structure and the associated decrease in the coordination number need not result in a reduction of the density because the interatomic spacings may decrease. The small deviation from Bachinskii's equation in the case of Mg_2Pb indicates that these changes are less marked in the melt of this compound for reasons already discussed.

Chapter 6

Physicochemical Properties of Molten $A^{IV}B^{VI}$ Compounds with the Galenite Structure

<u>§1. Crystal Structure and Chemical</u>

<u>Binding of $A^{IV}B^{VI}$ Compounds in the</u>

<u>Solid State</u>

The general nature of the chemical binding in the $A^{IV}-B^{VI}$ systems (A^{IV} = Ge, Sn, Pb; B^{VI} = S, Se, Te) is described by phase diagrams with one congruent-melting compound of the $A^{IV}B^{VI}$ type (cf. Figs. 131-133, Chapter 9) [203].

The nature of the chemical binding in these systems is affected by the nature of the "cation" and "anion" components. In the Ge–Te and Sn–Te systems, the composition of the $A^{IV}B^{VI}$ compounds is nonstoichiometric [49-51]. Until recently, it has not been clear whether germanium telluride melts congruently or whether it decomposes before melting by a peritectic transition. Klemm and Frischmuth [443] have reported that germanium telluride melts incongruently. However, McHugh and Tiller [47] have reported that GeTe melts with a definite maximum which corresponds to a composition with 50.61 at.% Te. A eutectic transition takes place 1° below the maximum in the liquidus curve; the composition at the eutectic point is 49.85 at.% Te. The congruent nature of the melting of germanium telluride has been confirmed by a differential thermal analysis, carried out by Abrikosov and Shelimova [51]. In their opinion, germanium telluride melts con-

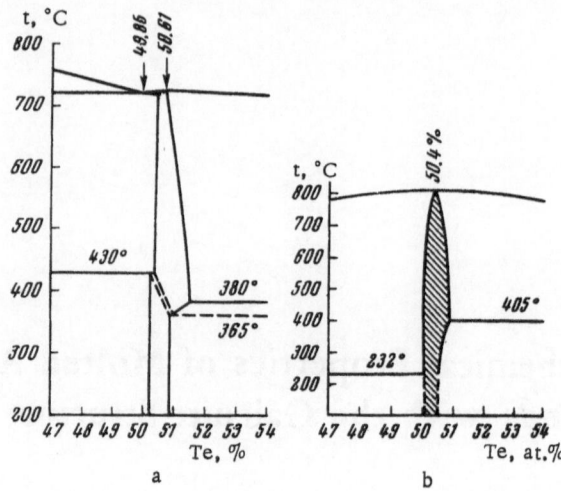

Fig. 72. Phase diagrams of two systems: a) Ge–Te [51]; b) Sn–Te [50].

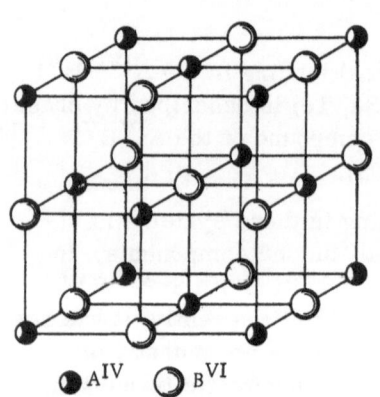

● A^{IV} ○ B^{VI}

Fig. 73.   Crystal structure of $A^{IV}B^{VI}$ compounds.

gruently and exhibits a considerable deviation from stoichiometry. However, the highest quenching temperature used by Abrikosov and Shelimova [51] was 600°C, which is almost 125° below the temperature of the invariant transition and therefore the limited solubility and solidus curves may be somewhat different than those reported. We shall consider this problem in more detail in connection with the analysis of the temperature dependences of the electrical conductivity of germanium and tin tellurides in the solid phase, but for the moment we shall bear in mind that the position of the maximum in the liquidus curve may be different than that reported by McHugh and Tiller [47]. The high-temperature modification of germanium telluride departs by 0.3 at.% from stoichiometry [51] and its homogeneity region extends about 1.2 at.% (at 430°C) along the composition axis. The homogeneity region of the low-temperature modification is somewhat narrower (50.2 at.% to 50.9 at.% Te). A eutectic transi-

tion is observed on the tellurium side at 380°C. The eutectic concentration is 85 at.% Te. Moreover, there have been reports of two horizontal lines, corresponding to invariant transitions in the solid state associated with an allotropic transformation, which is observed in GeTe-based solid solutions (Fig. 72a).

The Sn−Te system has one compound (SnTe), melting congruently at 790°C [203]. Tin telluride divides the phase diagram of this system into two partial diagrams of the eutectic type and the eutectic on the tin side is degenerate. The system near the SnTe region has been investigated in detail by Abrikosov and Shelimova [49, 50] and by Brebrick [444]. As in the case of germanium telluride, a deviation from stoichiometry has been observed. Brebrick [444] has carried out a thermal analysis and has shown that the melting-point maximum (805.9°C) does not coincide with the stoichiometric composition, but lies at 50.4 at.% Te. This differs somewhat from the results reported in [445], where this maximum has been found at 50.9 at.% Te. The limits of the homogeneity region of tin telluride, deduced by Brebrick from the electrical properties, are in fairly good agreement with the results reported in [50], which have been obtained by microscopic and x-ray diffraction analyses and by measurements of the microhardness. According to Abrikosov and Shelimova [50], the homogeneity region at 400°C extends from 50.1 to 50.9 at.% Te (Fig. 72b). The solidus curves at high temperatures may be modified by later measurements because the highest quenching temperature employed in [50] to determine these curves has been 100° lower than the melting point of tin telluride.

From an examination of the phase diagrams of the Pb−S, Pb−Se, and Pb−Te systems [203], it follows that only one compound PbB^{VI} is formed in these systems and that this compound melts congruently at much higher temperatures than its components.

The Pb−Te phase diagram, given by Hansen and Anderko, includes a wide range of solid solutions based on lead telluride (from 30 to 57 at.% Te). However, Miller, Komarek, and Cadoff [45, 446], as well as Brebrick, Allgaier, and Gubner [447, 448], who investigated the phase diagram near lead telluride, found that these results are incorrect and that the homogeneity region of lead telluride is narrow. Moreover, they found a small deviation from stoichiometry at high temperatures. Thus, according to Miller,

Komarek, and Cadoff [45], the maximum in the liquidus curve (923.9°C) does not coincide with the stoichiometric composition but is displaced toward tellurium and corresponds to 50.002 at.% Te and a hole density of $3.8 \cdot 10^{18}$ cm^{-3}.

According to Brebrick and Gubner [448], this shift is slightly larger (50.012 at.% Te, $7 \cdot 10^{18}$ holes per 1 cm^3). The deviation from stoichiometry decreases at lower temperatures and at 860°C the solidus line on the lead side intersects the stoichiometric composition (according to Miller, Komarek, and Cadoff [45], this intersection occurs 0.7° below the maximum melting point). The maximum extent of the homogeneity region of lead telluride is estimated in [446] to be 0.04 at.%. Brebrick, Allgaier, and Gubner [447, 448] have shown that the homogeneity region is widest at 800°C and that it represents 0.015 at.% ($1.5 \cdot 10^{18}$ cm^{-3} electron density for PbTe saturated with lead, and $7.7 \cdot 10^{18}$ cm^{-3} hole density for PbTe, saturated with tellurium). The homogeneity region decreases rapidly when the temperature is reduced.

The phase diagram of the Pb−Se system near lead selenide has been refined by Goldberg and Mitchell [449]. They have shown that the maximum in the liquidus curve (1065°C) corresponds to 50.005 at.% Se. Brebrick and Gubner [450], who have investigated this system near lead telluride in the temperature range 500-300°C, have found a narrow homogeneity region (0.012 at.% wide), which is symmetrical with respect to the stoichiometric composition.

The nature of the phase equilibrium in the Pb−S system has not yet been finally determined because of the high vapor pressure of sulfur. Nevertheless, Bloem and Kröger [451] have definitely established that the homogeneity region of lead sulfide is very narrow (PbS$_{0.9795}$ − PbS$_{1.0005}$) and that there is a small deviation from stoichiometry at high temperatures. The maximum in the liquidus curve (1127°C) is displaced from the stoichiometric composition toward lead by 0.015 at.% (this corresponds to $6 \cdot 10^{18}$ excess lead atoms per cm^3).

The A^{IV}B^{VI} compounds have the galenite-type structure. This structure is shown in Fig. 73.

The unit cell is basically of the fcc type, in which the chalcogen atoms are distributed in accordance with the closest packing law and the A^{IV} atoms fill all the octahedral vacancies. The co-

ordination number is 6 for atoms of both types, i.e., each A^{IV} atom is surrounded by six B^{VI} atoms and each B^{VI} atom is surrounded by six A^{IV} atoms. Each unit cell contains four A^{IV} atoms and four B^{VI} atoms, i.e., four $A^{IV}B^{VI}$ formula units. The lattice parameters† of the $A^{IV}B^{VI}$ compounds (NaCl-type structure, K = 6) are given below (in Å):

$$\begin{aligned}
&\text{GeTe} - 5.98; \; 6.02 \; (600°C) \\
&\text{SnTe} - 6.29 \\
&\text{PbTe} - 6.45 \\
&\text{PbSe} - 6.13 \\
&\text{PbS} \; - 5.94
\end{aligned}$$

The lattice period increases when the "cation" or "anion" is replaced with a heavier element. Germanium telluride has two modifications. According to Schubert and Fricke [457, 458], the low-temperature modification of germanium telluride has the rhombohedral structure with the parameters $a = 5.986$ Å and $\alpha = 88.35°$, whereas the high-temperature modification has the NaCl-type cubic structure. The temperature of the rhombohedral-to-cubic transition depends on the composition. In the presence of excess germanium, the transition takes place at 460°C, but in the presence of excess tellurium it occurs at 390°C. The lattice parameter of the high-temperature modification of GeTe on the germanium side is $a = 6.01$ Å at 460°C and on the tellurium side it is 5.992 Å at 390°C [458].

According to Shelimova, Abrikosov, and Zhdanova [51], the polymorphic transition temperature of GeTe-based solid solutions is 430°C on the germanium side and 365°C on the tellurium side; the lattice period of stoichiometric germanium telluride at 600°C is 6.020 ± 0.005 Å.

The nature of the chemical binding in the $A^{IV}B^{VI}$ compounds has not yet been finally resolved. Initially, it was assumed that lead chalcogenides are ionic semiconductors [455]. However, calculations of some properties, carried out on the assumption of an ionic interaction, have been found to disagree with the experimental data, as presented convincingly by Tsidil'kovskii [459, 460]. In-

---

†Average values taken from [203, 452-456]. The values given in these papers are in good agreement with one another.

vestigations of thermomagnetic properties, which are very sensitive to the nature of the interaction between carriers and the crystal lattice, have demonstrated that the covalent binding predominates in lead chalcogenides [459, 460]. A similar conclusion has been reported by Devyatkova and Smirnov [461], who investigated the thermal conductivity of these compounds.

There are two conflicting views about the nature of the chemical binding in tin and germanium tellurides. Sagar and Miller [462] investigated the electrical properties and found that tin telluride is a semimetal with overlapping valence and conduction bands. This conclusion has been confirmed by Damon, Martin, and Miller [463], who investigated the electrical conductivity, Hall coefficient, and the thermoelectric power in transverse magnetic fields. Similar views on the binding in germanium telluride have been expressed in [464, 465]. On the other hand, many dependences, in particular the anomalous dependence of the thermoelectric power on the effective hole density, cannot be explained satisfactorily by the semimetal model. Consequently, Allgaier and Scheie [466] have explained the electrical properties of tin telluride on the basis of their semiconductor model, in which the valence band has two sub-bands of different energies. Later, this model was confirmed by Brebrick et al. [467-469], who investigated the dependence of the Hall coefficient and thermoelectric power on the temperature and effective carrier density. Thus, it follows from [466-469] that tin telluride is a semiconductor with a high carrier density because of the deviations from stoichiometry. However, even if it is assumed that the valence band of tin telluride is complex, the published experimental data can be explained only qualitatively.

It has been suggested that germanium telluride is also a semiconductor with a complex band structure. Thus, Kolomoets et al. [470] showed that the carrier density can be altered by almost one order of magnitude by varying the tellurium concentration. This suggests that germanium telluride is a semiconductor. From these results, as well as from the dependences of the thermoelectric power and electrical conductivity on the composition and temperature, Kolomoets et al. [471] concluded that the valence band is complex and consists of two sub-bands with different densities of states, which suggests the presence of heavy and light holes. The band structure of germanium telluride proposed in [471]

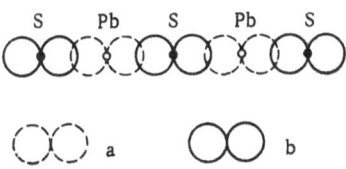

Fig. 74. Schematic representation of the overlap of the p eigenfunctions along the [100] direction in the PbS lattice [420]; a and b represent, respectively, weakly and strongly filled quantum states [420].

is practically identical with the band structure of tin telluride described in [468].

The present authors [472] have suggested that germanium and tin tellurides are degenerate semiconductors, in which the forbidden band has not been found so far because of the presence of equilibrium deviations from stoichiometry. The same conclusion was reached simultaneously by Abrikosov and Shelimova, who were the first to plot the phase diagrams in the range of compositions next to the compounds SnTe and GeTe [49-51]. The nature of the temperature dependences of the electrical conductivity and thermoelectric power of germanium and tin tellurides, reported in [473, 474], may be due to a reduction in the deviation from stoichiometry at higher temperatures (this should alter the density and nature of the carriers). This problem will be discussed in more detail when the relevant experimental data are presented.

Krebs [420] has analyzed the nature of the chemical binding in semiconducting compounds discussed in the present chapter and has concluded that the covalent binding is predominant. Electron pairs of the anion in compounds of the PbS-type have a tendency to fill the quantum states in the cation which has become vacant due to ionization. This results in a partial neutralization of the charge and an increase in the covalent component of the binding [316].

Such an analysis of the chemical binding in lead sulfide has enabled Krebs to explain the semiconducting nature of compounds with the rocksalt structure and the relatively narrow forbidden band of these compounds (the latter observation had not been explained before). The analysis carried out by Krebs can be summarized as follows. A Pb^{2+} ion has three vacant 6p states. A S^{2-} ion has one pair in the 3s state and three pairs in the 3p state. The 3p functions are oriented along the axes of an orthogonal system of coordinates and have, in contrast to the eigenfunctions of the sp^3 hybrid, large values along two opposite directions, i.e., along the negative and positive directions of each axis (cf. Fig. 74). This

Table 27. Properties of $A^{IV}B^{VI}$ Compounds in Solid State

Property	GeTe	SnTe	PbTe	PbSe	PbS	References
ΔX^*	—	0.4	0.5	0.8	0.9	[476]
t_{mp}, °C	725	790	917	1088	1119	[203, 424, 454, 477]†
Q_f, kcal/g-atom	5.65	4,0	3.75	4.25	4.35	[424, 480, 481]
$-\Delta H_{298}$, kcal/g-atom	4.0	7,3	8.4	10.5	11.25	[424, 477, 480, 485, 486]
ΔE, eV	—	0.26	0.29	0.22	0.34	[111, 496]‡
u_n, cm²·V⁻¹·sec⁻¹(4.2°K)	—	—	80·10⁴	14·10⁴	80·10⁴	⎱[463, 497]
u_p, cm²·V⁻¹·sec⁻¹ (77.4°K)	—	0.65·10³ (78°K)	17·10³	14·10³	15·10³	⎰

* Values of ΔX allow for the valence of the components in the $A^{IV}B^{VI}$ compounds. According to a model proposed by Krebs [418], the valence of the "cations" in the $A^{IV}B^{VI}$ compounds is two. However, the values of the electronegativity of divalent germanium is not known and therefore the difference of the electronegativities is not given for germanium telluride.

†Somewhat different values of the melting points are given in [45, 444, 449, 451, 478, 479]; also slightly different values of the heat of fusion are given in [482-484], and of the heat of formation in [452, 455, 484, 487].

‡There are many papers dealing with the determination of the forbidden band width of the $A^{IV}B^{VI}$ compounds. The published results given in these papers are in poor agreement (cf. [455, 479, 488-495]). Therefore, Table 27 lists the values of the forbidden band width at room temperature, which according to Smith[111], are the most reliable

orthogonal orientation is exhibited by the p functions directed toward six lead ions surrounding each sulfur ion. The three p eigenfunctions of each lead ion are directed in exactly the same way toward the six nearest sulfur ions. Consequently, the p eigenfunctions of sulfur and lead can overlap strongly so that electrons are pulled away from the sulfur ion to the lead ion and partly neutralize the charge of the latter. The covalent component of the binding is produced in this way. The p electrons are located mainly near the sulfur atom since it has a larger number (four) of the p electrons in the free state and is therefore much more electronegative.

The p states of lead are filled less completely. The appearance of the covalent binding along two directions (for each pair of the p electrons) is attributed by Krebs to resonance. The resonance binding is with the Pb^{2+} ions located to the left and right of S^-:

$$Pb^+ \leftarrow S^- \; Pb^{2+} \underset{\longleftrightarrow}{\text{resonance}} Pb^{2+} \; S^- \rightarrow Pb^+.$$

The resonance bond, formed by electrons in the p states, reduces considerably the energy required for the excitation of electrons to the conduction band and this explains the narrow forbidden band of compounds of this type. This is because when an electron-hole pair is formed, one electron is split off from a S^{2-} ion and, therefore, fewer electrons participate in the covalent bond between an S^- ion (formed by this loss of one electron) and the nearest Pb^{2+} ions. Therefore, the S^{2-} ions in the second coordination sphere can participate more strongly, via their electrons, in the formation of covalent bonds with Pb^{2+} ions in the first coordination sphere. As a result of this, the wave functions of the p electrons of S^{2-} ions in the second coordination sphere overlap the corresponding wave functions of the S^- ion, which results in some compensation of the lost negative charge. Thus, the loss of an electron in a system of valence bonds affects the crystal as a whole and not only the region next to one or two atoms. Consequently, the ionization in rocksalt-type crystals should be easier. The strong covalent component of the binding in lead sulfide is responsible for the relatively short distance between lead and sulfur atoms (2.97 Å), which is 3% smaller than the value calculated from the Gol'dshmidt values of the ionic radii (3.06 Å). The same covalent component is responsible also for the 4% higher experimental value of the lattice energy than the value calculated on the assumption of ionic binding (732 instead of 705 kcal/mole). In the case of lead selenide, this difference is 6% (727 instead of 684 kcal/mole) [475]. Hence, we may conclude that the covalent component makes a considerable contribution to the balance of forces which represent the chemical binding between atoms in the lattices of the $A^{IV}B^{VI}$ compounds. Table 27 lists some properties of the $A^{IV}B^{VI}$ compounds, which can be used to follow the changes in the nature of the chemical binding in these compounds due to the replacement of the "cation" and "anion" components.

It follows from Table 27 that the replacement of the "anion" with a lighter element enhances the ionic component of the binding, as indicated by an increase in the difference between the electronegativities. This replacement increases the strength of the chemical bonds (the heat of formation increases).

When the "cation" is replaced with a lighter element, the ionic component of the binding is not greatly affected but the metal-

lic component is enhanced. This reduces the strength of the bonds, as indicated by a decrease in the heat of formation.

Homologous series of analogs in the A^{IV}B^{VI} group, formed by the replacement of "cations" and "anions" with other elements, differ somewhat from the series of compounds in other groups (for example, A^{III}B^{V} and Mg$_2$B^{IV}).

The forbidden band width of A^{IV}B^{VI} compounds does not decrease (a decrease is expected because the binding energy of valence electrons decreases when the atomic number is increased) but varies anomalously. The forbidden band has not yet been discovered in germanium telluride and its width for lead telluride is greater than those for tin telluride and lead selenide.

Moreover, the melting point in the "cation" substitution series increases as the atomic number of the "cation" gets higher, whereas, in all other cases, it decreases. It follows that studies of the physicochemical properties of the A^{IV}B^{VI} compounds at the melting point and in the liquid phase are extremely important, since such studies should help in the solution of the problem of the appearance of semiconducting properties in compounds of this group in the solid state.

§2.   Changes in the Short-Range Order and the Chemical Binding at the Melting Points of A^{IV}B^{VI} Compounds

1. Changes in the Electrical and Magnetic Properties at the Melting Points of A^{IV}B^{VI} Compounds

These compounds were prepared from the following materials: germanium single crystals of nearly intrinsic resistivity; tin of OVCh-000 grade (99.9995% Sn); lead of S-00 grade (99.9992% Pb); tellurium of the TA-1 grade, subjected to additional purification by vacuum distillation (purity not less than 99.999%); selenium of analytic purity grade; high-purity sulfur of grade A-2. The compounds were synthesized by direct fusion of the components, taken in their stoichiometric ratios and placed in evacuated (to 10^{-3} mm Hg) and sealed quartz ampoules. Since slight oxidation of the lead was still possible in such a vacuum and the resultant lead monoxide

could react with the quartz, the lead chalcogenides were prepared
in quartz ampoules coated internally with a mirrorlike layer of
pyrocarbon, deposited using the method described in [98]. The
fusion was carried out for 4-5 h at temperatures 50-60° higher
than the melting point, and the melt was stirred vigorously from
time to time; cooling took place in the furnace at a rate of 20-30°
per hour. The compounds obtained were crushed and placed in
sharp-ended evacuated (10^{-3} mm Hg) quartz ampoules (the ampoules used for lead telluride were coated internally with graphite). These ampoules were placed in a tubular resistance furnace
and the temperature gradient along this furnace was measured
first before carrying out directional crystallization (horizontal
zone refining).

The conditions during this zone refining were:

	GeTe	GeTe	SnTe	PbTe
Rate of motion of the ampoule, mm/min	0.4	0.2	0.7	0.2
Temperature of hottest zone, °C	740	790	850	950—980

This treatment produced ingots of GeTe 90 mm long and 9 mm in
diameter, containing a second phase throughout the ingot. The
SnTe ingots were also of the double-phase type. In the case of
PbTe, we obtained single-phase ingots with the same microhardness along the length; these ingots were dense and coarse-grained,
9 mm in diameter, and 70 mm long.

Since the stoichiometric GeTe and SnTe contained (immediately after preparation and after their horizontal zone refining) a
small amount of a second phase, we used slightly nonstoichiometric
charges. The single-phase form of GeTe was obtained in the presence of 0.6 at.% excess tellurium; and of SnTe, in the presence of
0.3 at.% excess Te [50, 51].

Horizontal zone refining was not used in the preparation of
the lead sulfide and selenide. Investigations of the microstructure
indicated that the ingots prepared from components taken in the
stoichiometric ratio were of the single-phase type. Bearing in
mind the narrow region of the homogeneity of alloys based on lead
sulfide and selenide, we could assume that the composition of the
materials obtained was close to stoichiometric.

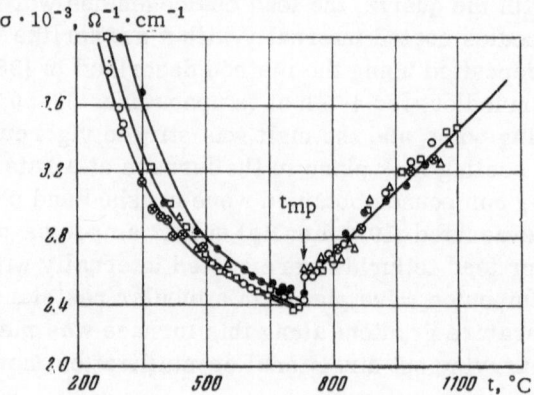

Fig. 75. Temperature dependence of the electrical conductivity of germanium telluride in the solid and liquid states (results of five experiments).

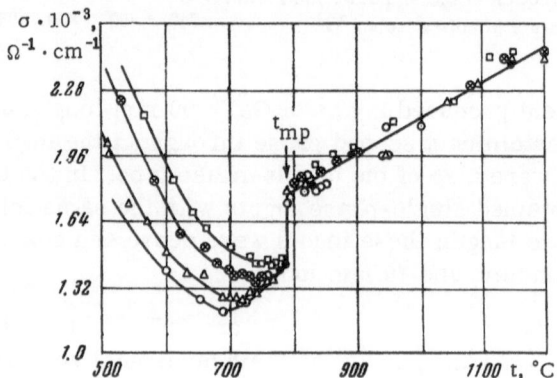

Fig. 76. Temperature dependence of the electrical conductivity of tin telluride in the solid and liquid states (results of five experiments).

Samples were cut from PbS, PbSe, SnTe, and GeTe stoichiometric ingots, prepared by direct fusion of the components, from GeTe ingots with an excess of 0.6 at.% Te, and from SnTe ingots with an excess of 0.3 at.% Te (the latter two materials were also prepared by direct fusion), as well as from the middle parts of the lead telluride ingots prepared by the horizontal zone melting method.

Samples of 10 mm diameter and 10 mm high were sealed into evacuated (10^{-3} mm Hg) cylindrical quartz ampoules of suitable dimensions. Since the lead telluride, selenide, and sulfide reacted with the quartz and, moreover, the lead selenide and sulfide were very volatile at high temperatures, these compounds were placed in corundum cans with ground corundum covers, which were sealed into evacuated quartz ampoules in the same manner as described in Chapter 5 for the Mg_2B^{IV} compounds. The measurements were carried out on 2-5 samples of each compound. The electrical conductivity and magnetic susceptibility were measured by a method described in Chapter 2 [472, 473, 498].

Published data on the electrical conductivity near the melting point and in the liquid state were available only for lead telluride and selenide [3, 8]. The temperature dependences of the electrical conductivity and magnetic susceptibility of other compounds have not yet been investigated in this region. The results of all the measurements are presented in Figs. 75-77.

When the temperature is increased the electrical conductivity of germanium telluride in the solid state falls rapidly at first and then more slowly. At the melting point, only a small rise is observed but in the liquid state the electrical conductivity rises quite rapidly. The results obtained for different samples in the solid state do not agree; this was probably due to the different degrees of deviation from stoichiometry. However, just before the melting point, all three curves practically merge.

A very similar dependence is also observed for tin telluride (Fig. 76) except that the conductivity rises slightly just before the melting point; this is probably due to a reduction in the deviations from stoichiometry and the appearance of the intrinsic conduction.

The results for two samples of lead chalcogenides, originating from different batches, are given in Fig. 77. Lead telluride exhibits a rise in the electrical conductivity before its melting point, which is associated with the intrinsic conduction. Lead sulfide and selenide exhibit some increase in the electrical conductivity, but just before the melting points the conductivity decreases appreciably. This decrease is barely noticeable for lead telluride. The electrical conductivity of lead telluride rises strongly at the

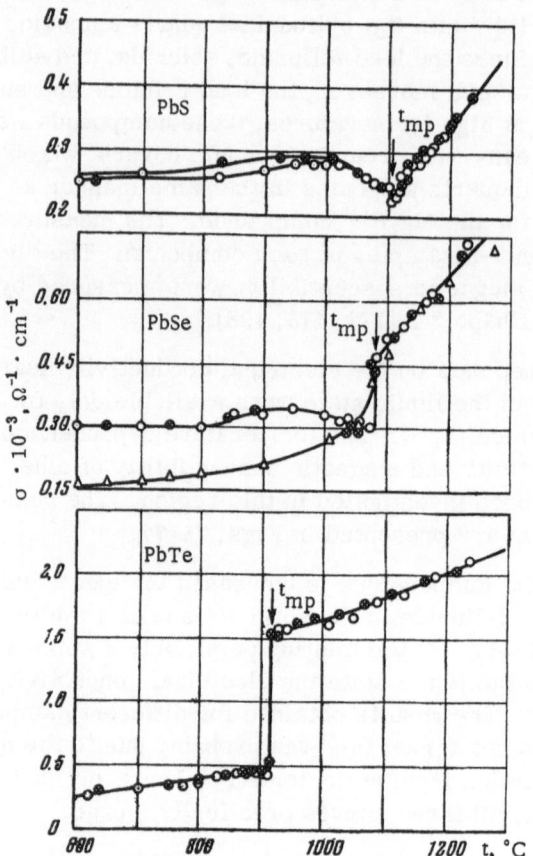

Fig. 77. Temperature dependence of the electrical conductivity of lead chalcogenides in the solid and liquid states. O, ⊗)
Results of the present authors; △) Regel' [3, 8].

melting point, but the rise is much less for lead selenide and the
conductivity of lead sulfide decreases at its melting point. The
electrical conductivity of all three compounds increases appreciably with temperature in the molten state.

The results for solid lead telluride are fairly close to those
obtained by A. R. Regel', but our values for the liquid state are appreciably higher. This is probably due to the fact that Regel' did
not allow for the change in the density at the melting point of lead
telluride. Our results for lead selenide below the melting point

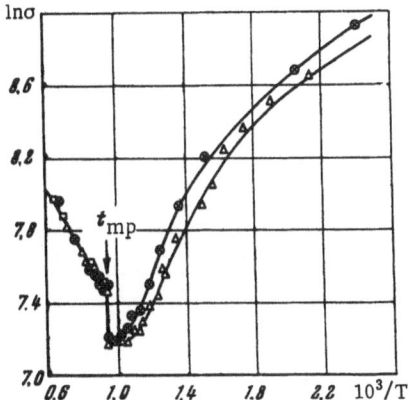

Fig. 78. Dependence $\ln\sigma \sim f(1/T)$ for tin telluride in the solid and liquid states (results for three samples).

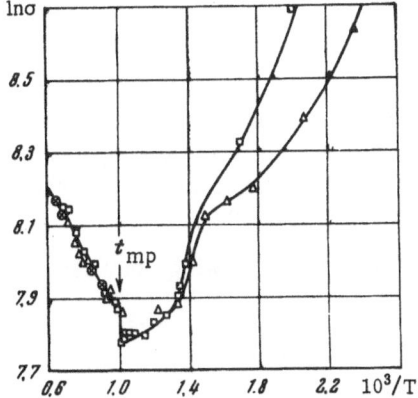

Fig. 79. Dependence $\ln\sigma \sim f(1/T)$ for germanium telluride in the solid and liquid states (results for three samples).

and in the liquid phase agree well with those of Regel' but they diverge at lower temperatures in the solid state. This divergence is most probably due to the different purities of the materials used to prepare the compound.

Table 28. Changes in Electrical Conductivity
at the Melting Points of the $A^{IV}B^{VI}$ Compounds

Compound	$\sigma_s \cdot 10^{-3}$, $\Omega^{-1} \cdot cm^{-1}$	$\sigma_l \cdot 10^{-3}$, $\Omega^{-1} \cdot cm^{-1}$	σ_s / σ_l
GeTe	~2.40	~2.60	~0.92
SnTe	~1.44	~1.80	~0.80
PbTe	~0.42	~1.52	~0.28
PbSe	~0.30	~0.45	~0.67
PbS	~0.25	~0.22	~1.14

These results were analyzed and plotted in semilogarithmic
coordinates in the form of dependences of the natural logarithm of
the electrical conductivity on the reciprocal of the absolute temperature (Figs. 78-80).

It is evident from Figs. 78 and 79 that in the solid state the
dependence changes markedly below the melting point; this is most
probably due to a rise in the carrier density because of the dissociation of the lattice (we shall discuss this point in detail in the
next subsection). It is worth noting the steep fall in the $\ln \sigma \propto f(1/T)$
curve for germanium telluride near the allotropic transition, which
is practically imperceptible in the temperature dependence of the
electrical conductivity plotted in the usual coordinates. An intrinsic conduction branch is observed fairly clearly for lead chalcogenides in the solid state (Fig. 80).

Calculations yield the following results:

Compound	PbTe	PbSe	PbS
Forbidden band width, eV	~0.3	~0.23	~0.36

These results are in fairly good agreement with the values given in
Table 27. The value of ΔE for lead sulfide is particularly close to
the result obtained by Scanlon [499] by optical and thermal methods
(0.37 eV). The results obtained in the determination of ΔE indicate
that our experimental temperature dependence of the electrical
conductivity of the $A^{IV}B^{VI}$ compounds in the solid state are quite
reliable. The dependences $\ln \sigma \propto f(1/T)$ for the liquid phases of
all five investigated compounds are strictly linear and consequently the temperature dependence of the electrical conductivity of

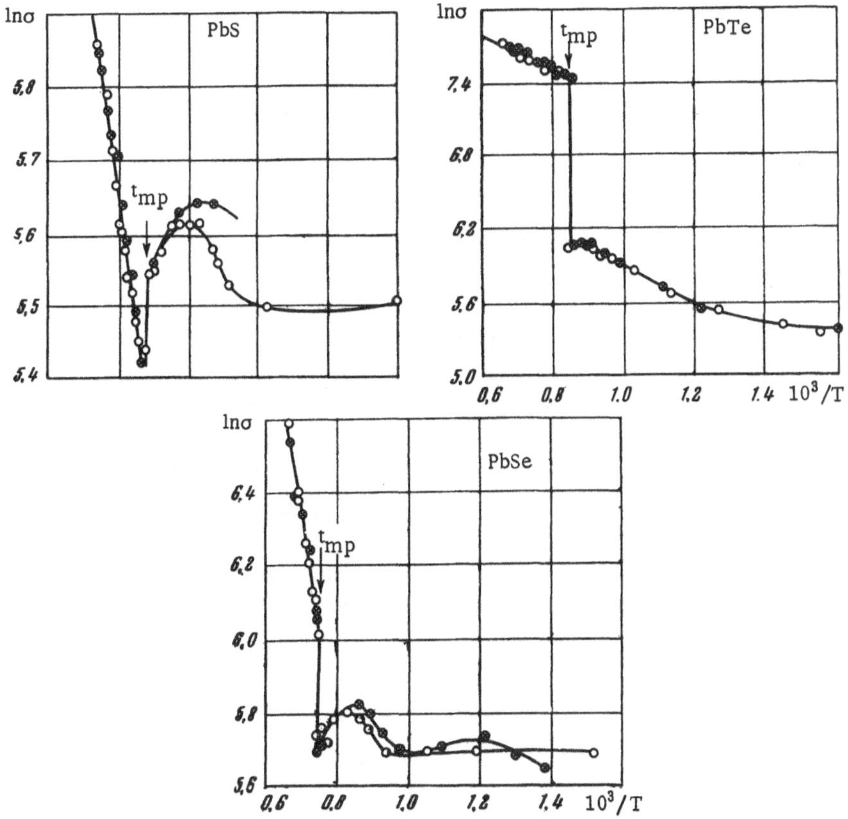

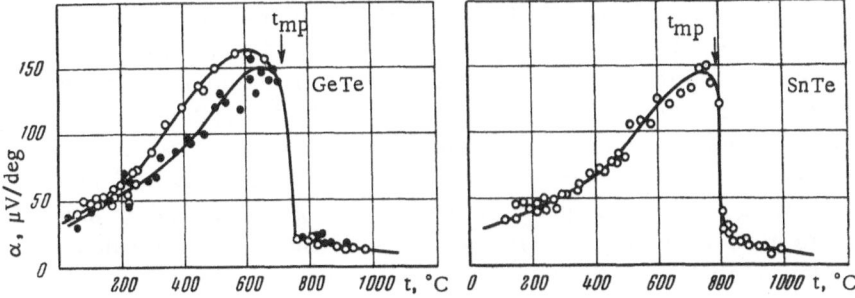

Fig. 80. Dependence $\ln \sigma \propto f(1/T)$ for lead chalcogenides in the solid and liquid states (results for two samples of each compound).

Fig. 81. Temperature dependences of the thermoelectric power of germanium and tin tellurides in the solid and liquid states. O and ● represent different samples.

molten $A^{IV}B^{VI}$ semiconductors can be described by means of the
equation

$$\sigma_{l(s)} = \sigma_0 \exp(-\Delta E/2kT), \tag{79}$$

where ΔE is the activation energy of the electrical conductivity in
the liquid phase.

Table 28 lists the values of the electrical conductivities of
the $A^{IV}B^{VI}$ compounds, in the solid and liquid phases, at their
melting points, as well as the values of the ratio σ_s/σ_l, which can
be regarded as a measure of the sudden change in the electrical
conductivity at the melting point. It is evident from this table that
the electrical conductivity of the solid and liquid phases increases
along the series PbS → GeTe. The absolute value of the electrical
conductivity of germanium telluride in the liquid phase does not
reach values typical of liquid metals (cf. Chapters 4 and 5) and occupies an intermediate position between metals and semiconductors (closer to semiconductors). The other compounds have electrical conductivities, which are typical of semiconductors.

Bearing in mind the absolute values of the electrical conductivity and its positive temperature coefficients in the liquid state
(cf. Figs. 75-80), we may conclude that the $A^{IV}B^{VI}$ compounds
undergo the "semiconductor—semiconductor" transition at the
melting point. Thus, the $A^{IV}B^{VI}$ compounds in the liquid state are
semiconductors whose values of the electrical conductivity decrease along the series GeTe → PbS.

Semiconducting properties are due to covalent or covalent—
ionic bonds between atoms (in the latter case, the covalent binding
predominates). Consequently, the $A^{IV}B^{VI}$ compounds in the liquid
phase have mainly covalent bonds. In the solid state covalent
crystals have three-dimensional networks of rigid bonds. Such
networks do not exist in the liquid state but are replaced by one-
dimensional molecular [3] or chainlike [65] structures. Hence,
we may conclude that one-dimensional structures may be formed
in molten $A^{IV}B^{VI}$ compounds. The nature of the packing of structural units in the liquid state may be such that the short-range
order is still close to the solid-phase structure either over a fairly wide range of temperatures or at least near the melting point.
In the latter case, heating should result in the rearrangement of

the short-range order, so as to produce a stable configuration which does not vary greatly when the temperature is increased. The fall in the electrical conductivity of lead sulfide at the melting point (Table 28) suggests that this compound has a molecular structure in the liquid state. The nature of this jump in the electrical conductivity of lead sulfide is similar to that observed by Regel' [3] for lead selenide. Other compounds of this group are more likely to form chainlike structures of the $-A-B-A-B-$ type, at the ends of which carriers are generated. This difference between the structure of molten lead sulfide and the structures of the other investigated $A^{IV}B^{VI}$ compounds may be due to an increase in the ionic component of the binding along the $A^{IV}Te \rightarrow PbSe \rightarrow PbS$ series, which is supported by the observed increase in the difference between the electronegativities (cf. Table 27). The rise in the electrical conductivity at the melting points of the $A^{IV}B^{VI}$ compounds (with the exception of PbS) indicates an increase in the carrier density because we can hardly expect an increase in the mobility in the transition to a more defective structure in the molten state.

The maximum change in the electrical conductivity is observed for lead telluride. This change decreases and even changes sign when the "anion" is replaced with a lighter element. The same is observed also in the case of the "cation" replacement. The value of the change decreases considerably along the series PbTe $\rightarrow$ SnTe $\rightarrow$ GeTe. However, we cannot determine exactly the cause of this decrease, because the value of the electrical conductivity changes simultaneously in the liquid and solid phases along this series. Nevertheless, we can definitely conclude that the main cause of the changes in the general nature of the electrical conductivity of lead chalcogenides, which are observed when the "anion" is replaced with a lighter element, is an increase in the ionic component of the binding. In the case of the tellurides of Group IV elements, the replacement of the "cation" with a lighter element increases the metallic component of the binding.

The temperature dependences of the thermoelectric power of these five compounds in the solid and liquid states are in full agreement with the results obtained from the investigations of the electrical conductivity.

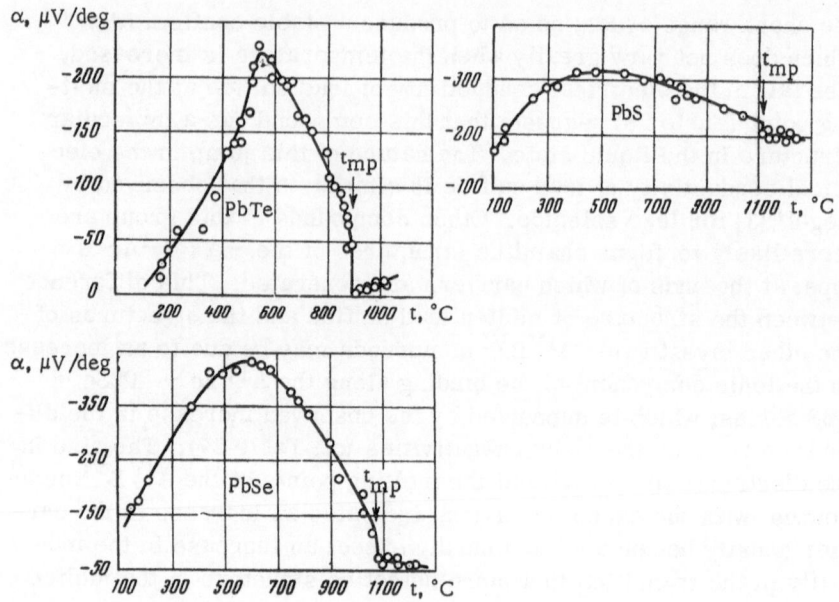

Fig. 82. Temperature dependences of the thermoelectric power of lead chalcogenides in the solid and liquid state.

The results of measurements on many samples of the A^{IV}B^{VI} compounds, obtained both during the heating and cooling of them, are collected in the temperature dependences of the thermoelectric power shown in Figs. 81 and 82.

The temperature dependences for germanium and lead tellurides [474, 500, 501] are given in Fig. 81. We can see that the thermoelectric power in the solid state increases rapidly when the temperature is increased and that only just before the melting point is there a small drop, which follows the maximum in the $\alpha \propto f(t)$ curves. At the melting point, the thermoelectric power falls sharply and it continues to fall, although more slowly, in the liquid state. The p-type of conduction is retained in the molten state. Both the nature of the temperature dependences of the thermoelectric power of these compounds in the solid state and the absolute values of this property are in good agreement with the results obtained in other investigations. The thermoelectric power in the molten state had not been determined before.

Table 29. Changes in Thermoelectric Power
at the Melting Points of the $A^{IV}B^{VI}$ Compounds

Compound	α_s, $\mu V/deg$	α_l, $\mu V/deg$	α_s/α_l
GeTe	$+130$	$+ 21$	6
SnTe	$+140$	$+ 28$	5
PbTe	$- 60$	$- 10$	6
PbSe	-120	$- 60$	2
PbS	-220	-200	1.1

Using germanium telluride, we also checked the influence of the purity of the tellurium used to prepare this compound on the nature of the temperature dependence of the thermoelectric power. We prepared samples of this compound from unpurified tellurium of the TA-1 grade. The results of measurements carried out on these samples are presented in Fig. 81 (black dots). We can see that the thermoelectric power of the less pure samples is slightly higher than that of pure samples in the solid state, but there is no difference in the liquid state.

Table 29 lists the values of the thermoelectric power of the $A^{IV}B^{VI}$ compounds at their melting points.

It follows from this table that the melting of germanium and tin tellurides reduces the thermoelectric power by a factor of 5-6 but it does not alter its positive sign. The constancy of the sign indicates that the semiconducting properties of these compounds are retained in the molten state. This is also supported by a fall of the thermoelectric power during heating in the molten state. The considerable drop of the thermoelectric power at the melting point may be due to the equalization of the electron and hole mobilities which follows from the general expressions for the thermoelectric power of a semiconductor in the intrinsic conduction region. The thermoelectric power should also decrease because of the rise in the carrier density, indicated by the rise in the electrical conductivity at the melting point.

Thus, the p-type conduction and the fall of the thermoelectric power when the temperature of the melt is increased both support the conclusion that germanium and tin tellurides remain semicon-

ductors in the molten state and are not semimetals in the solid state.

There is some correlation between the temperature dependences of the thermoelectric power and electrical conductivity. The correlation is observed in the molten and solid states of GeTe and SnTe. A maximum in the $\alpha \propto f(t)$ curves corresponds to a sharp change of the slope (a minimum in the case of SnTe) of the $\sigma \propto f(t)$ curves, which is evidently due to the appearance of intrinsic conduction (this point will be discussed in detail later).

Figure 82 shows the temperature dependences of the thermoelectric power of lead chalcogenides in the solid and liquid states. In the solid state, all three compounds exhibit maxima and have the p-type conduction. The fall of the thermoelectric power beyond the maximum in the solid state is evidently due to intrinsic conduction. At the melting point, the thermoelectric power falls suddenly and continues to fall when the liquid phase is heated. The absolute values of the thermoelectric power in the liquid state increase markedly when an "anion" is replaced with a lighter element, reaching a fairly large value (about 200 μV/deg) for lead sulfide. Table 29 gives the values of the thermoelectric power of lead chalcogenides at their melting points. It is evident from this table that the fall in the thermoelectric power at the melting point decreases regularly along the series PbTe → PbSe → PbS.

The value of this fall for lead telluride is almost exactly the same as that for germanium and tin tellurides.

These results indicate that semiconducting properties are retained by lead chalcogenides in the molten state and that these properties are enhanced when an "anion" is replaced with a lighter element. The replacement of a "cation" with a lighter element leaves the value of the fall at the melting point practically unaffected; and the low absolute values of the thermoelectric power also indicate a considerable increase in the metallic component of the binding in the molten state of the A^{IV}Te compounds.

Some additional data on the nature of the melting of the $A^{IV}B^{VI}$ compounds are provided by the magnetic susceptibility, which was investigated only for tellurides of Group IV elements [472, 473].

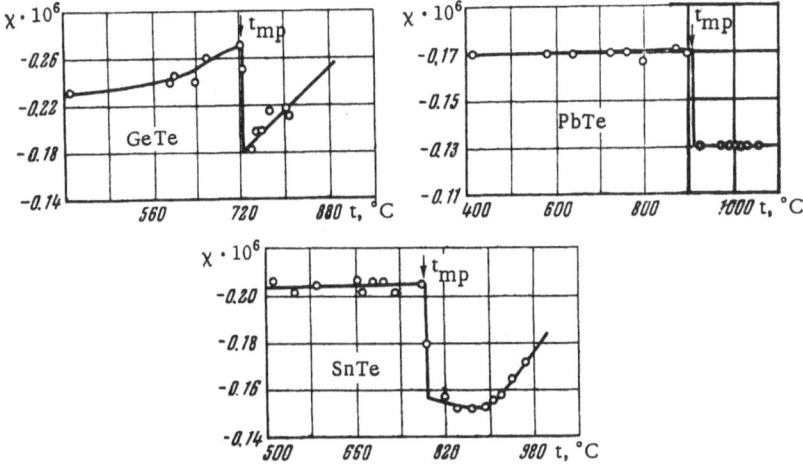

Fig. 83. Temperature dependences of the magnetic susceptibility of germanium, tin, and lead tellurides in the solid and liquid states.

This investigation of the magnetic susceptibility indicated that germanium, tin, and lead tellurides were diamagnetic in the solid and liquid states. The magnetic susceptibility of lead and tin tellurides (Fig. 83) in the solid state varied little with temperature. The magnetic susceptibility of germanium telluride increased considerably with temperature in the solid state. At the melting point, the susceptibility of all three compounds fell suddenly; in the molten state, the susceptibility of lead telluride was practically unaffected by temperature, but that of germanium and tin tellurides increased appreciably as the temperature rose. The change in the magnetic susceptibility at the melting point could be due to an increase in the carrier density, making an additional contribution to the paramagnetic component of the susceptibility because of the spin paramagnetism, or it could be due to an enhancement of the Van Vleck polarization paramagnetism, the role of which could be important [502]. In general, the magnetic susceptibility is a very complex property and it is difficult to identify the causes of its change at the melting point (we have indicated only some factors which may affect the susceptibility). However, it is worth stressing the complete similarity of the change in the magnetic susceptibility at the melting points of all three $A^{IV}Te$ compounds, which indicates a similarity in the changes of the chemical binding. These

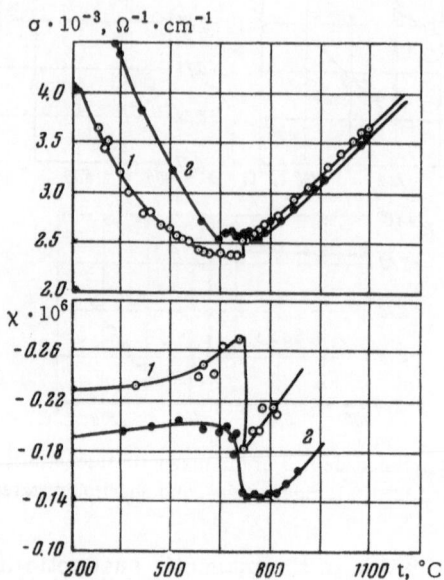

Fig. 84. Temperature dependences of the electrical conductivity and magnetic susceptibility of germanium telluride of stoichiometric composition (1) and with an excess
of 0.6 at.% Te (2) in the solid and liquid states.

similarities are the main reason for studies of the magnetic susceptibility, as pointed out by Professor Ya. G. Dorfman (private
communication).

2. Some Conclusions about the Physicochemical Nature
of Germanium and Tin Tellurides in the Solid State

Until recently, the views about the physicochemical nature
of the solid state of germanium and tin tellurides have been contradictory. However, our investigations of the electrical conductivity
and magnetic susceptibility in the solid and liquid states, which
have established that there is a rapid fall of the electrical conductivity when the temperature is increased followed by a slowing
down of this fall just before the melting point (this has also been
reported by Kolomoets et al. [470, 471]), indicate that germanium
and tin tellurides are semiconductors in the solid state [472]. In
this case, as in the case of Mg_2Pb (Chapter 5), the complete analogy

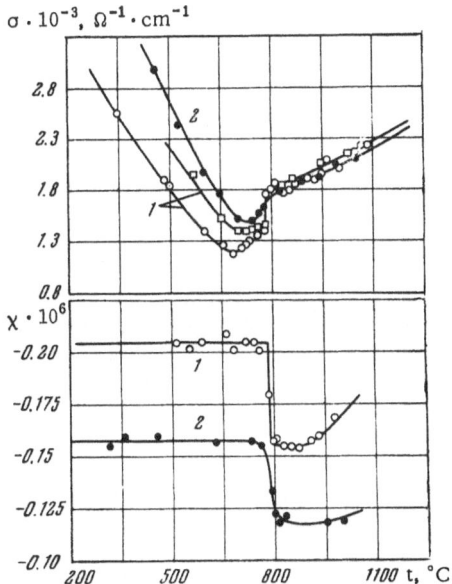

Fig. 85. Temperature dependences of the electrical conductivity and magnetic susceptibility of tin telluride of
stoichiometric composition (1) and with an excess of 0.3
at.% Te (2) in the solid and liquid states.

between the sudden changes in the electrical conductivity and the
magnetic susceptibility at the melting points of all three Group IV
tellurides (GeTe, SnTe, PbTe), as well as the indications that
these substances remain semiconductors in the liquid phase, show
clearly that these substances are semiconductors in the solid state.
The semimetallic nature of the properties of these compounds at
low temperatures is due to a high carrier density resulting from
nonstoichiometric compositions [47, 49, 50].

Thus, simply by analyzing changes in the electrical conductivity at the melting points of the $A^{IV}Te$ compounds, we have been
able to establish that germanium and tin tellurides are degenerate
semiconductors.

We shall now consider the temperature dependence of the
electrical conductivity in the solid state. We have already mentioned the slowing down of the rapid fall of the conductivity just before
the melting point in the case of germanium telluride and a change

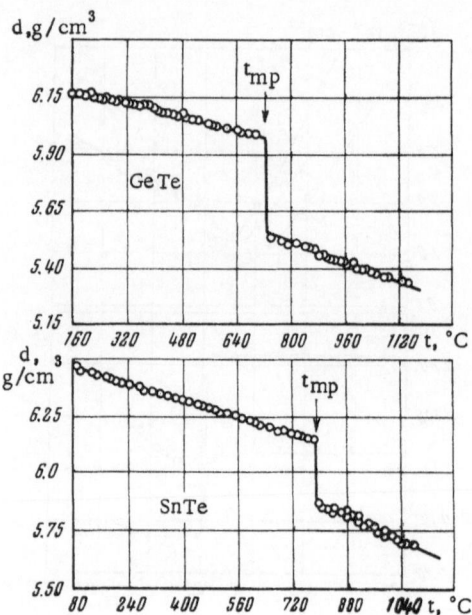

Fig. 86.   Temperature dependences of the density of germanium and tin tellurides in the solid and liquid states.

in the sign of the temperature coefficient in the case of tin telluride, which is most probably due to the appearance of intrinsic conduction. This conclusion is in agreement with the results obtained for the thermoelectric power, which begins to fall in the same range of temperatures [471, 474]. The appearance of intrinsic conduction at such high temperatures in practically degenerate (at room temperature) substances indicates a decrease in the degree of degeneracy. This decrease may be due to a shift of the composition at high temperatures in the direction of the stoichiometric ratio. To check qualitatively this hypothesis, we investigated alloys whose compositions lay in the homogeneous region, in which the changes attributed to stoichiometric compositions could not take place. We determined the temperature dependences of the electrical conductivity and magnetic susceptibility of alloys with an excess of tellurium: 0.6 at.% in the Ge−Te system and 0.3 at.% in the Sn−Te system. The results are presented in Figs. 84 and 85, where they are compared with the temperature dependences of stoichiometric compositions. It is evident from these figures that the electrical

conductivity of alloys with an excess of tellurium is higher than
that of the corresponding stoichiometric compounds, but the magnetic susceptibility of the alloys is lower than that of the compounds. Tellurium must be electrically active in these alloys.
Moreover, the alloys do not exhibit a sudden change in their electrical conductivity at the melting point and the transition from the
solid-to-liquid state is fairly broad. The same is true of the magnetic susceptibility. A definite change in the electrical conductivity and the magnetic susceptibility of the stoichiometric compounds indicates that deviations from stoichiometry probably decrease when the temperature is increased, as established by
Abrikosov et al. for the Sb — Te system [48]. The final solution of
this problem will be provided by a careful investigation of the
equilibrium deviations from stoichiometry at higher temperatures
in the Ge — Te and Sn — Te systems, and by the accurate plotting of
the solidus curves in the relevant range of temperatures.

3. Volume Changes Due to the Melting
of the $A^{IV}B^{VI}$ Compounds

We have already mentioned that investigations of the density
may yield information on changes in the short-range order which
take place at the melting point. For this reason, we shall consider
the nature of the volume changes due to the melting of the $A^{IV}B^{VI}$
compounds and to the heating of their melts. The temperature dependence of the density in the solid and liquid states and its change
at the melting point has not yet been investigated for these compounds. We used the method and apparatus described in Chapter 2.
Samples used in the investigation of the density in the solid state
were prepared by remelting synthesized compounds in quartz ampoules of suitable diameter. PbTe was subjected to a horizontal
zone melting. The initial values of the density at room temperature
were taken from the published data [50, 51, 455]:

Compound	GeTe	SnTe	PbTe	PbSe	PbS
Density, g/cm^3	6.19	6.45	8.16	8.1	7.5

The density in the solid and liquid state was measured using
several samples of each compound (several measurements were
carried out on each sample and these measurements indicated good
reproducibility. The results are presented in Figs. 86 and 87 [503].

Table 30. Volume Expansion Coefficients and Changes
in Density at the Melting Points of the $A^{IV}B^{VI}$ Compounds

Com-pound	$\beta_s \cdot 10^5$, deg^{-1}	d_s, g/cm^3	d_l, g/cm^3	Δd, g/cm^3	Δd, %
GeTe	6.9	5.97	5.57	0.40	~6.7
SnTe	6.9	6.15	5.85	0.30	~4.9
PbTe	6.4	7.69	7.45	0.24	~3.1
PbSe	5.7	7.57	7.10	0.47	~6.2
PbS	4.8	7.07	6.45	0.62	~8.8

When the temperature is increased the density of all the investigated compounds (with the exception of germanium telluride) decreases linearly right down to the melting point. In the case of germanium telluride, a polymorphic transition was observed at 430°C [51]. This transition is manifested by a slowing down of the rate of fall of the density with temperature, which increases again at higher temperatures. The values obtained for the volume expansion coefficients of these compounds are listed in Table 30 (the volume expansion coefficient of germanium telluride is given for the high-temperature modification with the NaCl-type structure). The volume expansion coefficient is practically independent of temperature and the listed values are averages for the investigated range of temperatures. At the melting points, the densities of all the investigated compounds decrease suddenly and continue to decrease practically linearly at still higher temperatures in the molten state. Table 30 lists the densities of these compounds in the solid and liquid states at the melting points.

It follows from Table 30 that the sudden change in the density at the melting point is reduced when "cations" and "anions" are replaced with heavier elements in the two investigated series of chemical analogs: GeTe → SnTe → PbTe and PbS → PbSe → PbTe. The fall of the density at the melting point, particularly if its value is small, can be regarded as an indication that the general nature of the short-range order and of the chemical binding is retained in the liquid state, which is in good agreement with the results obtained by investigating the electrical conductivity.

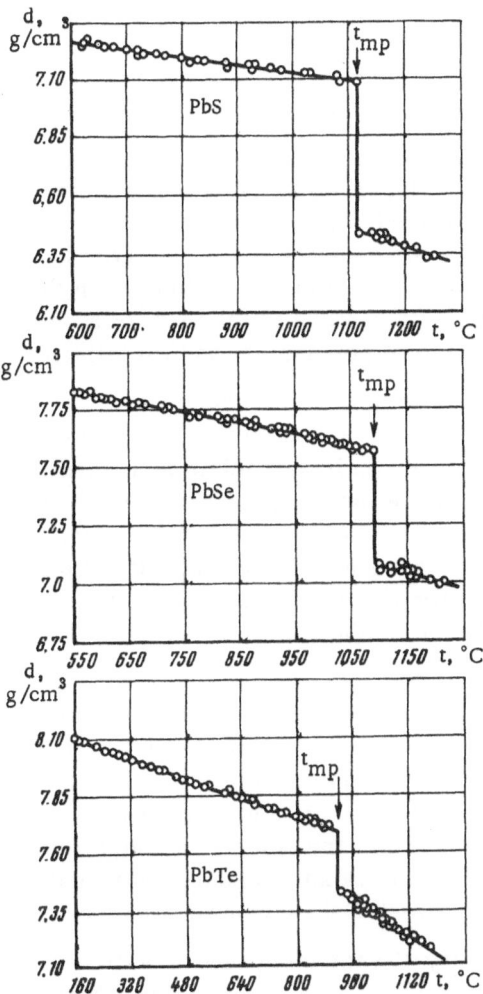

Fig. 87. Temperature dependences of the density of lead
chalcogenides in the solid and liquid states.

§3. Changes in the Short-Range Order during the Heating of Molten $A^{IV}B^{VI}$ Compounds

The investigations of the electrical conductivity, magnetic susceptibility, and density of the $A^{IV}B^{VI}$ compounds near the melting point and in the liquid state indicate that covalent bonds are most probably retained above the melting point and that a short-range structure of the chain or molecular type is formed in the liquid state (the liquids are likely also to contain the dissociation products, which could even be in the form of atoms or ions, particularly in the case of germanium and tin tellurides).

It seemed of interest to determine how the short-range order varied with temperature above the melting point. In particular, it was not clear whether an intermediate structure existed over a range of temperatures (as in the case of the $A^{III}B^V$ or Mg_2B^{IV} compounds). To obtain information on this point, we investigated the temperature dependence of the viscosity of our $A^{IV}B^{VI}$ compounds in the molten state. The viscosity was measured by the torsional oscillation method (Chapter 2) and calculated using the formulas for weakly viscous liquids.

In the investigation of the lead chalcogenides, the quartz ampoules were coated internally with a mirrorlike layer of pyrocarbon using a method described in [98]. These ampoules were placed in cylindrical graphite containers with screw-on lids. This prevented deformation of the samples (at high temperatures, the quartz softened and the vapor pressure deformed the ampoules), which could have affected the absolute values of the viscosity.

The viscosity of each compound was measured using two or three different samples. The results of measurements of the kinematic viscosity are given in Figs. 88-92. The same figures include the temperature dependences of the dynamic viscosity (η), calculated from the density measured earlier (preceding subsection). The viscosity of these $A^{IV}B^{VI}$ compounds has not yet been investigated before.

The results presented in Figs. 88-92 indicate that the viscosity of these compounds decreases smoothly when the temperature is decreased and the fall of the viscosity is somewhat faster near

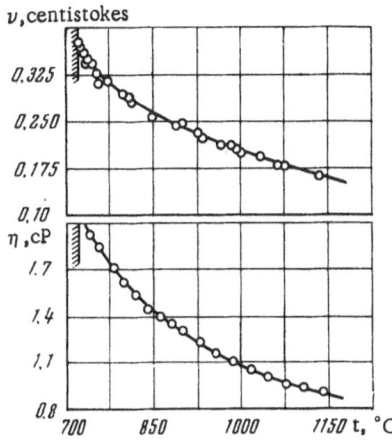

Fig. 88. Temperature dependences of the kinematic and dynamic viscosities of molten germanium telluride.

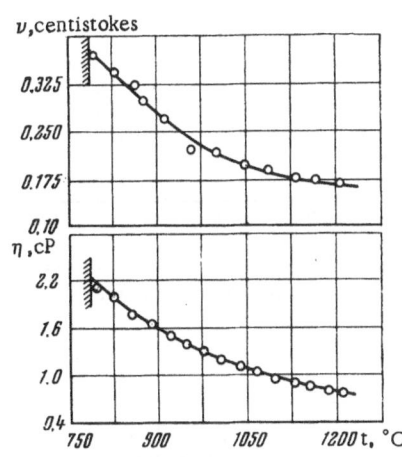

Fig. 89. Temperature dependences of the kinematic and dynamic viscosities of molten tin telluride.

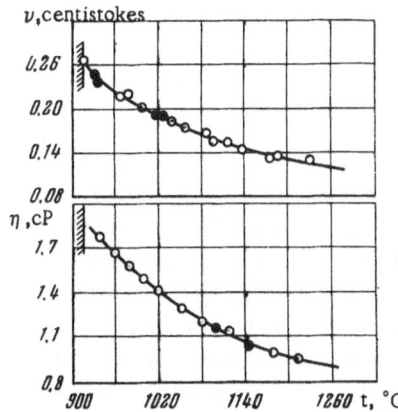

Fig. 90. Temperature dependences of the kinematic and dynamic viscosities of molten lead telluride.

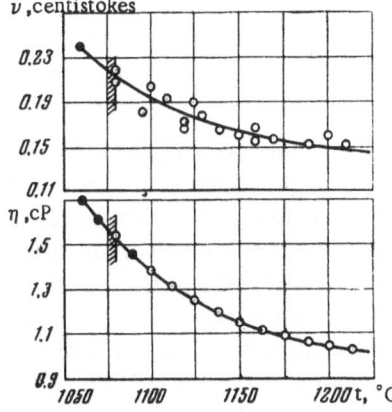

Fig. 91. Temperature dependences of the kinematic and dynamic viscosities of molten lead selenide. The black dots represent supercooling.

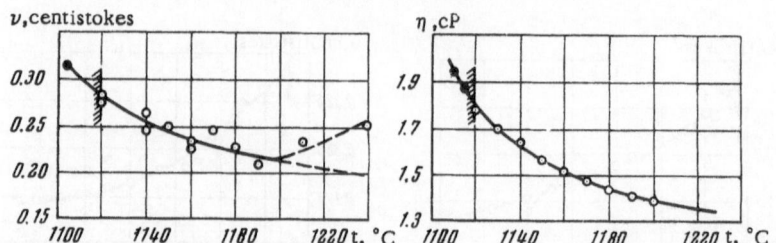

Fig. 92. Temperature dependences of the kinematic and dynamic viscosities of molten lead sulfide. The black dots represent supercooling.

the melting point, particularly in the case of tin telluride (Fig. 89) and, to a lesser degree, in the case of germanium and lead tellurides (Figs. 88 and 90). The scatter of the values of the viscosity of lead selenide and sulfide is much greater than the scatter for the other three compounds, and in the case of lead sulfide reproducible results are not obtained above 1200°C, but the viscosity tends to increase with increasing temperature (this is due to the proximity of the boiling of lead sulfide at 1280°C [424]).

The results obtained yield some information on the changes in the short-range order when they are compared with theoretical relationships in the same manner as in Chapter 3.

Using the data on the kinematic viscosity of the molten $A^{IV}B^{VI}$ compounds, we calculated the free activation energy of viscous flow. The results of these calculations are presented in Fig. 93. It is evident from this figure that at high temperatures, considerably above the melting point, this energy increases linearly with temperature. The temperature interval (measured from the melting point) above which this linear dependence is observed increases along the series GeTe → SnTe → PbTe, i.e., when a "cation" is replaced with a heavier element, and it decreases along the series PbTe → PbSe → PnS, i.e., when an "anion" is replaced with a lighter element.

Table 31 gives the values of the activation entropy of viscous flow, calculated from the tangents of the slopes of the straight lines $F_b \propto f(t)$. The constancy of the activation entropy at temperatures Δt above the melting point (Table 31) shows that the

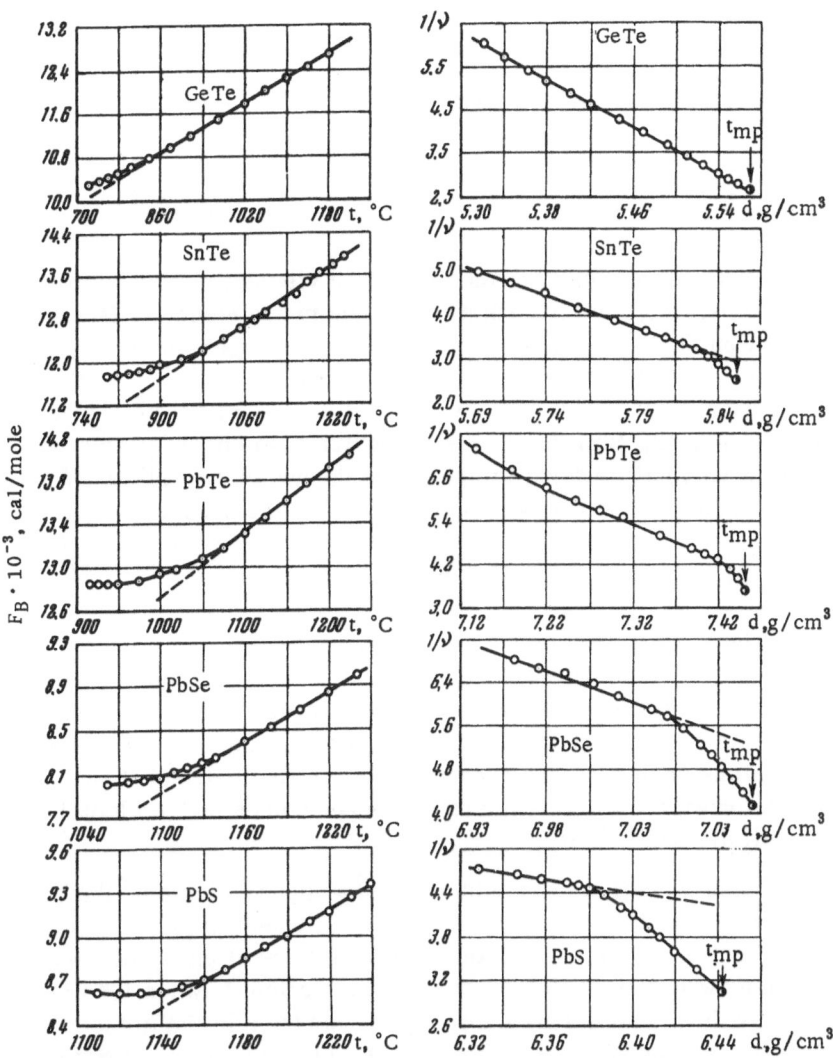

Fig. 93. Temperature dependences of the free activation energy of viscous flow of the $A^{IV}B^{VI}$ compounds in the molten state.

Fig. 94. Dependence of the reciprocal of the kinematic viscosity on the density of the $A^{IV}B^{VI}$ melts at various temperatures (dashed lines represent results calculated from Bachinskii's equation).

Table 31. Values of E_b and S_b of $A^{IV}B^{VI}$ Melts

Compound	t_{mp}, °C	Δt, °C	$E_b \cdot 10^{-3}$, cal/mole	$-S_b$, cal $\cdot$ mole$^{-1} \cdot$ deg^{-1}
GeTe	725	~80	4.70	5.8
SnTe	790	~140	4.90	6.1
PbTe	917	~140	6.85	6.0
PbSe	1088	~50	6.85	7.5
PbS	1119	~50	9.80	7.5

temperature dependence of the kinematic viscosity obeys an exponential law, i.e., the logarithm of the kinematic viscosity is a linear function of the reciprocal of the absolute temperature. This exponential dependence has been checked, at temperatures Δt above the melting point (Table 31), by plotting the logarithm of ν as a function of $1/T$. Similar dependences are also observed for the dynamic viscosity, whose temperature dependences well above the melting point are well described by Frenkel's equation

$$\eta = \exp(E_b/RT). \tag{80}$$

The results presented in Table 31 indicate that the activation energies of viscous flow of the molten $A^{IV}B^{VI}$ compounds are relatively low. This may be due to a weak interaction between the structural units taking part in the viscous motion, in agreement with the conclusions about the short-range order in the melts of these compounds drawn from the temperature dependences of the electrical conductivity and density.

When a "cation" is replaced with a lighter element, the activation energy of viscous flow decreases, which is probably due to a reduction in the dimensions of the structural units, because of both a reduction in the dimensions of "cations" (when lead is replaced with tin and germanium) and an increase in the dissociation along the series PbTe → SnTe → GeTe (this has been confirmed by special investigations, which will be discussed later). When an "anion" is replaced with a lighter element, the following behavior is observed. The activation energies of viscous flow of molten lead telluride and selenide remain practically constant, while the corresponding energy of lead sulfide increases by a factor of almost 1.5. This indicates changes in the structure of activated

complexes and an increase in the height of the potential barriers which they have to overcome. The latter may be due to an en-hancement of the interaction between the structural units in the melt resulting from an increase in their polarity. This conclusion is in agreement with our earlier conclusion on the formation of a molecular structure in molten lead sulfide, which has been de-duced by analyzing the temperature dependence of the electrical conductivity in the molten state.

The fairly large values of the activation entropy of viscous flow of the molten $A^{IV}B^{VI}$ compounds indicate that the structural units, participating in the process of viscous flow in the melts, have a lower entropy than the structural units which are in the state of stable equilibrium. Thus, an analysis of the process of viscous flow in the $A^{IV}B^{VI}$ melts gives at least a partial explana-tion of the changes in the various properties (in particular, the electrical conductivity) which take place when lead sulfide and the other four compounds are melted.

We shall now consider the deviations of $F_b \propto f(t)$ and $\log \nu \propto f(1/T)$ from linearity, which are observed immediately above the melting points in the temperature intervals listed in Table 31. It is evident from Fig. 93 that all the investigated $A^{IV}B^{VI}$ compounds, with the exception of germanium telluride, ex-hibit considerable deviations from Eqs. (67) and (74). This indi-cates that appreciable changes in the short-range order take place in these temperature intervals Δt. In the case of germanium tel-luride, these deviations are very small due to a higher degree of dissociation immediately above the melting point.

Consequently, within a temperature interval Δt (cf. Table 31) the activation entropy of viscous flow does not remain constant but can even change its sign (this happens in the case of lead sulfide). Consequently, we may definitely conclude that when the tempera-ture is increased above the melting point, considerable changes take place in the short-range order in the melts of the investigated group of compounds. Bearing in mind the results obtained on the electrical conductivity and density, we may conclude that these changes are associated with a transformation of the structure from a three-dimensional system of covalent bonds to a linear structure, which is more stable under conditions of high mobility in the liquid. Dissociation of the compounds also takes place in the same

temperature interval. In the case of germanium telluride, the dissociation occurs mainly near the melting point and changes in the short-range order are weaker. In the case of tin and lead telluride, the dissociation takes place over a range of temperatures (above the melting point) and this may account for the large values of Δt of these compounds. In the case of lead sulfide, the deviations are stronger than for the other compounds but they occur in a narrower range of temperatures, which may indicate a different transformation of the short-range order (as already mentioned in the section dealing with the electrical conductivity).

Analysis of these results by means of the Bachinskii equation shows (Fig. 94) that all these compounds exhibit regular deviations of the type observed for water when the dependence of the reciprocal of the kinematic viscosity is plotted as a function of the density. These deviations are due to a gradual transformation (during cooling) of the short-range order from a linear to a three-dimensional form (this applies to the distribution of the covalent bonds).

Thus, an analysis of the temperature dependence of the viscosity, together with the data on the electrical conductivity and density in the molten state, shows that when the molten $A^{IV}B^{VI}$ compounds are heated, certain structural changes take place: these changes involve a transformation of the three-dimensional distribution of the covalent bonds to a linear configuration as well as the dissociation of the compounds in the molten state. The conclusion about the dissociation is supported most convincingly by the physicochemical analysis of the corresponding binary liquid systems, which will be discussed later.

Chapter 7

Physicochemical Properties of Molten $A_2^V B_3^{VI}$ Compounds with the Tetradymite Structure

Tetradymite-type compounds have a layered structure in the solid state and valuable thermoelectric properties. Therefore, investigations of the physicochemical properties of these compounds at the melting point and in the liquid state are of scientific and practical interest.

We investigated the following three $A_2^V B_3^{VI}$ tetradymite-type compounds: Bi_2Te_3, Bi_2Se_3, Sb_2Te_3. We shall consider the structure and chemical binding of these compounds in the solid state as well as the changes which take place at their melting points and during heating of their melts.

§1. Crystal Structure and Chemical Binding of $A_2^V B_3^{VI}$ Compounds in the Solid State

The $A_2^V B_3^{VI}$-type compounds are formed from the elements of Groups V and VI in Mendeleev's periodic table. The phase equilibria in the $A^V - B^{VI}$ systems are described by phase diagrams with one congruently melting compound and several intermediate phases, formed as a result of peritectic transformations. The published data for the $Bi-Te$, $Bi-Se$, and $Sb-Te$ systems have been reviewed by Hansen and Anderko [203]. However, the diagrams in Hansen and Anderko's handbook do not give the true pictures of the phase equilibria in these systems.

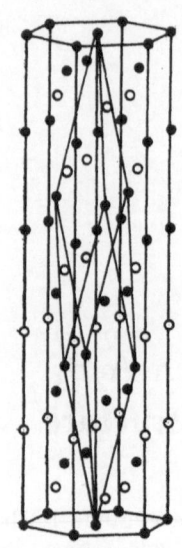

Fig. 95. Crystal structure of the $A_2^V B_3^{VI}$ compounds. ◯) A atoms; ●) B atoms.

The fullest investigation of these systems was carried out by Abrikosov, Bankina, and Poretskaya [44, 48, 504, 505]. They plotted the phase diagrams of the Bi—Te, Bi—Se, Sb—Te systems (Figs. 130-132) from the thermal, microstructure, x-ray diffraction, and thermoelectric data on equilibrium alloys.

We shall concentrate our attention on the compounds Bi_2Te_3, Bi_2Se_3, and Sb_2Te_3 rather than on other phases and transitions in these systems.

According to Abrikosov et al. [504], Bi_2Te_3 dissolves about 1 wt.% Te, but hardly any bismuth. A detailed investigation of the phase diagram in the region of Bi_2Te_3 has been carried out by Satterthwaite and Ure [506]. They have established that the highest melting point occurs on the bismuth side of the stoichiometric composition. Stoichiometric melts always yield p-type Bi_2Te_3 with an excess of bismuth. In order to obtain n-type Bi_2Te_3, the excess of tellurium in the liquid phase must not be less than 2.5 at.%. Offergeld and Van Cakenberghe [46] have also found a slight deviation from stoichiometry by the differential thermal analysis method. According to their data, the congruently melting composition corresponds to 59.935 at.% Te. According to Abrikosov et al. [505], the solubility of Bi and Se in Bi_2Se_3 is slight, which is also supported by measurements of the electrical properties. Abrikosov et al. maintain that bismuth selenide single crystals are always n-type with a high carrier density of the order of 10^{19} cm^{-3}. Other workers [506, 507] assume that, as in the case of bismuth telluride, this is due to a deviation from the stoichiometric composition. According to Offergeld and Van Cakenberghe [46], the composition with the maximum liquidus temperature contains 59.98 at.% Se.

The deviation reported in [44] in the direction of excess antimony in the Sb—Te system has been confirmed by Offergeld and Van Cakenberghe [46], who found that the composition with the maximum liquidus temperature contains 59.6 at.% Te. In a later

Table 32. Lattice Parameters of $A_2^V B_3^{VI}$ Compounds
[457, 508, 510, 512, 514, 516-521]

| Compound | Lattice parameter | | | |
| | hexagonal cell | | rhombohedral cell | |
	a, Å	c, Å	a, Å	α, deg
Bi_2Te_3	4.38	30.45	10.45	24.8
Bi_2Se_3	4.14	28.7	9.82	24.4
Sb_2Te_3	4.25	30.3	—	—

Table 33. Interatomic Spacings in $A_2^V B_3^{VI}$ Compounds
(in Å)

| Compound | Interatomic spacings | | | Ref. |
	$Bi - Te^1 (Se^1)$ $(Sb - Te^1)$	$Bi - Te^2 (Se^2)$ $(Sb - Te^2)$	$Te^1 - Te^1$ $(Se^1 - Se^1)$	
Bi_2Te_3	3.12	3.22	3.57	[510]
	3.04	3.24	3.72	[508]
Bi_2Se_3	2.99	3.07	3.30	[508]
	2.99	3.05	3.31	[513]
Sb_2Te_3	3.03	3.16	3.62	[521]

investigation [48], the temperature dependence of the deviation of Sb_2Te_3 from stoichiometry was examined. It was found that the composition with 59.2 at.% Te is of the single-phase type at 400°C, but when the temperature is increased the deviation from stoichiometry decreases and it practically disappears at the melting point.

According to x-ray diffraction investigations [508-516], Bi_2Te_3, Bi_2Se_3, and Sb_2Te_3 have the structure of the mineral tetradymite (Bi_2Te_2S), which has been investigated in detail by Harker [517]. Its structure consists of nine closely packed layers of tellurium (selenium) atoms, in which the layer sequence is as fol-
                            chhchhchhc
lows: ABABCBCACA ("c" and "h" denote, respectively, the cubic and hexagonal configurations of a given layer). In this packing, two out of three octahedral vacancies between tellurium (selenium) atoms are occupied by bismuth (antimony). The structure has

rhombohedral symmetry (space group $D_{3d}^5 = R\bar{3}m$, structure type SZZ). The unit rhombohedral cell contains one $A_2^V B_3^{VI}$ molecule. The unit hexagonal cell contains three $A_2^V B_3^{VI}$ molecules. The crystal structure of these compounds is shown in Fig. 95, and the average lattice parameters are listed in Table 32.

The tetradymite-type compounds have a layered structure. A unit hexagonal cell in these compounds consists of three five-layer packets with the following distribution of layers in each packet: $-B$, $B-A-B-A-B$, $B-$, where A represents an element of Group V (Bi or Sb) and B represents an element of Group VI (Te or Se). Each layer contains only atoms of one kind. Atoms in a given layer are located at the corners and in the center of a regular hexagon. The relative positions of the layers are such that each atom in a given layer has three nearest neighbors in the next layer and the bond symmetry is almost octahedral. The chemical binding in these $A_2^V B_3^{VI}$ compounds indicates the presence of two types of tellurium (selenium) atom, which have different binding with the atoms in the neighboring layers. According to Drabble and Goodman [522], layers in each packet of Bi_2Te_3 are distributed as follows:

$$Te^1 - Bi - Te^2 - Bi - Te^1.$$

According to Lange [510], Te^2 atoms have an almost octahedral environment of six bismuth atoms and the angles between the bonds differ only by 4.5° from the true octahedral symmetry. Bismuth atoms are bound to three Te^2 atoms on one side, and three Te^1 atoms on the other. Again the bond configuration is nearly octahedral. The nearest neighbors of each of Te^1 atom are three bismuth atoms in the same packet and three Te^1 atoms in the next five-layer packet. The angles between the $Bi-Te^1$ bonds are approximately 89°21', and between the Te^1-Te^1 bonds, 75°42'. Table 33 lists the interatomic spacings in the investigated $A_2^V B_3^{VI}$ compounds, taken from various sources.

The published information on the nature of the binding within and between the layers is contradictory. Assuming that a given layer contains atoms of one type only, Konorov [523] concludes that the binding within the layers is of the covalent type with some admixture of the metallic binding in layers containing bismuth. Krebs [524] suggests the metallic binding within each atomic layer. According to Airapetyants and Efimova [525], the distance between

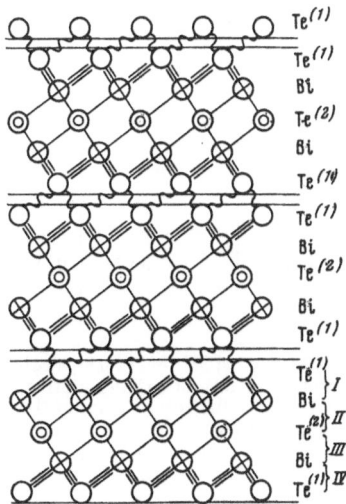

Fig. 96. Chemical bonds in bismuth telluride [514].
I, IV) Covalent + ionic; II, III) covalent.

atoms in one layer (4.38 Å) is greater than the sum of two van der Waals radii of tellurium (2.0 Å) or bismuth (1.69 Å) [476], and, therefore, they conclude that there is practically no binding between the atoms within one layer.

The distance between the five-layer packets, i.e., the length of the $Te^1 - Te^1$ bonds (cf. Table 33) is of the same order as the length of the bonds between the tellurium atoms in neighboring tellurium chains (3.46-3.74 Å) [195, 522]. It is usually assumed that van der Waals forces act between the tellurium chains. It has been suggested [522, 525, 526] that the five-layer packets are bound by weak van der Waals forces, which explains the easy cleavage of these crystals along the {0001} planes. Thus, according to the Drabble–Goodman bond scheme for bismuth telluride [522], the valence electrons of the Te^1 atoms are used only to form bonds with the nearest three bismuth atoms, i.e., to form three almost orthogonal (with respect to one another) bonds. This agrees with the conclusion of Lagrenaudie [527] that only the p orbitals and p electrons of Te^1 take part in the bond formation, while the two 5s electrons form a pair. However, Mooser and Pearson [195] conclude that the length of the $Te^1 - Te^1$ ($Se^1 - Se^1$) bonds indicates that the binding between the packets is stronger than that expected from the sole presence of the van der Waals forces. Therefore,

Table 34. Properties of $A_2^V B_3^{VI}$ Compounds in Solid State

Property	Bi_2Te_3	Bi_2Se_3	Sb_2Te_3	Reference
ΔX	0.3	0.6	0.3	[476]
t_{mp}, °C	585	706	622	[203.531—533]
Q_f, cal·(g-atom)$^{-1}$	5.67	—	4.73	[531.532]
$-\Delta H$, kcal·(g-atom)$^{-1}$	4.88 (400°)	6.69 (0°)	5.60 (25°)	[528.531.534]
ΔE^*, eV	0.21(therm)	0.35 (opt)	0.3 (opt)	[506.535.536.539]
u_n†, cm^2·V^{-1}·sec^{-1}	1140	600	—	
u_p†, cm^2·V^{-1}·sec^{-1}	680	—	360	

* A value of 0.16 eV has been obtained in [537-541] for Bi_2Te_3 by a thermal method. According to [542, 543], the value of the forbidden band width of bismuth telluride, determined by an optical method, is 0.13 or 0.15 eV. Values of the forbidden band width of bismuth selenide and antimony telluride, obtained by an optical method, are given in accordance with the results reported in [539]. A value of 0.28 eV is reported in [544] for bismuth selenide and 0.21 eV is given in [545] for antimony telluride.
† This table lists the maximum values of the mobilities reported in [506, 535, 539, 541, 546-550].

they suggest the possibility of some covalent binding between the packets due to the pd hybridization.

There are also several viewpoints about the nature of the binding between layers within packets. According to some authors [514, 522, 526], covalent–ionic bonds appear between bismuth and Te^1 layers and purely covalent bonds are formed between bismuth and Te^2. These conclusions are drawn from the observation that the bismuth–Te^1 bonds are shorter than the bismuth–Te^2 bonds, and, consequently, the former are stronger. Other workers [519, 525, 528, 529] conclude that the Bi–Te^1 bonds are mainly covalent and that the Bi–Te^2 bonds are mainly ionic, because the distance between the bismuth and Te^1 layers is approximately equal to the sum of the covalent octahedral radii, and the distance Bi–Te^2 is greater than the sum of the covalent radii and approximately equal to the sum of the ionic radii.

Figure 96 shows schematically the system of chemical bonds in bismuth telluride, proposed by Drabble and Goodman [522], which also applies to other $A_2^V B_3^{VI}$ compounds with the tetradymite structure.

Drabble and Goodman [522] have assumed that the Te^2 atoms are surrounded almost octahedrally by six bismuth atoms and that the binding orbitals between these atoms are sp^{3}d^2 hybrids, i.e.,

the chemical bonds between these atoms are formed by the s and p electrons of the bismuth and Te^2 atoms. The shorter distance between Te^1 and bismuth is attributed to an ionic component in the binding.

Mooser and Pearson [195] have proposed a different system of bonds. They assume a bond resonance between the Te^2 and bismuth atoms in the presence of a p^3d^3 hybridization. According to Shtrum [530], the Drabble−Goodman scheme is more acceptable because it agrees better with the crystal structure and because it explains well the difference between the behavior of the two types of tellurium atom as well as the change in the forbidden band width when tellurium is replaced with selenium.

This description of the chemical binding in bismuth telluride applies also to bismuth selenide, as well as to antimony telluride. Table 34 lists some properties of the $A_2^V B_3^{VI}$ compounds with the tetradymite structure.

Examination of the data presented in this table shows that bismuth selenide has a stronger ionic component of the binding than the other compounds. This somewhat strengthens the chemical bonds, a phenomenon which is manifested by higher values of the melting point and heat of formation, as well as a larger forbidden band width.

Thus, the $A_2^V B_3^{VI}$ compounds are distinguished by the covalent−ionic nature of the chemical binding in the solid state and, with their layered structure, are interesting materials because of the change in their properties at the melting point.

§ 2. Changes in the Short-Range Order and the Chemical Binding at the Melting Points of $A_2^V B_3^{VI}$ Compounds

1. Changes in the Electrical Conductivity and Thermoelectric Power at the Melting Points of the $A_2^V B_3^{VI}$ Compounds

The electrical conductivity is very sensitive to changes in the type of binding. Therefore, in order to obtain information on the general nature of the changes in the chemical binding at the

melting points of the compounds considered here, we investigated the electrical conductivity changes due to the transition from the solid to the liquid state.

The published results of Regel' on the electrical conductivity of bismuth telluride are inconsistent. Thus, according to [3, 8], the electrical conductivity of this compound in the liquid state is 2800 $\Omega^{-1} \cdot cm^{-1}$ at the melting point, but according to other papers by the same author [9, 144] the conductivity is 6200 $\Omega^{-1} \cdot cm^{-1}$. There are no published data on the temperature dependence of the electrical conductivity of antimony telluride.

We investigated Bi_2Se_3, Bi_2Te_3, and Sb_2Te_3. The compounds were synthesized from V-000 grade bismuth (<3 $\cdot 10^{-4}\%$ impurity content), "Su Extra" grade antimony ($\sim 10^{-5}\%$ impurity content), TA-1 grade tellurium, which was subjected to double distillation in vacuum (this made it spectroscopically pure), and selenium of analytic purity.

The compounds were synthesized in quartz ampoules evacuated to 10^{-3} mm Hg and kept at temperatures 80-100° above the corresponding melting point. The process took 3-4 h and the melt was stirred periodically (by shaking the ampoule). Cooling took place in a furnace at a rate of 30-40° per hour. The quality of the compounds obtained was checked by metallographic analysis. Bi_2Se_3 and Bi_2Te_3 had a single-phase structure. Bearing in mind the very narrow range of homogeneity of alloys based on these compounds, we could conclude that the single-phase structure indicated a fairly close approach to stoichiometry. Samples of Sb_2Te_3 of stoichiometric composition contained a small amount of the second phase, which indicated an equilibrium deviation from stoichiometry reported earlier for this compound [44, 48]. The ingots thus obtained were used to cut samples of 10 mm diameter and 10 mm high. The bismuth selenide was subjected to horizontal zone melting at a rate of 0.2 mm/min and the temperature in the hottest part of the furnace was 750°C. This produced ingots of 10 mm diameter and 60 mm high. The middle parts of these ingots were used to cut samples employed in the measurements of the electrical conductivity.

The electrical conductivity was measured by the electrode-less method in a rotating magnetic field using apparatus whose construction has been described in Chapter 2.

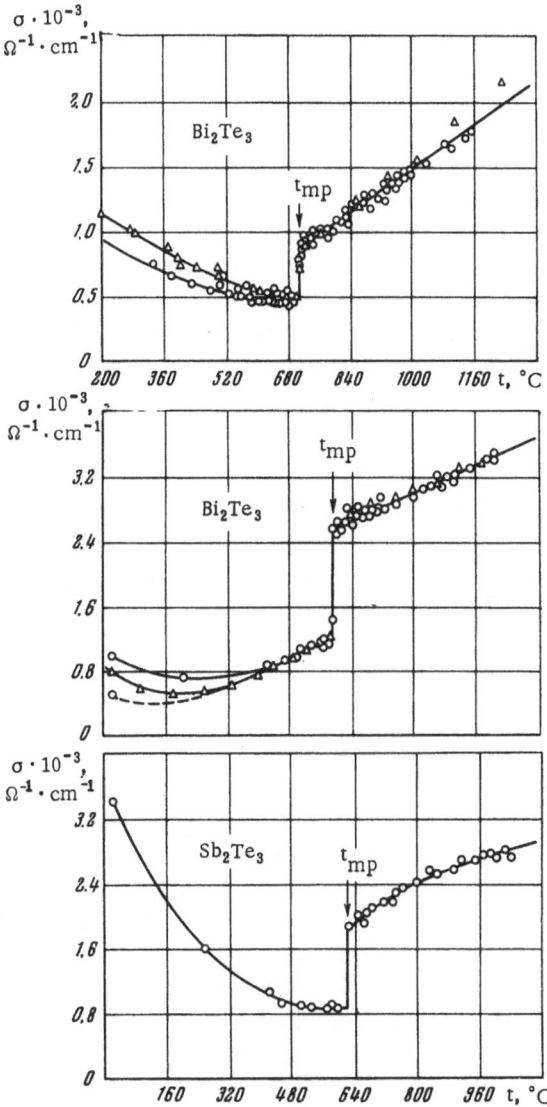

Fig. 97. Temperature dependences of the electrical conductivity
of the $A_2^V B_3^{VI}$ compounds in the liquid and solid states. O) Results
of the present authors obtained for five (Bi_2Se_3), five (Bi_2Te_3), and
two (Sb_2Te_3) different samples; $\triangle$) Regel' [3, 8].

The results obtained for Bi_2Se_3, Bi_2Te_3, and Sb_2Te_3 in the solid and liquid states are presented in Fig. 97 [552]. It follows from this figure that the electrical conductivity of bismuth selenide in the solid state decreases right down to the melting point. About 100-120°C before the melting point, the rate of fall of the conductivity slows down considerably and just before the melting point the conductivity is practically independent of temperature. This is undoubtedly due to the appearance of intrinsic conduction, but we have been unable to observe an intrinsic conduction branch, probably because of the insufficient purity of the selenium used to synthesize this compound. At the melting point, the conductivity of bismuth selenide increases suddenly and continued to increase during further heating in the molten state.

Similar results were obtained also for bismuth telluride (Fig. 97), but in this case an intrinsic conduction branch was observed clearly for all the investigated samples. The data in Fig. 97 include the values of the electrical conductivity found by Regel' [3, 8]. It is evident from Fig. 97 that both in the liquid and solid states our high-temperature results agree with those of Regel' for Bi_2Se_3 and Bi_2Te_3. At lower temperatures, our results differ somewhat from those of Regel' because of the different purities of the initial materials used to synthesize the compounds.

The general nature of the temperature dependence of the electrical conductivity of antimony telluride (Fig. 97) in the solid and liquid states is similar to the temperature dependence observed for bismuth selenide, except that in the solid state the conductivity of antimony telluride decreases more rapidly; however, just before the melting point, the rate of decrease also slows down. These results are typical of semiconductors whose deviations from the stoichiometric composition decrease when the temperature is increased; they are also in good agreement with the nature of the phase equilibria in the investigated systems [44, 48]. At its melting point, the conductivity of antimony telluride increases suddenly.

Table 35 gives the values of the electrical conductivity of these three compounds at their melting points in the solid and liquid states, as well as the values of the sudden change in the electrical conductivity in the form of the ratio σ_S/σ_l. Table 35 shows that for all three compounds the ratio σ_S/σ_l is ≈ 0.5. The small changes in the electrical conductivity at the melting point indicate

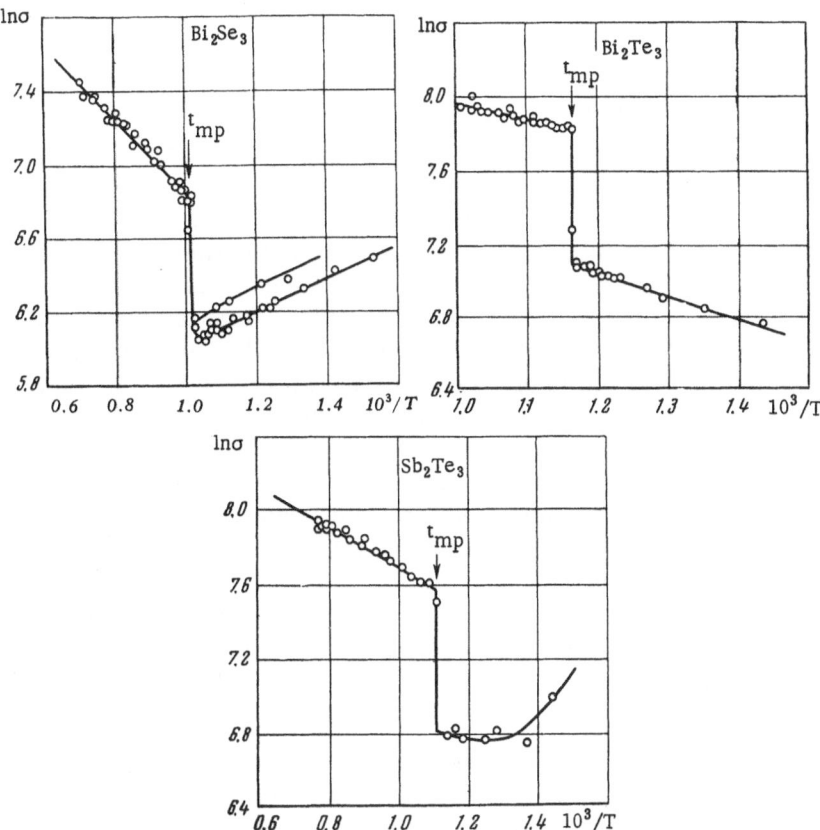

Fig. 98. Dependences $\ln \sigma \propto f\,(1/T)$ for the $A_2^V B_3^{VI}$ compounds in the solid and liquid states.

Table 35. Electrical Conductivity of Bi_2Se_3,
Bi_2Te_3, and Sb_2Te_3 at Their Melting Points in
the Solid and Liquid States

Com-pound	$\sigma_s \cdot 10^{-3}$, $\Omega^{-1} \cdot cm^{-1}$	$\sigma_l \cdot 10^{-3}$, $\Omega^{-1} \cdot cm^{-1}$	σ_s / σ_l
Bi_2Se_3	0.45	0.90	
Bi_2Te_3	1.25	2.58	~0.5
Sb_2Te_3	0.90	1.85	

small changes in the nature of the chemical binding, and the same value of the conductivity ratio for all three compounds shows that these changes are identical in all of them.

The electrical conductivity data for Bi_2Se_3, Bi_2Te_3, and Sb_2Te_3 are presented in Fig. 98 on a semilogarithmic scale in the form of a dependence of the natural logarithm of the electrical conductivity on the reciprocal of the absolute temperature. It follows from Fig. 98 that the temperature dependence of the electrical conductivity of bismuth telluride in the solid state is strictly exponential. The forbidden band width of bismuth telluride, found from the tangent of the slope of the straight line $\ln \sigma \propto f(1/T)$, is 0.206 eV, which is in good agreement with $\Delta E = 0.21$ eV given in Table 34 (the latter value has also been deduced from the temperature dependence of the electrical conductivity). It follows from Fig. 98 that the semilogarithmic dependences $\ln \sigma \propto f(1/T)$ of all three compounds in the molten state are linear. Consequently, the temperature dependences of the electrical conductivity of the $A_2^V B_3^{VI}$ melts can be described by an exponential equation, which applies also to solid semiconductors in the intrinsic conduction region.

Thus, the changes in the electrical conductivity at the melting points of Bi_2Se_3, Bi_2Te_3, and Sb_2Te_3 and the positive temperature coefficients of the electrical conductivity in the liquid phase indicate that the predominantly covalent chemical binding in the solid state is retained after the melting of these compounds. However, we must point out the considerable degree of metallization after melting, which increases along the series $Bi_2Se_3 \rightarrow Sb_2Te_3 \rightarrow Bi_2Te_3$, as indicated by an increase in the absolute values of the electrical conductivity of the melts (Table 35). This is due to a general increase in the "metallic" component of the binding due to "anion" and "cation" replacement with heavier elements.

The retention of the covalent bonds under high-mobility conditions in the liquid is probable only if a one-dimensional chain or molecular structure is formed [65]. We are of the opinion that the formation of a chainlike structure in the liquid state is more likely in antimony and bismuth chalcogenides, because, in the case of a molecular structure the electrical conductivity would have changed in the opposite direction and the conductivity itself would have been much lower. Even in the case of a dissociating com-

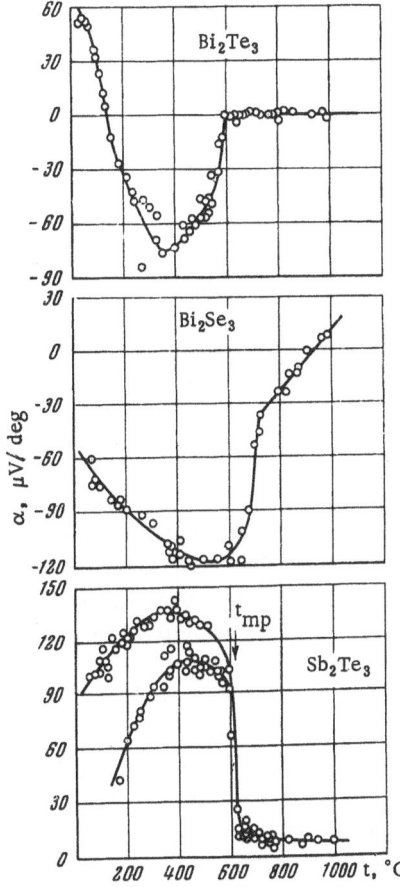

Fig. 99. Temperature dependences of the thermoelectric power of the $A_2^V B_3^{VI}$ in the solid and liquid states.

pound, the general nature of the change in the electrical conductivity in the presence of a molecular structure should be different from that observed, as indicated by the published data on mercury chalcogenides [8].

The dissociation of bismuth and antimony tellurides (indicated by the physicochemical analysis of these systems in the liquid state [552]) gives rise to a considerable increase in the carrier density so that the electrical conductivity in the liquid state is fairly high (its value is typical of the transition region between metals and semiconductors). Nevertheless, the results obtained indicate that the three compounds undergo, according to the classification due to Regel', a "semiconductor — semiconductor" transition at the melting point (in the liquid state, bismuth and antimony tellurides are degenerate semiconductors).

The results obtained on the thermoelectric power are in agreement with those which follow from the electrical conductivity. Figure 99 shows the temperature dependences of the thermoelectric power of all three compounds in the solid and liquid states.

It follows from Fig. 99 that the absolute values of the thermoelectric power of the investigated compounds have maxima at temperatures below their melting points. The fall of the absolute values of the thermoelectric power at temperatures beyond these maxima is apparently due to the appearance of intrinsic conduction. In this respect, Sb_2Te_3 is interesting because its intrinsic

Table 36. Thermoelectric Power of Sb_2Te_3,
Bi_2Se_3, and Bi_2Te_3 at Their Melting Points in
the Solid and Liquid States

Com- pound	$\sim\alpha_s$, $\mu V/deg$	$\sim\alpha_l$, $\mu V/deg$	α_s/α_l
Sb_2Te_3	$+90$	$+11$	8.2
Bi_2Se_3	-90	-35	2.5
Bi_2Te_3	-45	-3	15.0

conduction at high temperatures appears due to a reduction in the deviation from stoichiometry.

We used two different batches of Sb_2Te_3 samples so that we obtained two curves in the solid state. However, these curves merged before the melting point and the results for both branches were practically identical in the liquid state. The two curves were evidently due to somewhat different deviations from stoichiometry in samples originating from different batches. The nature of the temperature dependence of the thermoelectric power of antimony telluride in the solid state is in agreement with the electrical conductivity data and with the results for germanium and tin tellurides, which also exhibit deviations from stoichiometry. Thus, the general nature of the temperature dependences of the thermoelectric power provided an additional proof of a reduction in the deviations from stoichiometry in the Sb—Te system when the temperature is increased.

The results obtained for different samples of bismuth telluride or selenide were identical, but the scatter of the results was somewhat greater in the solid than in the liquid state.

The absolute values of the thermoelectric power of all three compounds decrease strongly at the melting point. The thermoelectric power of molten bismuth telluride is close to zero in the investigated range of temperatures (up to 1000°C). In the case of the other two compounds, the heating of the melt reduces the thermoelectric power and the negative sign for antimony and bismuth selenides changes to positive above a certain temperature.

This reversal of the sign of the thermoelectric power probably occurs also in the case of bismuth telluride, but the thermoelectric power of this compound in the whole investigated range of
temperatures is very low and lies within the limits of the experimental error.

Table 36 lists the values of the thermoelectric power of the
investigated compounds at their melting points in the solid and liquid states; they were obtained by extrapolation of the temperature dependences of the thermoelectric power in the solid and liquid states to the melting point.

It follows from the results in Table 36 that the change in the
thermoelectric power at the melting point, represented by the
ratio α_s / α_l, is very considerable, but different for each of the
three compounds. The value of this change is reduced by the replacement of an "anion" with a lighter element, which may be attributed to a tendency of the thermoelectric power in the liquid
state to increase when tellurium is replaced by selenium. Nothing
definite can be said about the effect of the "cation" replacement.

The nature of the temperature dependence of the thermoelectric power in the liquid state and particularly the transition from
n- to p-type conduction (after heating above a certain temperature)
confirms, to some extent, the conclusion that semiconducting
properties are exhibited by these compounds in the liquid state.
The strong fall of the thermoelectric power at the melting point is
obviously due to a considerable rise in the carrier density (as indicated by an increase in the electrical conductivity at the melting
point) and a reduction in the difference between the values of the
electron and hole mobilities. The change in the sign of the conduction above a certain temperature, observed for bismuth selenide,
may be associated with a change in the ratio of the electron and
hole mobilities.

On the whole, the results obtained by measuring the thermoelectric power of molten antimony and bismuth chalcogenides indicate some metallization of the bonds due to melting, which is particularly important in the case of bismuth telluride (this is in good
agreement with the results obtained in the investigation of the electrical conductivity).

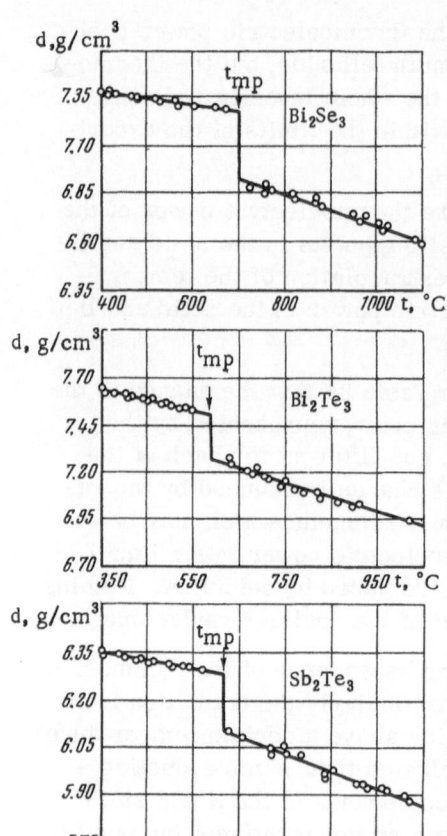

Fig. 100. Temperature dependences of the density of the $A_2^V B_3^{VI}$ compounds in the solid and liquid states.

2. Volume Changes at the Melting Points of the $A_2^V B_3^{VI}$ Compounds

To determine the general direction of changes in the short-range order caused by the melting of $A_2^V B_3^{VI}$ compounds, we carried out a detailed investigation of the temperature dependence of the density of these compounds in the solid and liquid states. Such investigations had not been carried out before. The published results simply indicate that the density of bismuth and antimony tellurides decreases at the melting point [247, 553].

The method used in this investigation is described in Chapter 2. The samples used to determine the density in the solid state were 5 mm in diameter and 25–30 mm high; they were prepared by the remelting of synthesized compounds in quartz tubes of suitable dimensions. The initial values of the density at room temperature were taken from [520]:

Compound	Bi_2Se_3	Bi_2Te_3	Sb_2Te_3
Density, g/cm^3	7.51	7.73	6.44

The density was measured using three or four different samples of each compound; measurements repeated on a given sample indicated good reproducibility of the results.

The temperature dependences of the density are presented in Fig. 100 [544]. We can see from that figure that the results obtained for different samples of the same compound are in good agreement. When the temperature is increased, the density of all

Table 37. Volume Expansion Coefficients and Density
of the $A_2^V B_3^{VI}$ Compounds at Their Melting Points in the
Solid and Liquid States

Compound	$\beta_s \cdot 10^5$, deg^{-1}	d_s, g/cm^3	d_l, g/cm^3	Δd, g/cm^3	Δd, %
Bi_2Se_3	4.2	7.27	6.97	0.30	4.1
Bi_2Te_3	4.2	7.50	7.26	0.24	3.2
Sb_2Te_3	3.9	6.29	6.09	0.20	3.2

three compounds decreases practically linearly to the melting
point. Consequently, the thermal expansion coefficient is virtually
independent of temperature. Table 37 lists the values of the volume expansion coefficients for all three compounds.

At their melting points, the density of all three compounds
decreases suddenly. The values of the density in the solid and liquid states at the melting point, as well as the changes in the density due to fusion are also given in Table 37.

It follows from this table that the melting of the $A_2^V B_3^{VI}$ compounds is accompanied by very small volume changes, which indicate qualitatively that the general nature of the short-range order
is retained when these compounds melt. The reduction in the density at the melting point is typical of the retention of the nature of
chemical binding in the liquid state. Thus, the general nature of
the changes in the density at the melting points leads to the same
conclusions about the short-range order and the chemical binding
as the investigation of the electrical conductivity.

No definite conclusions can be drawn about the influence of
the replacement of the components, because the number of the
$A_2^V B_3^{VI}$ compounds investigated is too small.

§3. Changes in the Short-Range Order during the Heating of Molten $A_2^V B_3^{VI}$ Compounds

Investigations of the electrical conductivity and density of
the $A_2^V B_3^{VI}$ compounds in the solid and liquid states have indicated
that the covalent bonds may be retained in the liquid phase. To

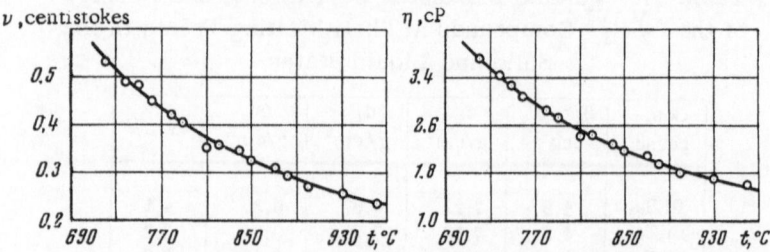

Fig. 101. Temperature dependences of the kinematic and dynamic viscosities of molten bismuth selenide.

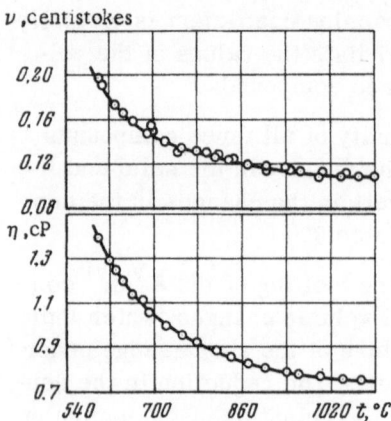

Fig. 102. Temperature dependences of the kinematic and dynamic viscosities of molten bismuth telluride.

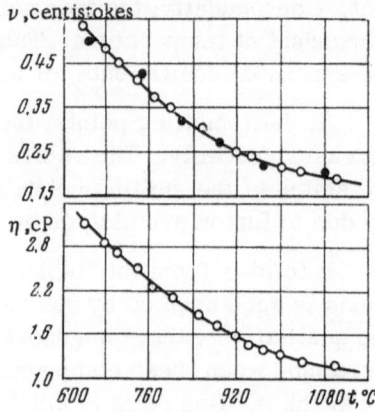

Fig. 103. Temperature dependences of the kinematic and dynamic viscosities of molten antimony telluride. ●) Results reported in [48].

follow changes in the short-range order during the heating of the melts, we investigated the temperature dependence of the viscosity of these compounds in the molten state [552].

The kinematic viscosity of the $A_2^V B_3^{VI}$ compounds was investigated using a method described in Chapter 2. The viscosity was calculated from the formulas for weakly viscous liquids. At high temperatures, the tellurium reacted with the quartz and therefore all the measurements on the tellurides were carried out in quartz ampoules coated internally with a mirrorlike layer of pyrocarbon (this layer was deposited by a method described in [98]). The viscosity was measured using two or three different samples of each

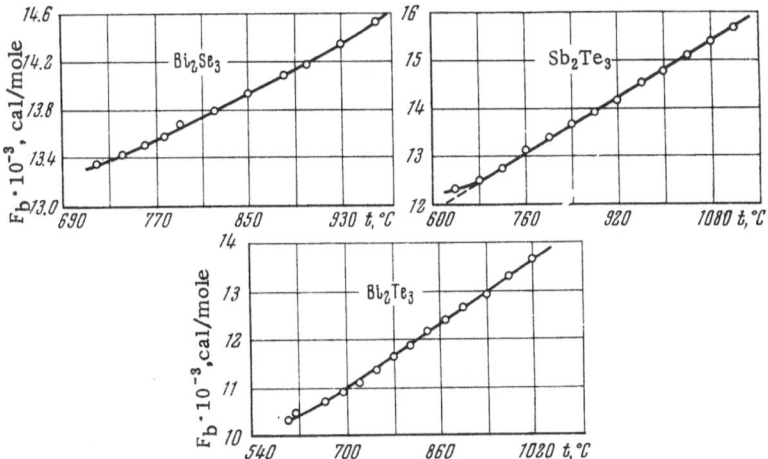

Fig. 104. Temperature dependences of the free activation energy of the viscous flow of the molten $A_2^V B_3^{VI}$ compounds.

compound. The results of the measurements are presented in Figs. 101-103. The same figures include the temperature dependences of the dynamic viscosity, calculated using the results on the temperature dependence of the density. There are no published data on the viscosity of bismuth chalcogenides. It follows from Figs. 101-103 that the viscosity of the $A_2^V B_3^{VI}$ compounds decreases smoothly when the temperature is increased. The results obtained for different samples are in good mutual agreement.

In order to find changes in the short-range order which take place during heating, the results obtained were analyzed using Eyring's theory of activated complexes [148] as well as Bachinskii's equation [62, 63]. The free activation energy of viscous flow was calculated from the kinematic viscosity of the molten $A_2^V B_3^{VI}$ compounds. The results of these calculations are presented in Fig. 104 in the form of a temperature dependence of this free activation energy. It follows from this figure that the free activation energy of viscous flow increases almost linearly (with a slight upward curvature) when the temperature is increased. No characteristic changes can be seen in the investigated range of temperatures (above the melting point); in this respect these compounds are similar to the $A^{IV}B^{VI}$ materials. Consequently, the activation entropy of viscous flow remains practically constant over the whole

Table 38. Activation Energy and Entropy
of the Viscous Flow of the Molten $A_2^V B_3^{VI}$
Compounds

Compound	$E_b \cdot 10^{-3}$, cal/mole	$-S_b$, cal $\cdot$ mole$^{-1} \cdot$ deg^{-1}
Bi_2Se_3	9.7	5.05
Bi_2Te_3	2.7	7.80
Sb_2Te_3	6.1	7.35

investigated range of temperatures, indicating the lack of any appreciable changes in the short-range order above the melting
point. The average values of the activation entropy of viscous flow
of molten $A_2^V B_3^{VI}$ compounds, calculated from the tangents of the
slopes of the curves $F_b \propto f$ (t) in accordance with Eq. (67), are
listed in Table 38.

In agreement with the constancy of the activation entropy of
the viscous flow of the $A_2^V B_3^{VI}$ melts, the dependence of the kinematic viscosity on the reciprocal of the absolute temperature is
described satisfactorily by the exponential equation (74) (a slight
deviation is observed only in the case of bismuth telluride). Exactly the same behavior is observed in the case of the dynamic
viscosity, which obeys the Frenkel' equation.

The average values of the activation energy of viscous flow,
calculated from $\log \nu \propto f(1/T)$, are given in Table 38.

The values of the activation energy of viscous flow of the
$A_2^V B_3^{VI}$ compounds in the molten state (Table 38) are relatively
small. As in the case of the $A^{IV} B^{VI}$ compounds, this indicates a
weak interaction between the structural units and, consequently, it
means that the process of viscous flow is relatively easy (this conclusion follows also from the study of the electrical conductivity,
which indicates the possibility of the retention of covalent bonds).
Considerable differences between the values of E_b for bismuth telluride and selenide may be due to the different degree of their dissociation in the melt, beginning from temperatures immediately
above the melting point. Bismuth selenide is more stable than
bismuth telluride, as indicated by the liquidus curves and direct

investigations of the thermal stability of these compounds by the physicochemical analysis of the corresponding binary liquid systems (cf. Chapter 9). Consequently, the structural units in molten bismuth selenide should be considerably larger than in the telluride and this should reduce the potential barriers which have to be overcome when these structural units are set in motion during viscous flow. Antimony telluride occupies an intermediate position. The absence of appreciable deviations of the dependences $F_b \propto f(t)$ and $\log \nu \propto f(1/T)$ from linearity in the whole investigated range of temperatures indicates the absence of appreciable changes in the short-range order above the melting point. Hence, it follows that the melting of compounds with a layered structure of the tetradymite type causes no basic changes in the short-range order. Consequently, we may conclude that the nature of the short-range order retained in the liquid state is basically the same as in the solid state. Thus, the results of an analysis of the viscosity of the molten $A_2^V B_3^{VI}$ compounds indicate qualitatively that covalent bonds may be retained in the liquid state, in agreement with the electrical conductivity and density data.

Further heating of the melts produces slight changes in the short-range order which are likely to be due to slight changes in the degree of dissociation, resulting in some variation in the size of structural units. This conclusion is confirmed by an analysis of the viscosity data by means of Bachinskii's equation.

Chapter 8

Changes in the Physicochemical Properties at the Melting Points of Semiconducting Compounds Belonging to Different Structure Groups. Some General Problems in the Physical Chemistry of Melts

We have considered several groups of compounds with the zinc-blende, antifluorite, galenite, and tetradymite structure in the solid phase. The compounds belonging to the first three groups have the fcc lattice; the compounds in the last group have a layered structure with complex binding between the atoms.

Thus, we have sufficient data to enable us to compare the nature of the changes which take place during the melting of semiconductors with different structures [555, 556]. We shall consider the results of these comparisons and some general relationships, employing the experimental data reported in the present monograph. We shall also consider some general problems in the physical chemistry of the melts which have been touched upon in this book.

§1. Relationships Governing Changes in the Physicochemical Properties at the Melting Points of Semiconductors Belonging to Different Structure Groups and the Role of the Initial Structure

We shall consider first the data obtained for the large group of compounds which have the zinc-blende structure. Changes in the electrical conductivity, which reflect changes in the nature of the chemical binding during the melting of compounds belonging to this group show that in the case of the $A^{III}B^V$ compounds ($A^{III}Sb$ and $A^{III}As$), the transition from the solid to the liquid state is accompanied by a strong increase in the electrical conductivity to values of the order of $10^4 \ \Omega^{-1} \cdot cm^{-1}$, which are typical of metals. When the melts of these compounds are heated, their electrical conductivity decreases like that of metals. From these results, we have concluded that the $A^{III}B^V$ group of compounds undergoes, like germanium and silicon, transitions to a metallike state in the liquid phase, i.e., according to the Regel' classification the fusion of these substances is of the "semiconductor—metal" type.

Investigations of the electrical conductivity of zinc and cadmium tellurides have shown that their electrical conductivity also increases strongly at the melting point. However, the electrical conductivity of molten zinc and cadmium tellurides at the melting points is about 1.5-2 orders of magnitude less than the conductivity of the molten $A^{III}B^V$ compounds at their melting points. Heating molten zinc and cadmium tellurides increases their electrical conductivity. The magnetic susceptibility of these compounds changes suddenly at the melting point but the absolute value of this change is considerably less than that observed for the $A^{III}B^V$ compounds – it is only 8-10%, indicating much weaker changes in the nature of the chemical binding. The results quoted suggest that the covalent type of binding is retained in zinc and cadmium tellurides in the liquid state, so that they remain semiconductors after melting.

Thus, according to the Regel' classification, zinc and cadmium tellurides undergo transitions of the "semiconductor—semiconductor" type. The same type of transition is observed also at the melting points of gallium and indium tellurides (Ga_2Te_3 and In_2Te_3).

The electrical conductivity of CuI increases somewhat after melting. However, the absolute value of the electrical conductivity of molten CuI is approximately two orders of magnitude less than that of the $A^{II}B^{VI}$ compounds and four orders of magnitude less than the conductivity of the $A^{III}B^{V}$ group. These results indicate that the melting of cuprous iodide produces an ionic liquid in which the structure units are Cu^{+} and I^{-} ions.

The general nature of the changes in the properties at the melting point is similar in series of chemical analogs in which "cations" and "anions" are replaced with heavier or lighter elements.

Changes in the metallic component of the binding in series of chemical analogs in which a "cation" is replaced do not affect the general nature of the changes in the properties at the melting point and during the subsequent heating of the melt.

In series of analogs in which an "anion" is replaced, the ionic component of the binding varies considerably. Nevertheless, the changes in the properties caused by the fusion of these compounds are in general similar to the changes observed in series of analogs in which a "cation" is replaced.

A completely different relationship governs the changes in the isoelectronic series, such as

$$InSb - (In_2Te_3) - CdTe,$$
$$GaSb - (Ga_2Te_3) - ZnTe - CuI.$$

As already mentioned, the first members of these series (InSb and GaSb) become metallic after melting, the middle members $(In_2Te_3, Ga_2Te_3, CdTe, ZnTe)$ retain their covalent binding and remain semiconductors after melting, while the last member of the isoelectronic series of the $A^{I}B^{VII}$ compounds (cuprous iodide) becomes an ionic liquid in the molten state.

We thus observe a very interesting pattern: members of an isoelectronic series which have similar chemical binding, the same structure, practically identical interatomic spacings and density in the solid state, differ basically in the nature of the changes in their physicochemical properties at the melting point. Similar behavior is observed also in other structure groups.

Series of chemical analogs in which "cations" and "anions" are replaced exhibit basically the same pattern in changes of the physicochemical properties at the solid-to-liquid transition. This can be seen most clearly in the following three series:

$$Mg_2Si - Mg_2Ge - Mg_2Sn - Mg_2Pb,$$
$$GeTe - SnTe - PbTe,$$
$$PbTe - PbSe - PbS, \text{ etc.}$$

Some conclusions can be drawn about the role of the initial structure. The results for the compounds with the zinc-blende structure indicate that compounds with the same initial structure can exhibit completely different types of transition at the melting point. The same is true of other groups of compounds. The Mg_2B^{IV} and $A^{IV}B^{VI}$ compounds, which have similar fcc structures and the same average value of the coordination number (K = 6), differ greatly in the nature of their transition to the molten state: the former (Mg_2B^{IV}) exhibit the "semiconductor−metal" transition, whereas the latter ($A^{IV}B^{VI}$) undergo the "semiconductor− semiconductor" transition.

Thus, the retention of semiconducting properties in the molten state is not simply related to the structure although in some cases a prediction can be made. Thus, the melting of elements or compounds with chainlike or layered structures (for example, selenium, $A^{III}B^{VI}$ and $A_2^VB_3^{VI}$ compounds), in which weak van der Waals forces act between the chains or layers, should not destroy the semiconducting properties because the melting process consists mainly in the breaking of weak bonds and it can increase the covalent binding within chains, layers, or molecules.

Zhuze and Shelykh [400] attribute the retention of semiconducting properties in the liquid state to the influence of the ionic component of the binding and the associated different degree of packing of the crystal lattice. They assume that the differences observed in the nature of melting of the $A^{III}B^V$ and $A_2^{III}B_3^{VI}$ compounds are due to the considerably higher degree of the ionicity of the chemical bonds in gallium and indium tellurides (this applies even much more to other chalcogenides, which have the zinc-blende or wurtzite structure) than in the case of the $A^{III}B^V$ compounds; the consequence of this is a much stronger packing of the

lattice volume with "anions," which is retained after melting so that the type of binding and the conduction mechanism are conserved in the molten state [400]. Using the results of Batsanov [320], whose view is that 70-90% of the electropositive atoms are "ions" when the ionicity of the binding is only 30-50%, Zhuze and Shelykh conclude that a relatively small increase in the degree of the ionicity (in the transition from one class of compounds to another) greatly increases the packing of the unit cell volume. In the zinc-blende structure, the packing increases from 34% for the covalent binding to 74% for the purely ionic binding [400]. Moreover, Zhuze and Shelykh assume that the melting does not alter basically the closest packing of "anions" or the conduction mechanism in substances with a strongly ionic binding, and that the short-range order is either conserved or changes slightly in the molten state [400].

However, estimates of the degree of ionicity of the binding are still very inaccurate, because there is no reliable calculation method, and even Zhuze and Shelykh [400] admit to "some arbitrariness" in such calculations. Moreover, it is not clear why the packing of "anions" should be retained after melting when the crystal structure is destroyed. The point is that, in spite of a certain contribution from the ionic binding, the dominant binding of the $A_2^{III}B_3^{VI}$ and $A^{II}B^{VI}$ compounds is covalent. Consequently, if we assume the retention of the "anion" sublattice after melting, we must also assume the possibility of the retention of the three-dimensional system of covalent bonds, which — in our opinion — is impossible because of the large amplitudes of the lattice vibrations and the high mobility of the structure units in the melt. Covalent bonds are more likely to be retained as one-dimensional chainlike or molecular structures. In some cases, two-dimensional units may exist in a certain range of temperatures above the melting point and the covalent bonds may be retained in such units.

In the case of the melting of the $A_2^{III}B_3^{VI}$ and $A^{II}B^{VI}$ compounds with the zinc-blende lattice such a change in the structure is possible after the breaking of the bonds along planes with the maximum reticular density ({110} planes) but further heating should give rise to a chainlike or molecular structure, which is more stable under the conditions of rapid thermal motion in the liquid. This is supported by the anomalous changes in the viscos-

ity in the molten $A_2^{III}B_3^{VI}$ and $A^{II}B^{VI}$ compounds, as well as by the increase in the density during heating above the melting point, which is observed in Ga_2Te_3 and In_2Te_3 [24, 28, 29].

The degree of the ionicity in the GeTe → PbS series in the $A^{IV}B^{VI}$ group varies within fairly wide limits, but nevertheless all the $A^{IV}B^{VI}$ compounds exhibit similar changes in their physicochemical properties at their melting points and remain semiconductors in the liquid state. It is interesting to compare, for example, the A^{III}Sb compounds, for which the difference between the electronegativities is 0.6 (on the scale used in Bokii's book [476]), and the A^{IV}Te compounds, for which the difference of the electronegativities is 0.4-0.5 (both groups have similar fcc lattices). However, compounds belonging to the former group $(A^{III}Sb)$ melt in accordance with the "semiconductor−metal" transition, while the compounds in the latter group undergo the "semiconductor−semiconductor" transition.

Another example illustrating the deficiency of the concepts suggested at the beginning of this section is provided by the $A^{II}B^{VI}$ and Mg_2B^{IV} groups. In both cases, the difference between the electronegativities is 0.6 [476], but the compounds in the former group $(A^{II}B^{VI})$ melt, as reported in [26], in accordance with the "semiconductor−semiconductor" transition, while the members of the latter group (Mg_2B^{IV}) undergo the "semiconductor−metal" transformation [433]. In this case, the compounds in both groups have fcc structures which are even more similar than those in the case of the $A^{III}Sb$ and A^{IV}Te groups.

Thus, the problem of the retention of semiconducting properties by some compounds and not by others cannot be solved by a "geometrical" theory and by considering the degree of the ionicity of the binding, as suggested by Zhuze and Shelykh [400].

We shall now consider again the role of the initial structure: it does not play a decisive role in the formation of the short-range order in the molten state, with the exception of those cases when the presence of weak bonds between structure units in chainlike, molecular, or layered lattices seems to "prepare" a crystal for the formation of a melt consisting of the same structure units (selenium, the $A^{III}B^{VI}$, $A_2^{V}B_3^{VI}$ compounds, etc.).

It follows that the main factor which influences the nature of the changes in the properties of a solid semiconductor at its melting point is the structure of the outer electron shells of the atoms which form the compound. Here, we can use the idea, suggested earlier, that the possibility of the retention of the covalent bonds after melting is related to the possibility of the formation of a chainlike or molecular structure, which is stable under the conditions of high mobility of particles in a liquid. One of the methods for deducing probable melt structure from an analysis of the structure of the outer electron shells of the components of a compound is to construct probable electron—valence schemes. Unfortunately, such schemes are often constructed in a purely formal manner because it is difficult to predict a priori the most probable and energetically favored state of electrons in the outer electron shells of the components. However, in some cases this method provides us with useful information. Thus, in the molten $A^{III}B^V$ compounds the retention of the covalent bonds is possible only if they form an arsenic-like three-dimensional system with a coordination number of 3. However, this process is not very likely. Therefore, as confirmed by experiments, the melting of the $A^{III}B^V$ compounds produces liquids with high coordination and some of the valence electrons form an electron gas. However, bearing in mind the insufficient reliability of this method, we shall not construct electron—valence schemes for the compounds discussed in this book. We shall mention only that in the case of the $A^{IV}B^{VI}$ and $A_2^V B_3^{VI}$ compounds, which retain their semiconducting properties after melting, we can predict formally the possibility of the formation of a molecular or a chainlike structure in the molten state.

§2. Influence of the Initial Structure on the "After-Melting" Phenomena. Comparison with the Values of the Entropy of Fusion

The role of the initial structure and the degree of the retention of its main features after melting can be discussed, as shown by Regel' [7], by considering the entropy of fusion (S_f). The entropy of fusion is found from the expression

$$S_f = \frac{Q_f \cdot 10^3}{T_{mp}}, \tag{81}$$

where Q_f is the heat of fusion (in kcal/g-atom, and T_{mp} is the melting point (in °K).

The entropy of fusion represents the change in the degree of ordering after melting, due to changes in the structure and in the nature of the chemical binding. Large values of the entropy of fusion indicate considerable changes at the melting point.

Bochvar and Kuznetsov [557] estimated the entropy of fusion of chemical elements and demonstrated that this characteristic is a periodic function of the atomic number. Regel' [7] compared changes in the entropy of fusion of chemical elements having different structures and different chemical binding. He found that close-packed metals, which suffer little change on transition to the liquid state, have very small values of the entropy of fusion: about $2 \text{ cal} \cdot (\text{g-atom})^{-1} \cdot \text{deg}^{-1}$. Semimetals (Sb, Bi), which undergo much greater changes at the melting point (mainly changes in the coordination number), have higher values of the entropy of fusion: about $5 \text{ cal} \cdot (\text{g-atom})^{-1} \cdot \text{deg}^{-1}$. Finally, semiconductors (Ge, Si), which undergo very great changes in their structure and chemical binding after melting, have even larger values of the entropy of fusion: about $7 \text{ cal} \cdot (\text{g-atom})^{-1} \cdot \text{deg}^{-1}$. On the other hand, a semiconducting element such as selenium, which retains its chainlike structure and covalent binding after melting, has a small entropy

Table 39. Entropies of Fusion of Semiconductors Belonging to Different Structure Groups

Element	S_f, cal·g-atom^{-1}·deg^{-1}	$A^{III}B^V$	S_f, cal·g-atom^{-1}·deg^{-1}	Mg_2B^{IV}	S_f, cal·g-atom^{-1}·deg^{-1}	$A^{IV}B^{VI}$	S_f, cal·g-atom^{-1}·deg^{-1}	$A_2^V B_3^{VI}$	S_f, cal·g-atom^{-1}·deg^{-1}
Si	~7.1	AlSb	~5.2	Mg$_2$Si	~5.0	GeTe	~5.7	Bi$_2$Se$_3$	—
Ge	~6.9	GaSb	~6.1	Mg$_2$Ge	—	SnTe	~3.7 (5.1)	Bi$_2$Te$_3$	~6.6
Se	~3	InSb	~7.2	Mg$_2$Sn	~3.6	PbTe	~3.1	Sb$_2$Te$_3$	~5.3
Te	~5.7	GaAs	~7.7	Mg$_2$Pb	~3.8	PbSe	~3.1		
		InAs	~5.2			PbS	~3.1		

of fusion: ~3.06 cal · (g-atom)$^{-1}$ · deg^{-1}. The entropy of fusion of
tellurium, whose physicochemical properties change somewhat
more strongly than those of selenium, is correspondingly higher:
~5.7 cal · (g-atom)$^{-1}$ · deg^{-1}.

Comparative estimates of the entropies of fusion of the
$A^{III}B^{V}$ compounds have been carried out by us and by Zhuze and
Shelykh [400]. It has been found that the entropies of fusion of
these compounds are similar to the corresponding entropies of
germanium and silicon, which indicates considerable changes in
the structure and nature of the chemical binding in the transition
from the solid-to-liquid state. On the other hand, the $A_2^{III}B_3^{VI}$ compounds have much lower values of the entropy of fusion and their
semiconducting properties are retained above the melting point
[400].

Thus, some predictions about changes in the short-range
order at the melting point can be made on the basis of the entropy
of fusion. We may assume that this entropy is a measure of the
difference between the short-range structures in the solid and liquid states. Consequently, in order to obtain fuller information
about the investigated compounds, we shall estimate their entropies of fusion and compare these entropies with changes in the
short-range order and the nature of the chemical binding, which
have been deduced by analyzing the changes in the physicochemical
properties at the melting point. The results of estimates based on
Eq. (81) are given in Table 39. The values of the heats of fusion
and of the melting points required in these calculations were taken
from Tables 6, 13, 17, 23, 27, and 34.

The results given in Table 39 indicate that the correlation
between the entropies of fusion and the changes in the physicochemical properties at the melting points of the Mg_2B^{IV}, $A^{IV}B^{VI}$,
and $A_2^{V}B_3^{VI}$ compounds is not as definite as the correlation for elements and for the $A^{III}B^{V}$ compounds. Thus, the entropies of fusion
of the Mg_2B^{VI} compounds are smaller than those of the $A^{III}B^{V}$
group (particularly in the case of Mg_2Sn and Mg_2Pb). This somewhat unexpected result (from the point of view of analogy between
changes in the properties) indicates that the short-range order
formed after melting retains, to a considerable degree, its characteristic features observed in the solid phase. The similarity in
the distribution of the structure units in the Mg_2B^{IV} compounds is

much greater than in the case of the $A^{III}B^V$ compounds. This may be due to the fact that elements of the antifluorite structure with the coordination number 8 are generally retained in Mg_2B^{IV} compounds after melting (although the nature of the chemical binding changes) and only the packing of the structure increases because of a more compact arrangement of these elements resulting from the rapid thermal motion of the atoms. The cause of this contradiction between the entropy of fusion and the nature of changes in the properties at the melting point and the short-range order in molten Mg_2B^{IV} compounds may be determined by direct investigations using diffraction methods.

A definite correlation between changes in the physicochemical properties at the melting point and the entropy of fusion is observed for the $A^{IV}B^{VI}$ and $A_2^VB_3^{VI}$ compounds.

The $A^{IV}B^{VI}$ compounds which retain their chemical binding after melting have relatively small values of the entropy of fusion (with the exception of germanium telluride). The higher value of the entropy of germanium telluride may be due to its dissociation at the melting point. In the case of tin telluride, there is a published value of the heat of fusion (5.1 kcal/g-atom) [482]. The entropy of fusion calculated from this heat of fusion is given in Table 39 in parentheses. If the value in parentheses is more correct, it obviously indicates the dissociation of tin telluride at the melting point.

The higher values of the entropy of fusion of antimony and bismuth tellurides may also be associated with changes in the short-range structure, which are either due to the dissociation of these compounds at the melting point or due to differences between the solid-state structure and the final covalent structure of the melt (the latter is less likely).

Summarizing our discussion, we must conclude that there is no unambiguous correlation of the entropy of fusion with the nature of changes in the chemical binding at the melting point or with the "after-melting" phenomena. The definite correlation which is observed for elements and the $A^{III}B^V$ compounds does not apply to other materials.

§3. Activation of Electrical Conduction

in Liquid Phases Retaining Semiconducting

Properties after Melting

When semiconducting properties are retained after melting, the temperature dependence of the electrical conductivity is exponential and can be described by the equation which applies to solid crystalline semiconductors in the intrinsic conduction region. However, the physical meaning of the quantity ΔE in the index of the exponent and of the pre-exponential factor is not very clear.

Many of the compounds described in the present book are semiconductors in the liquid state. Their values of ΔE (in eV), calculated from the tangents of the slopes of the straight lines $\ln \sigma \propto f(1/T)$ (Figs. 78-80 and 98) are as follows:

GeTe — 0.15	PbS — 1.09
SnTe — 0.32	Bi_2Se_3 — 0.37
PbTe — 0.29	Bi_2Te_3 — 0.13
PbSe — 1.15	Sb_2Te_3 — 0.17

It follows from these values that there is no regular variation of ΔE when "cations" and "anions" are replaced with other elements in compounds of the $A^{IV}B^{VI}$ group. ΔE for lead selenide and sulfide is several times larger than for lead telluride and other tellurides. This is difficult to explain, since this difference is not observed in the solid state and there are no basic differences in the nature of the fusion of these compounds (as judged by changes in their physicochemical properties). In view of this, the value of ΔE, calculated in this way for liquid semiconductors, can hardly be the analog of the forbidden band width of solid semiconductors and consequently it cannot represent the energy necessary for the breaking of a single bond, because the energy spectrum of a liquid semiconductor is very complex. Moreover, we cannot ignore changes in the mobility in the liquid state when the temperature is raised, because the scattering by thermal vibrations in the melt should increase considerably when the temperature is raised. Therefore, the rise of the electrical conductivity in the liquid state cannot be simply the consequence of an increase in the carrier density. Therefore, the quantity ΔE for a liquid semiconductor, calculated from the tangents of the slopes of the straight lines

$\log \sigma \propto f(1/T)$ should be regarded as some complex integral activation characteristic, representing many processes which take
place during the heating of a liquid semiconductor. Nevertheless,
the temperature dependence of the electrical conductivity of liquid
semiconductors is described satisfactorily by Eq. (79), and this is
supported also by the results of other investigators. Therefore,
the meaning of the quantity ΔE, and of the processes which it represents, require further study.

§4. Relationships Governing Volume Changes at the Melting Points of Semiconductors Belonging to Different Structure Groups

The determination of the general relationships governing
volume changes during the melting of semiconducting elements and
compounds belonging to various groups is of great interest because
an analysis of such changes and a comparison with the results of
investigations of the electrical properties can give us additional information on the changes in the short-range order at the melting
point.

The results of investigations of changes in the density at the
melting points of semiconductors belonging to different groups are
summarized in Table 40.

The results presented in Table 40 show that the nature of the
volume changes can be used to divide the investigated substances
into two groups.

The first group includes germanium, silicon, the $A^{III}B^V$
compounds with the zinc-blende structure, and the Mg_2B^{IV} compounds with the antifluorite structure. The volume of all these
substances decreases after melting.

The second group comprises the $A^{IV}B^{VI}$ compounds with the
galenite structure, the $A_2^{III}B_3^{VI}$ compounds with the zinc-blende
structure, the $A^{III}B^{VI}$ compounds with layered structures, and the
$A_2^V B_3^{VI}$ compounds with the tetradymite structure. The volume of
all these compounds increases after melting.

Table 40. Volume Changes Due to Melting and Values of dT/dP
for Semiconductors Belonging to Various Groups

Substance	Type of structure	T_{mp}, °K	Q_f, kcal per mole	d_s, g/cm³	d_l, g/cm³	$\Delta d =$ $d_l = d_s$; g/cm³	$\dfrac{d_s-d_l}{d_s}$, %	dT/dP, deg/atom
Si	Diamond	1693	12.1	2.30	2.53	+0.23	+10.0	−0.0038
Ge		1210	8.35	5.26	5.51	+0.25	+4.75	−0.0020
AlSb	ZnS	1353	14.2	4.18	4.72	+0.54	+12.9	−0.0093
GaSb		985	12.0	5.60	6.06	+0.46	+8.2	−0.0053
InSb		809	11.6	5.76	6.48	+0.72	+12.5	−0.0076
GaAs		1511	23.2	5.16	5.71	+0.55	+10.7	−0.0041
InAs		1215	12.6	5.50	5.89	+0.39	+7.1	−0.0049
Mg₂Si	CaF₂	1375	20.4	1.84	2.27	+0.43	+23.3	−0.0129
Mg₂Sn		1051	11.4	3.45	3.52	+0.07	+2.03	−0.0022
Mg₂Pb		823	9.3	5.00	5.20	+0.20	+4.0	−0.0041
CuI	ZnS	875	2.6	5.36	4.84	−0.48	−9.0	+0.0295
Ga₂Te₃		1063	—	5.35	5.086	−0.264	−4.93	—
In₂Te₃		940	—	5.77	5.54	−0.23	−3.98	—
GaTe	Layered	1097	—	5.32	5.17	−0.15	−2.82	—
InTe		969	—	5.88	5.65	−0.23	−3.91	—
GeTe	Galenite	998	11.3	5.97	5.57	−0.40	−6.7	+0.0051
SnTe		1063	8.0	6.15	5.85	−0.30	−4.9	+0.0067
PbTe		1190	7.5	7.69	7.45	−0.24	−3.1	+0.0054
PbSe		1361	8.5	7.57	7.10	−0.47	−6.2	+0.0099
PbS		1392	8.7	7.07	6.45	−0.62	−8.8	+0.0130
Bi₂Se₃	Tetra- dymite	979	—	7.27	6.97	−0.30	−4.1	—
Bi₂Te₃		858	28.35	7.5	7.26	−0.24	−3.2	+0.0026
Sb₂Te₃		895	23.65	6.29	6.09	−0.20	−3.2	+0.0031
Se	Hexag- onal	493	1.5	4.69	3.975	−0.715	−15.2	+0.0243
Te		725	4.17	6.1	5.775	−0.325	−5.3	+0.0048

A comparison of the nature of the changes in the volume at
the melting points of semiconductors with different structures in
the solid state shows that the direction of the volume changes is
not related to the degree of packing of the initial crystal structure.

In fact, those $A^{III}B^V$ compounds which have the loose-packed
structure of the zinc-blende type with a coordination number of 4
(like germanium and silicon) exhibit a contraction after melting.

On the other hand, the $A_2^{III}Te_3$ compounds which have an even looser structure of the defect zinc-blende type, exhibit an expansion at the melting point.

Changes in the density at the melting points of series of analog compounds in which "cations" and "anions" are replaced within a given isostructural group show clearly that only in the case of the galenite group does the sudden change in the density decrease in a regular manner when a "cation" is replaced with a heavier element or an "anion" is replaced with a lighter element. In this case, we have the following series:

$$GeTe \rightarrow SnTe \rightarrow PbTe, \ PbTe \rightarrow PbSe \rightarrow PbS.$$

In the $A^{III}B^V$ group with the sphalerite structure, there is no correlation between volume changes at the melting point in a series of compounds in which "cations" and "anions" are replaced with other elements. The same behavior is observed also in isoelectronic series: in this case, the compounds with the same structure and practically the same density in the solid phase differ not only in the magnitude but even in the sign of the changes in the density at the melting point (cf. GaSb and CuI).

However, we must point out a very definite correlation between the sign of the change in the density and changes in the nature of the chemical binding at the melting point.

Germanium, silicon, the $A^{III}B^V$ and Mg_2B^{IV} compounds, all of which become metallic in the liquid state (cf. Chapters 3-5), exhibit a contraction at the melting point, while all the other compounds, which retain (at least partly) the covalent binding in the liquid state, exhibit an expansion.

Thus, an analysis of the volume changes of a large number of substances, listed in Table 40, shows that the substances which retain their chemical binding in the liquid state exhibit an expansion at the melting point, which is obviously due to an increase in the average interatomic spacings, increase in the number of vacancies, etc.

This conclusion follows from our analysis of substances with mainly covalent type of chemical binding. Analysis of the data for metals shows that the same applies to substances with the metallic type of binding.

The nature of the volume change observed during the melting of bismuth, which is regarded as anomalous, is in fact due to the destruction of covalent bonds and the transition from a mixed covalent–metallic type of binding in the solid state to a purely metallic binding in the liquid phase; this is reflected also in the sign of the change in the electrical conductivity (in contrast to true metals, the electrical conductivity of bismuth increases in the molten state).

The electrical conductivity of all substances in the first group (i.e., those whose chemical binding is altered by melting) increases at the melting point, while in the case of substances belonging to the second group (whose chemical binding does not change at the melting point) changes in the electrical conductivity are slight.

Thus, a correlation between the nature of the volume changes and the changes in the electrical properties at the melting point found by Regel', is exhibited by a large number of substances.

We shall now analyze the volume changes observed at the melting points of semiconductors belonging to various groups using the Clausius–Clapeyron equation [59, 60] and we shall calculate the coefficient dT/dP, which represents the pressure dependence of the melting point

$$\frac{dT}{dP} = \frac{v_l - v_s}{S_f} = \frac{T_{mp}}{Q_f}\left(\frac{1}{d_l} - \frac{1}{d_s}\right), \tag{82}$$

where $(v_l - v_s)$ is the volume change at the melting point, Q_f is the heat of fusion, S_f is the entropy of fusion, and T_{mp} is the melting point.

The results of these calculations are given in Table 40. An analysis of these results indicates that when the pressure is increased the melting points of the substances in the first group (germanium, silicon, $A^{III}B^V$, Mg_2B^{VI} compounds) should decrease. The melting points of substances belonging to the second group should increase when the pressure is increased. There are no definite relationships between the coefficients dT/dP in series of analog compounds (when "cations" and "anions" are replaced with other elements) or in isoelectronic series. The values of the dT/dP coefficients can be compared with the experimental data on the influence of pressure on the melting point. The melting points

Table 41. Influence of Pressure on Melting Points of Some Semiconductors

Substance	Melting point, °C, at pressures of				
	1 atm	5000 atm		10,000 atm	
		calc.	exp.	calc.	exp.
Si	1420	1400	1390	1381	1360
Ge	937	927	920	916	900
Se	220	341	330	465	417
Te	452	474	470	499	482
InSb	536	497	483	458	440
InAs	942	917	930	892	900
GaSb	712	685	680	658	660
GaAs	1238	1216	1215	1196	1200
Bi_2Te_3	585	598	600	612	610

of silicon, germanium, and the $A^{III}B^V$ compounds have been investigated [558-563] and it has been found that the melting point is reduced by the application of pressure and the dependence $T_{mp} \propto f(P)$ is almost linear (we are not discussing here the influence of pressure on the melting points of phases which may exist at higher pressures). On the other hand, selenium, tellurium, and bismuth telluride exhibit an increase in the melting point when pressure is increased [564-566].

Using an approximate relationship

$$\frac{dT}{dP} \simeq \frac{\Delta T}{\Delta P}, \tag{83}$$

we can write

$$T_{mp}^{(P)} = T_{mp}^{(1)} + (P_2 - P_1)\frac{dT}{dP}, \tag{84}$$

where $T_{mp}^{(P)}$ is the melting point at a pressure P, and $T_{mp}^{(1)}$ is the melting point at the atmospheric pressure.

We shall now compare the melting points (Table 41) calculated from Eq. (84) using the values given in Table 40 with the values found experimentally [558-566].

It is evident from Table 41 that, up to 10,000 atm, the experimentally determined values of the melting points of many semiconductors are in good agreement with those calculated from Eq. (84). Hence, we may conclude that the values of the coefficient dT/dP given in Table 40 may be useful.

§ 5. Thermoelectric Effects at the Liquid − Solid Boundary in Semiconductors Belonging to Different Structure Groups

Cascaded (multistage) thermoelectric converters are a promising development in the direct conversion of thermal into electrical energy. The efficiency of the Carnot cycle can be increased by increasing the working temperature difference. One possible way of constructing a multistage thermoelectric converter is to use a liquid semiconductor for the high-temperature stage.

In this case, a liquid semiconductor would be in contact with a solid and the thermoelectric effects would be observed at the solid–liquid boundary.

In the calculation of the energy characteristics of such a thermoelectric converter, these effects–above all the Peltier effect–would have to be allowed for because they may be considerable and may make an important contribution to the heat balance of the converter. In view of this, we have estimated and compared the thermoelectric effects (the Seebeck and Peltier effects at the solid–liquid phase boundary) using the experimental data for semiconducting elements and compounds belonging to various groups.

We used the results of investigations of the thermoelectric power of various semiconductors near the melting point and in the liquid state.

For the sake of completeness, we calculated these data not only for the substances which retained their semiconducting properties in the liquid state, but also for those substances which became metallic above the melting point.

Table 42 lists the values of the thermoelectric power of the liquid and solid phases at the melting points of the substances considered in the present book.

Table 42. Thermoelectric Effects at Solid—Liquid Phase
Boundary for Semiconductors Belonging to Different Groups

Substance	α_s, μV/deg	α_l, μV/deg	$\Delta\alpha$, μV/deg	Π_s, V	Π_l, V	$\Delta\Pi$, V	$\Delta S \cdot 10^{23}$, J/deg
Ge	—90	0	90	—0.110	0	0.11	1.40
AlSb	—160	—60	80	—0.216	—0.108	0.108	1 28
GaSb	—60	0	60	—0.059	0	0.059	0.96
InSb	—120	—20	100	—0.095	—0.016	0.079	1.60
In$_2$Te$_3$	—50	+30	80	—0.047	+0.028	0.075	1.30
Ga$_2$Te$_3$	—290	—85	205	—0.310	—0.096	0.214	3.28
CuI	+550	+490	60	+0.485	+0.430	0.055	0 97
GeTe	+130	+21	109	+0.130	+0.021	0.109	1.75
SnTe	+140	+28	112	+0.149	+0.030	0.119	1.76
PbTe	—60	—10	50	—0.071	—0.012	0.059	0.80
PbSe	—120	—60	60	—0.163	—0.082	0.081	0.98
PbS	—220	—200	20	—0.306	—0.278	0.028	0.25
Sb$_2$Te$_3$	+90	+11	79	+0.080	+0.010	0.070	1.26
Bi$_2$Te$_3$	—45	—3	42	—0.039	—0.003	0.036	0.67
Bi$_2$Se$_3$	—90	—35	55	—0.088	—0.034	0.054	0.88
Cu$_2$Te	+130	+113	17	+0.182	+0.158	0.024	0.27
Cu$_2$Se	+177	+160	17	+0.246	+0.222	0.024	0.27
Cu$_2$S	+400	+325	75	+0 560	+0 460	0.100	1.20

Estimates of the sudden changes in the thermoelectric power
at the melting point show that in the majority of cases the thermoelectric power decreases considerably at the melting point. The
exceptions are copper chalcogenides and lead sulfide. This is evidently due to an increase in the density of carriers liberated by the
destruction of the crystal lattice and due to the equalization of the
electron and hole mobilities. The slight changes in the thermoelectric power, observed when the Cu_2B^{VI} compounds and PbS are
melted, can be explained by a nearly complete retention of the nature of the chemical binding and of the short-range order at the
solid-to-liquid transition. We calculated the Peltier coefficients
of the investigated substances in the solid and liquid phases (Π_s
and Π_l). We used Thomson's second thermoelectric relationship:

$$\Pi = \alpha T, \tag{85}$$

where Π is the Peltier coefficient, α is the Seebeck coefficient, and T is the temperature. The results of the calculations are given in Table 42. The change in the Peltier coefficient at the melting point is similar to the change in the thermoelectric power.

A large difference between the values of the Peltier coefficient in the solid and liquid states results in an evolution of a considerable amount of the Peltier heat at the solid–liquid boundary. This reduces the amount of heat passing through the thermoelectric converter and consequently reduces its power and efficiency.

Thus, it would be undesirable to use substances whose Seebeck and Peltier coefficients decrease markedly at the solid-to-liquid transition, if a solid–liquid phase boundary were to be included in the thermoelectric circuit.

It would be more desirable to use those semiconductors whose values of α and Π either change very little or increase in their transition from the solid-to-liquid state (an increase is possible when this transition is accompanied by an increase in the intramolecular forces).

To determine the influence of the change in the Peltier coefficient at the melting point on the heat balance in a thermoelectric converter, we calculated the Peltier heat Q_Π and compared it with the main term in the heat balance Q_T, representing the heat transmitted on to the thermal conductivity from the hot end to the cold:

$$Q_\Pi = \pi I, \tag{86}$$

$$I = \frac{\Delta T \alpha \sigma}{2}\left(\frac{s}{l}\right), \tag{87}$$

where I is the current, ΔT is the temperature difference between the ends of the converter (we assumed this difference to be 100°), σ is the electrical conductivity, s is the cross-sectional area of the converter, and l is its length.

Assuming a given value of the current (we took it to be 1 A) and knowing the changes in the thermoelectric power and in the electrical conductivity at the melting point, we could determine the ratio s/l, which was required in the calculation of the main term in the heat balance, Q_T:

$$Q_T = \Delta T \varkappa \left(\frac{s}{l}\right), \tag{88}$$

where $\varkappa$ is the thermal conductivity.

The values of Q_T were calculated for those compounds whose thermal conductivity was investigated by us at the melting point and in the liquid phase [88, 567].

These calculations showed that materials exhibiting the largest changes in the Peltier coefficient (Table 42) the amount of heat transmitted through a thermoelectric converter decreased by 10-15%. However, in the case of the smallest values of $\Delta\Pi$ the heat losses due to the evolution of the Peltier heat at the solid–liquid phase boundary were 2-3%.

Thus, the Peltier heat evolved at a solid–liquid phase boundary, in the case when the Seebeck coefficient changed at the melting point, could affect considerably the heat balance of a thermoelectric converter, particularly if $\Delta\alpha > \sim100\ \mu V/deg$. If this Peltier heat were ignored, considerable errors would be made in the design and construction of multistage thermoelectric units with a high-temperature stage in the form of a liquid semiconductor.

Analysis of the thermoelectric effects at a solid–liquid boundary allows us to calculate the change in the entropy per one electron in the transition from the solid-to-liquid state; this change is of interest in connection with estimates of the electronic component of the entropy of fusion. In fact, the amount of heat evolved at a solid–liquid phase boundary in a semiconductor, through which n electrons pass at a constant temperature (melting point) irrespective of how weak may be the electric current, is given by

$$dQ = T_{mp}(S_s - S_l)\,dn = T_{mp}\Delta S dn, \tag{89}$$

where S_S and S_l is the entropy per one electron in the solid and liquid states.

On the other hand,

$$dQ = \Delta\Pi\,edn, \tag{90}$$

where e is the electron charge. Comparison of Eqs. (86) and (87) shows that

$$\frac{\Delta S}{l} = \frac{\Delta \Pi}{T} = \Delta \alpha. \tag{91}$$

The results of calculations based on Eq. (91) are listed in Table 42. In those cases when the change in the carrier density at the melting point is known, we can estimate approximately the electronic component of the heat of fusion.

Unfortunately, there are as yet no data on the Peltier coefficient in the liquid state.

Using the results of Busch and Tieche [142] on the carrier density in molten germanium and estimates of the carrier density in the solid phase at the melting point (cf. Chapter 3), we obtain a reasonable value (about 40% of Q_f) which indicates that the contribution of the change in the free carrier density to the heat of fusion is considerable.

This confirms the conclusion (cf. Chapter 3) that germanium becomes metallic in the molten state.

§ 6. Application of the Viscosity Method in Investigations of Slow Reactions in Melts

1. Possible Application of the Viscosity Method in Investigations of the Kinetics of Slow Reactions in Melts

Investigations of the kinetics of the formation of chemical compounds in melts yield information on important aspects of chemical processes and led to a better understanding of the mechanism of interaction between substances. Investigations of the kinetics of chemical reactions in solutions are widely used to study chemical reactions. However, these investigations can hardly be applied to the melts of intermetallic compounds because the concentrations of the reacting substances usually cannot be determined by chemical means. For this reason, the reactions of formation and dissociation of intermetallic compounds have hardly been investigated; only qualitative conclusions on the completeness of a given reaction can be drawn from a microscopic analysis, and then the time necessary to complete a reaction between the molten components can be estimated.

We would like to point out the possibility of using the viscosity method in quantitative studies of the slow formation of compounds in melts [568].

The viscosity is a structure-sensitive property which reacts to small changes in a liquid. It is known that when the dimensions of the structure units in a liquid are increased, the viscosity increases in direct proportion to the size of such units [67]. The viscosity of the melt should increase with time, during the formation of a chemical compound and when the reaction is complete the viscosity should remain constant.

Thus, if we know the law governing the dependence of the viscosity of a melt on the concentration of a chemical compound in it, we can use the viscosity to follow the time dependence of the concentration of this chemical compound or the concentration of the initial mixture in the melt during the formation of this compound.

In 1887, Arrhenius proposed a formula for binary solutions:

$$\log \eta = x \log \eta_1 + (1 - x)\log \eta_2, \tag{92}$$

where x is the volume fraction of one of the components.

Arrhenius showed that this formula was in fairly good agreement with the experimental data over a relatively narrow range of concentrations (up to 10% of one of the components) [569].

Later, Kendall and Monroe [570] attributed the meaning of the total concentration of one of the components to the quantity x in the Arrhenius equation and investigated several tens of binary solutions. They found that this modification of the Arrhenius equation agreed better with the experimental data but full agreement over the whole range of concentrations was still not obtained.

Later, Kendall and Wright [571] reported very good agreement between the experimental and calculated values (maximum deviation 0.2%) for the phenetole−diphenyl ether system for values of x from 0 to 1.

Relatively recently, Pospekhov [572] analyzed the data on the viscosity reported in [573, 574] and showed that the logarithm of the viscosity was, to a high accuracy, additive in the following systems: acetophenone-1,1,2,2-tetrachloroethane, acetone−carbon tetrachloride, and ethyl acetate−diethyl succinate.

The simplest structure in the liquid state is exhibited by molten metals. Therefore, one can expect the Arrhenius equation to be obeyed by metals even more closely than by organic liquids, and the viscosity of a molten intermetallic compound to vary during a reaction in such a way as to obey fairly accurately the additivity rule for the logarithms of the viscosity. Thus, the viscosity can be used as a criterion of the concentration of a chemical compound in a melt.

We can use the values of the viscosity at the beginning and end of the investigated reaction. We can assume that the logarithm for each of these values is the sum of the logarithms of the viscosities of the substances present at the beginning and end of the reaction $(\log \eta = \sum_1^i x_i \log \eta_i)$. Then, a melt can be regarded as a quasi-binary system and we can apply Eq. (92) to it. The viscosity is found from the equation

$$\log \eta = c_i \log \eta_i + (1 - c_i)\log \eta_f, \tag{93}$$

where η is the viscosity of the system, which varies during a chemical reaction; c_i and c_f are the "molar" fractions of the initial and final products of the reaction; and η_i and η_f are the viscosities of the initial and final substances in the melt. Hence, we can determine c_i and c_f:

$$c_i = \frac{\log \eta_f - \log \eta}{\log \eta_f - \log \eta_i} \tag{94}$$

$$c_f = \frac{\log \eta_i - \log \eta}{\log \eta_i - \log \eta_f}. \tag{95}$$

If we assume, for the sake of simplicity, that the stoichiometric coefficients of the reacting components in the chemical reaction equation are all the same (the more general case of unequal coefficients results in much more cumbersome equations, without adding anything basically different), the determination of the molar fraction of each of the reacting substances presents no difficulties. The molar fractions of individual substances are the corresponding parts of c_i and c_f, which are inversely proportional to the number of the interacting substances.

First-Order Reactions. In general, a first-order reaction can be represented by an equation which symbolizes a typical dissociation reaction, for example, the dissociation of an intermetallic compound which cannot be investigated by the standard methods:

$$AB \rightarrow A + B. \qquad (A)$$

The reaction in a homogeneous melt is irreversible and we can apply to it the homogeneous kinetics equation. For a first-order equation, we have

$$K = \frac{2.303}{\tau} \log \frac{c}{c_0}, \qquad (96)$$

where K is the reaction rate constant; c and c_0 are, respectively, the concentration at a given moment and the initial concentration. Since c is given by Eq. (94) and $c_0 = 1$, we find that

$$K = \frac{2.303}{\tau} \log \frac{\log \eta_f - \log \eta_i}{\log \eta_f - \log \eta}, \qquad (97)$$

which shows that a criterion of a first-order reaction is the linearity of the function

$$\log \eta = f(\tau). \qquad (98)$$

Second-Order Reactions. A second-order reaction can be represented symbolically as follows:

$$A + B \rightarrow AB. \qquad (B)$$

This represents the usual formation of compounds. For a second-order reaction, we have

$$K = \frac{1}{\tau} \left(\frac{1}{c} - \frac{1}{c_0} \right). \qquad (99)$$

Since

$$c = \frac{1}{2} \frac{\log \eta_f - \log \eta}{\log \eta_f - \log \eta_i},$$

$$c_0 = \frac{1}{2}, \qquad (100)$$

we obtain

$$K = \frac{2}{\tau} \frac{\log \eta - \log \eta_i}{\log \eta_f - \log \eta}.$$ (101)

A criterion of a second-order reaction, when investigated by the viscosity method, is provided by the linearity of the function

$$\frac{a}{b - \log \eta} = f(\tau),$$ (102)

where a and b are constants.

Third-Order Reactions. A third-order reaction can be represented symbolically in the form

$$A + B + C \rightarrow ABC$$ (C)

Such reactions may take place also in other ways but the basic treatment is still the same. For third-order reactions, we have

$$K = \frac{1}{2\tau} \left(\frac{1}{c^2} - \frac{1}{c_0^2} \right),$$ (103)

bearing in mind that

$$c = \frac{1}{3} \frac{\log \eta_f - \log \eta}{\log \eta_f - \log \eta_i},$$ (104)

and

$$c_0 = \frac{1}{3},$$ (105)

we obtain

$$K = \frac{9}{2\tau} \left[\left(\frac{\log \eta_f - \log \eta_i}{\log \eta_f - \log \eta} \right)^2 - 1 \right].$$ (106)

In this case, a criterion of a third-order reaction, expressed in terms of the measured viscosity, is the linearity of the function

$$\frac{a}{b - (\log \eta)^2} = f(\tau).$$ (107)

General Form of Equations for n-th Order Reactions. Examination of the equations for the first-, second-, and third-order reactions yields the following expression for an n-th order reaction:

$$c = \frac{1}{n} \frac{\log \eta_f - \log \eta}{\log \eta_f - \log \eta_i} \tag{108}$$

and

$$K = \frac{1}{\tau} \frac{n^{n-1}}{n-1} \left[\left(\frac{\log \eta_f - \log \eta_i'}{\log \eta_f - \log \eta} \right)^{n-1} - 1 \right]. \tag{109}$$

It follows that the order of the reaction can be found from the linear function

$$\frac{a}{b - (\log \eta)^{n-1}} = f(\tau). \tag{110}$$

These relationships apply to irreversible reactions in homogeneous systems.

If changes in the density during the formation of a compound are small, the dynamic viscosity can be replaced with the kinematic viscosity.

The use of the viscosity method in qualitative investigations of the formation of intermetallic compounds has the following limitations.

1. The investigated reactions must take place sufficiently slowly because a single measurement of the viscosity requires 30–40 sec. If the reaction is fast, the viscosity will change during the measurement itself.

2. The degree of dispersion of the particles in the mixture may affect strongly the reaction and, consequently, the results of the measurements. In order to avoid this influence, the surfaces of the components should be carefully cleaned and the investigation should be carried out in high vacuum.

3. We must bear in mind that the assumption about the additivity of the logarithms of the viscosity of the initial mixture and of the synthesized compound is an approximation and is obviously not satisfied at all the stages of reaction. Therefore, in investigating any particular reaction it is necessary to determine specially the dependence of the viscosity on the concentration of the chemical compound in the melt by quenching at intermediate stages during the reaction and determining microscopically the concentration in a suitably prepared section.

In spite of these limitations, the method can be used successfully in studies of many intermetallic compounds and it can yield
information on details of chemical processes which take place during the formation of such compounds.

2. Investigation of the Kinetics of the Formation of Aluminum Antimonide by Viscosity Method

We determined only the kinematic viscosity.

The viscosity was measured in a viscometer whose construction and method of use have been described in Chapter 2.

The initial materials — aluminum and antimony — were washed
in alcohol directly after crushing; then they were dried and placed
in a corundum can of 12 mm diameter and 30 mm high, which was
heated first to 1200°C. The can containing the initial materials
was placed in a quartz ampoule, which was evacuated down to 10^{-4}
mm Hg and sealed.

A sample prepared in this way was placed in a heater and
the required temperature was reached in 2-3 min. The first read-
ing of the viscosity was made 2 or 3 min after reaching the isothermal state. Thereafter, the readings were taken after the system came to rest (in each measurement the system was set in
oscillation).

The viscosity was calculated using the formulas for weakly
viscous liquids (cf. Chapter 2).

The dependence of the viscosity on the duration of isothermal
treatment was determined at 1090°C (a), 1120°C (b), 1150°C (c), and
1200°C (d). The results of the measurements are presented in Fig.
105. It is evident from this figure that the curves are of the typical kinetic type.

The constancy of the viscosity after a certain time interval,
which had a definite value at a given temperature, indicated that
the process of formation of the compound in the melt had been completed.

Table 43 compares the values of the viscosity of the compound
synthesized under these conditions (ν'_{AlSb}) with the viscosity found
earlier for the same compound synthesized and purified by the pulling method (ν''_{AlSb}).

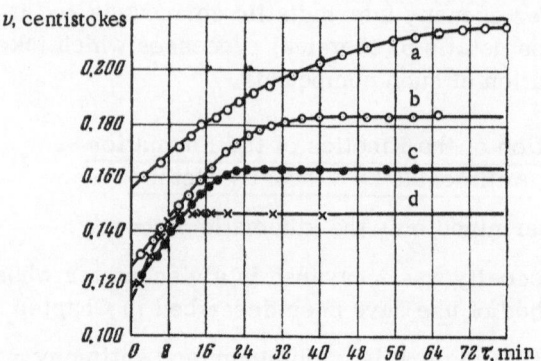

Fig. 105. Dependence of the viscosity of an aluminum—
antimony melt (the two components are present in equi-
atomic ratio) on the duration of exposure to a given temperature.

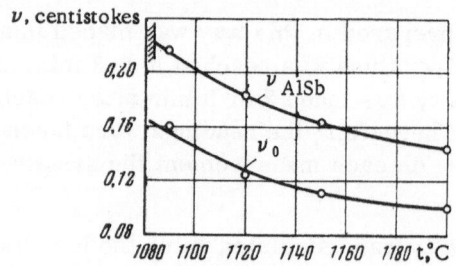

Fig. 106. Temperature dependence of the viscosity ν_0 of
the initial mixture of aluminum and antimony (taken in
proportions necessary to form a compound between them),
and the temperature dependence of the viscosity of the
chemical compound ν_{AlSb} obtained as a result of a reaction between these components.

We can see that the values of ν'_{AlSb} and ν''_{AlSb} are in good
agreement. Bearing in mind that we are dealing with the reaction
of formation of a compound whose molecule consists of two atoms,
we may assume that the reaction is bimolecular. Analysis of the
results presented in Fig. 105, using Eqs. (98), (102), and (107)
shows that the experimental results fit best a linear law corresponding to Eq. (102).

Table 43. Viscosity of AlSb at Various Temperatures

Temperature, °C	ν'_{AlSb}, centistokes	ν''_{AlSb}, centistokes	$\dfrac{\nu'_{AlSb} - \nu''_{AlSb}}{\nu'_{AlSb}} \times 100\%$
1090	0.219	0.226	3.20
1120	0.185	0.196	5.90
1150	0.165	0.176	6.70
1200	0.146	0.154	5.48

Table 44. Viscosity of Initial Mixture of Components (ν_0)
and of Synthesized Compound AlSb (ν_{AlSb})
at Various Temperatures

Temperature, °C	Viscosity, centistokes	
	ν_0	ν_{AlSb}
1090	0.160	0.219
1120	0.125	0.185
1150	0.117	0.165
1200	0.110	0.146

Hence, we may conclude that the reaction of formation of aluminum antimonide is a second-order reaction and that it can be written as follows: Al + Sb → AlSb, which is in agreement with case (B).

To estimate the order of the reaction by the standard criteria, it is necessary to transform the curves of Fig. 105 to show the dependence of the instantaneous concentration of aluminum (C_{Al}) or antimony (which is equivalent) on time at a given temperature. To do this, we shall rewrite Eq. (100) to make it applicable to the present case:

$$c_{Al} = \frac{\log(\nu_{AlSb}/\nu)}{2\log(\nu_{AlSb}/\nu_0)}. \tag{111}$$

The values of ν_0 and ν_{AlSb} for the investigated temperatures are listed in Table 44.

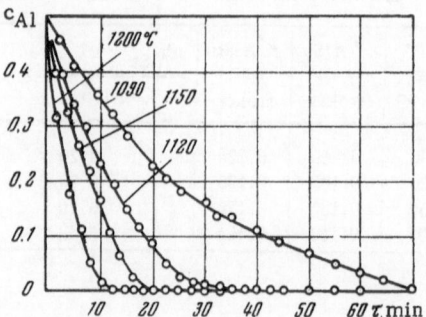

Fig. 107. Dependence of the concentration of aluminum in the melt on the duration of exposure to a given temperature during the formation of aluminum antimonide.

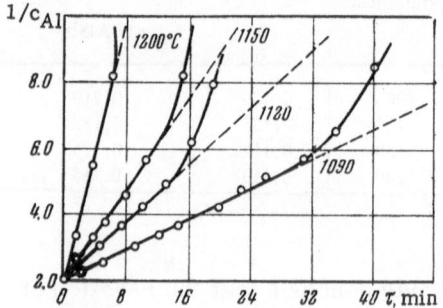

Fig. 108. Dependence of $1/c_{Al}$ on the duration of exposure to a given temperature.

The temperature dependences of ν_{AlSb} and ν_0 are shown in Fig. 106.

We can now use Eq. (111) to calculate the dependence of the concentration of aluminum in the melt during the formation of aluminum antimonide at these temperatures. The results of the calculation are shown in Fig. 107.

To determine the order of the reaction, we shall use a graphical method. Analysis of the dependences $\ln c_{Al} = f_1(\tau)$, $1/c_{Al} = f_2(\tau)$, and $1/c_{Al}^2 = f_3(\tau)$ shows that the experimental points fit best a straight line in the $1/c_{Al} = f_2(\tau)$ case (cf. Fig. 108).

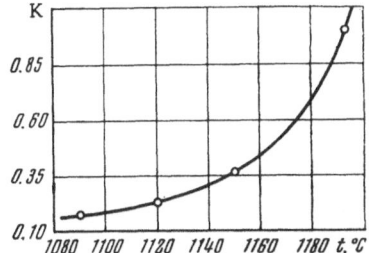
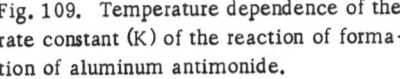

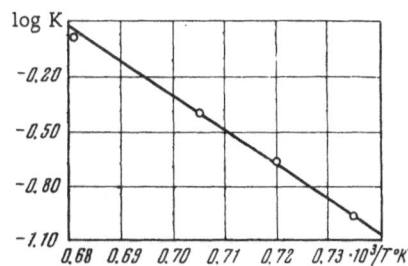

Fig. 109. Temperature dependence of the rate constant (K) of the reaction of formation of aluminum antimonide.

Fig. 110. Dependence of the logarithm of the rate constant of the reaction of formation of aluminum antimonide on the reciprocal of the absolute temperature.

On the basis of these results we may assume that the reaction of formation of aluminum antimonide can be regarded as a second-order reaction.

Thus, the analysis of the results obtained using the standard criteria confirms that the reaction of formation of aluminum antimonide is of the second order. The fact that the reaction is of the second order agrees with the suggestion that it is bimolecular; this indicates the correctness of the conclusion that the viscosity varies with the concentration of the compound in the melt and that the replacement of the dynamic viscosity with the kinematic value in Eq. (95) does not result in any significant errors.

The dependence $1/c_{Al} = f_2(\tau)$ is linear in the initial and middle stages of the reaction of formation of AlSb. In the final stage, an appreciable deviation is observed, which is obviously due to a departure from Eq. (93) and possibly to a greater change in the density in the final stages of the formation of AlSb.

Moreover, it is possible that in the final stage the reaction is more complex than a second-order bimolecular reaction.

Using the results obtained, we can calculate the reaction rate constant at each temperature from the formula

$$K = \frac{1 - 2c_{Al}}{c_{Al}\,\tau}. \qquad (112)$$

The results of such a calculation are:

Temperature, °C	K
1090	0.115
1120	0.222
1150	0.355
1200	1.02

The temperature dependence of the reaction rate constant is presented in Fig. 109. When this dependence is plotted using semilogarithmic coordinates we clearly obtain a linear dependence of the logarithm of the rate constant on the reciprocal of the absolute temperature during the reaction (Fig. 110), i.e., the dependence satisfies the Arrhenius equation:

$$\log K = \frac{A}{T} + B, \tag{113}$$

where A and B are constants typical of a given reaction.

From these results, we can calculate the activation energy of the reaction:

$$E = -4.57 \tan\alpha, \tag{114}$$

where α is the slope of the straight line representing the dependence of the logarithm of the reaction rate constant on the reciprocal of the absolute temperature (Fig. 110). The reaction activation energy determined in this way is $91,500 \pm 200$ cal/mole for the formation of aluminum antimonide.

This value of the activation energy is fairly large and typical of slow reactions.

Knowing the activation energy, we can calculate the relative number of active centers, i.e., particles which have a higher kinetic energy (collisions with these particles result in a chemical reaction). It is known [575] that the number of active centers or the relative number of active particles can be calculated from the equation

$$\log \frac{N^*}{N} = -\frac{E}{RT}, \tag{115}$$

where $N*/N$ is the relative number of active particles, and E is the reaction activation energy.

The application of Eq. (115) to the reaction considered gives the following values:

Temperature, °K	$N*/N$	Temperature, °K	$N*/N$
1353	$5 \cdot 10^{-15}$	1500	$8 \cdot 10^{-14}$
1400	$9 \cdot 10^{-15}$	1550	$2 \cdot 10^{-13}$
1450	$5 \cdot 10^{-14}$	1600	$8 \cdot 10^{-13}$

It follows from these results that the number of active centers increases rapidly with increasing temperature.

The temperature dependence of the reaction rate, which can be deduced from these results, is in fairly good agreement with the conclusions drawn from the degree of completeness of the reaction at various temperatures (Fig. 105).

Thus, measurements of the viscosity can be used successfully in investigations of the reaction kinetics in the melts.

§ 7. Application of the Viscosity Method in the Determination of the Crystallization Onset Temperature of Melts

This problem is important because the use of the thermographic method in the determination of the liquidus temperature is often difficult and may give rise to considerable errors. For example, even a very small error in the determination of the composition of a binary alloy (ΔC, Fig. 111) consisting of components with very different melting points (Fig. 111), when the addition of a more refractory component often raises strongly the liquidus line, may result in a large error (Δt) in the determination of the crystallization onset temperature of the alloy (t_1 or t_2, Fig. 111) [576].

The use of thermal analysis methods or the recording of polytherms of any other property (including the viscosity) for the purpose of determination of the phase transition temperature is methodologically incorrect because of the unavoidably large errors, which are the result of the inaccurate determination of the alloy composition.

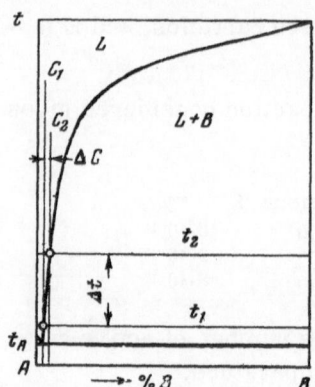

Fig. 111. Determination of the crystallization onset temperature by the polytherm method.

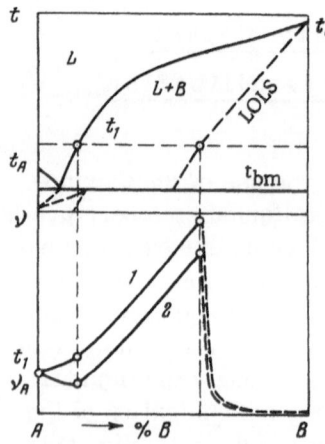

Fig. 112. Typical composition—viscosity dependences at a given temperature for the transition from a one-phase to a two-phase region in a phase diagram. LOLS — line of onset of linear shrinkage; t_{bm} — temperature of the beginning of melting.

In those cases when the transition from a one-phase to a two-phase region in the phase diagram is accompanied by a very small thermal effect the use of the thermographic method also meets with difficulties and may give inaccurate results. In such cases, we can use the viscosity method.

The use of the viscosity method is based on the Kurnakov correspondence principle [1] and on the assumption that the viscosity is a structure-sensitive property. The formation of a surface due to the precipitation of a second phase makes a liquid heterogeneous and should result in a large rise in the viscosity.

Shvidkovskii and Priss [74] have investigated the temperature dependence of the viscosity of Bi−Pb alloys and have found that when the temperature is reduced by not more than 20° below the liquidus line, the viscosity increases by a factor of 500.

Shvidkovskii and Goryaga [74, 188, 577] have shown that when liquid metals become heterogeneous due to the formation of the corresponding oxides the viscosity also increases considerably.

When we go over from a one-phase to a two-phase region in the phase diagram, the rate at which the viscosity increases due to the

lowering of temperature (or the increase in the concentration of the second component at a fixed temperature) should depend mainly on the nature and rate of crystallization (for details of the concept of the "crystallization rate," the reader is referred to a book by Academician A. A. Bochvar [578]). The precipitation of a second phase in the form of branched dendrites should result in a greater increase of the viscosity than in the case of the formation of regular crystals. In both cases, an increase in the rate of crystallization should increase the rate of rise of the viscosity in the two-phase region.

Kurnakov's correspondence principle can be used to deduce easily typical composition–viscosity diagrams at a given temperature (Fig. 112). In the one-phase region L, an increase in the concentration of the second component may result either in an increase (curve 1) or a decrease (curve 2) in the viscosity of the alloy. Within the limits of the homogeneity (one-phase) region, the viscosity may increase when the second component is added but this rise is not very strong, particularly when the change in the concentration is not very large. However, in the two-phase region (L + B) of the phase diagram, the viscosity should increase more rapidly because the liquid is heterogeneous (the increase should be much greater than in the homogeneous region). The viscosity isotherm has a kink, corresponding to the intersection of two concentration dependences of the viscosity in the one-phase and two-phase regions of the phase diagram. The coordinates of the phase transition point are determined from the position of this kink (Fig. 112). The viscosity increases in the two-phase region solely due to an increase in the relative concentration of the second phase in accordance with the lever rule, since the viscous properties of the remaining liquid should not vary because its concentration is constant (as indicated by a point on the liquidus line at a given temperature).

It should be mentioned that in the heterogeneous region of the phase diagram the viscosity increases rapidly when the concentration of the second component is increased at a given temperature; this applies within the liquid–solid mixed region, contained between the liquidus line and the line of the onset of linear shrinkage (LOLS in Fig. 112) [579, 580].

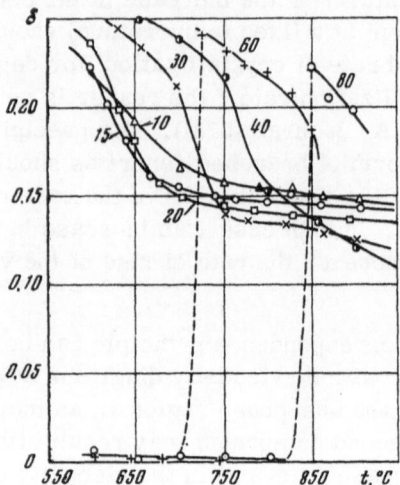

Fig. 113. Temperature dependences of the logarithmic damping decrement. Alloys of the Sb—Ge system; the numbers alongside the curves give the Ge concentration in at.%. The dashed curves indicate the sudden jump in the logarithmic decrement when the temperature of the sample crosses the LOLS.

Beyond the LOLS in the solid—liquid state, which is characterized by the presence of a crystalline core, the damping of oscillations of the suspension system in the determination of the viscosity should decrease strongly, since the residual liquid is split by growing crystals into isolated regions so that the system as a whole is closer to the solid state.

The decrease in the damping of oscillations in the determination of the viscosity is shown arbitrarily in Fig. 112 as a rapid fall of the viscosity. Thus, a rapid decrease in the damping of oscillations in the viscosity measurements, which is due to the transition from the liquid—solid to the solid—liquid state, can be used to plot the LOLS.

We shall illustrate the possibilities of this method by considering the well-known Sb—Ge system [203].

We prepared Sb—Ge alloys containing 10, 15, 20, 30, 40, 60, and 80% of Ge by weight.

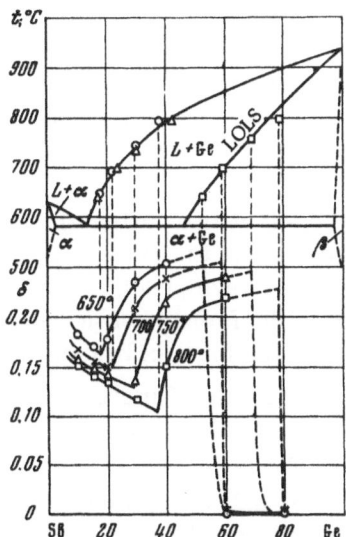

Fig. 114. Isotherms of the logarithmic decrement and the liquidus line for the Sb—Ge system. □, ○) Results obtained by the viscosity method; △) results obtained by thermal analysis [203].

The initial materials were 99.99% pure antimony and 99.9999% pure germanium.

The ratio of the height of the liquid column in the cylinder to the radius of the cylinder was ~4.5. The viscosity was measured in the 600-900°C range. The system was kept for 15 min at a given temperature before any measurements were carried out. The results of the measurements are presented in Fig. 113 in the form of the temperature dependence of the logarithmic decrement of damped oscillations (δ) for alloys of various compositions.

It follows from Fig. 113 that the values of δ for alloys in the two-phase region (L + B) are considerably greater than those for alloys in the homogeneous part (L) of the phase diagram.

Figure 114 shows the isotherms of δ, plotted from the results given in Fig. 113. The kinks in these isotherms were used to determine the corresponding points on the liquidus line.

It is evident from Fig. 114 that the liquidus line determined by the viscosity method practically coincides with the line plotted

by the thermal analysis method. The same figure also includes the LOLS, plotted from the results obtained in this investigation.

This system does not exhibit such a strong rise of the viscosity in the transition from the one-phase to the two-phase region as that reported by Shvidkovskii and Priss [74] for Bi − Pb alloys.

This is due to the fact that, under the experimental conditions employed in our study, the primary germanium nuclei, which made the liquid heterogeneous, were precipitated in the form of regular crystals rather than in the form of branched dendrites (crystallization of germanium in the form of branched dendrites is untypical).

We can see that the rate of increase of the viscosity in the two-phase region can be used to draw some conclusions about the shape of primary second phase precipitates.

In conclusion, we shall consider the liquidus line of a system whose components have strongly differing melting points.

As an example of such a system, we investigated the tin − silicon carbide system. Tin melts at 232 °C [203], whereas silicon carbide decomposes without melting at 2700 °C [203]. The tin − silicon carbide system has not been investigated before.

To determine the solubility of silicon carbide in liquid tin, we prepared alloys containing 0.02, 0.04, 0.06, 0.10, 0.20, and 0.30 mol.% SiC.

Tin of 99.99% purity and silicon carbide crystals of 99.9 % purity were used as the initial materials. These materials (each charge weighed 30 g) were placed in evacuated quartz ampoules of 14 mm diameter and were kept in a furnace at 1250-1300 °C for 5-7 h; during this period, the alloy was vigorously stirred from time to time by shaking the ampoule. Next, the ampoule containing the alloy was placed in the viscometer furnace, where the viscosity was determined during cooling from 1200-1400 °C. The alloy was held at each temperature for 30 min before the measurement of the viscosity. The results of the measurements are presented in Fig. 115. The temperature dependences of the viscosity of the alloys containing 0.06, 0.1, 0.2, and 0.3% SiC were smooth over the whole investigated range of temperatures and the viscosity of these alloys was approximately 10-12 times higher than the viscosity of pure tin.

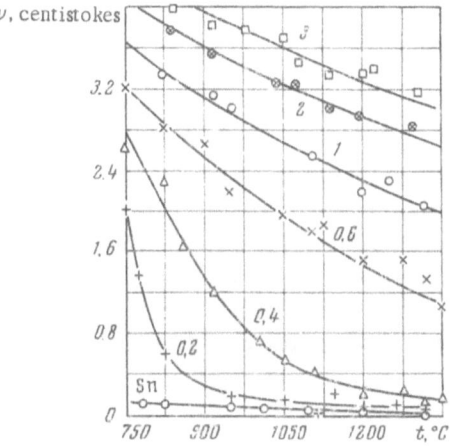

Fig. 115. Temperature dependences of the viscosity for alloys of the Sn−SiC system. The numbers alongside the curves give the SiC concentration in at.%.

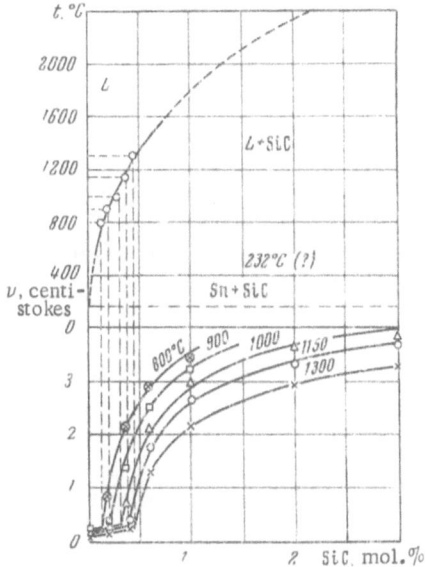

Fig. 116. Viscosity isotherms of the Sn−SiC system and the solubility of silicon carbide in tin.

The temperature dependences of the viscosity of the alloys with 0.02 and 0.06% SiC differed considerably from the dependences for alloys containing more SiC.

At high temperatures, the viscosity of the alloys with 0.02 and 0.04% SiC was slightly higher than the viscosity of pure tin, but when the temperature was reduced (below 950°C for the 0.02% SiC alloy and below 1100°C for the 0.04% SiC) the viscosity increased. This was due to a transition from the one-phase to the two-phase region. However, it was difficult to deduce the liquidus temperatures directly from these curves, since the transition from low to high values of viscosity was spread out over a range of temperatures.

When the results obtained were replotted in the form of viscosity isotherms (Fig. 116), we were able to construct a liquidus curve in the 800-1300°C range.

In this case, in contrast to the Bi – Pb system [74], we did not observe a strong increase in the viscosity; this was obviously due to the same reasons as those pointed out in the case of the Sb – Ge system. Moreover, when the ampoules containing the alloys with higher concentrations of SiC (0.1, 0.2, 0.3%) were opened, it was found that some silicon carbide was deposited on the inner surfaces, i.e., some fraction of the carbide did not take part in making the liquid inhomogeneous and, therefore, it did not affect the viscosity.

From these experimental data, we may conclude that the method of plotting phase transition lines and of the LOLS from the viscosity data can be recommended in those cases which have been listed at the beginning of this section.

Chapter 9

Physicochemical Analysis of Liquid Systems

§1. Physicochemical Analysis of the A^{III} – Sb Binary Liquid Systems [69]

In order to study the A^{III}–Sb systems, we prepared a series of binary alloys with compositions to the left and right of the A^{III}Sb compounds (27 alloys in all). The original materials were aluminum, gallium, indium, and antimony of purity not less than 99.999%.

The viscosity and electrical conductivity of alloys in the Al–Sb system were investigated by placing the samples in corundum cans of 11-12 mm diameter, which were put into evacuated quartz containers. The Ga–Sb and In–Sb alloys were placed in evauated cylindrical quartz ampoules, made of calibrated quartz tubes, whose internal diameter was 11-12 mm; the ampoules used for alloys of a given system were all made from the same tube.

In the viscosity measurements, we used Al–Sb samples weighing 8-10 g; for Ga–Sb and In–Sb systems we used samples weighing 12-15 g. In the measurements of the electrical conductivity, we used cylinders of identical size; they all were 10 mm high and of 11 mm diameter. A correction was made for the change in the volume due to melting on the assumption that the mixing rule was valid; we regarded each system as a mixture of two binary systems: A^{III} –A^{III}Sb and A^{III}Sb–Sb.

Each value of the viscosity and electrical conductivity represented the average of three to five measurements.

Analysis of the temperature dependences of the viscosity and electrical conductivity of the alloys give the following results.

At temperatures slightly higher than the melting points of the chemical compounds, the viscosities of these compounds were much higher than those of the other alloys. At higher temperatures, this difference gradually disappeared. The temperature coefficients of the viscosity ($d\nu/dt$) of the chemical compounds were considerably greater than those of the other alloys.

The values of the electrical conductivity of the chemical compounds were lower than those of the other alloys. The temperature coefficients of the electrical conductivity ($d\sigma/dt$) of the chemical compounds were smaller than those of the other alloys.

Let us now consider the concentration dependences of the viscosity and electrical conductivity at various temperatures. Figure 117 shows the isotherms of these properties side by side with the equilibrium phase diagrams of the Al−Sb (a), Ga−Sb (b), and In−Sb (c) systems.

We can see that pronounced extrema (a maximum of the viscosity and a minimum of the electrical conductivity) are found opposite the chemical compounds.

These extrema in the "composition−property" isotherms indicate the retention of a strong chemical interaction between the components in the transition from the solid-to-liquid state.

At high temperatures, these pronounced maxima in the viscosity isotherms and the minima in the conductivity isotherms become less marked, indicating a decrease in the chemical interaction forces between the components. When the temperature is increased, the maxima in the viscosity isotherms become broader and shift in the direction of the component whose viscosity is higher (aluminum in the Al−Sb system and antimony in the Ga−Sb system). According to Kurnakov's classification, these systems should be regarded as irrational, i.e., systems with a dissociating compound. The maximum in the viscosity isotherms of the In−Sb system also broadens and therefore indium antimonide should also be regarded as a compound which dissociates at higher temperatures, but in this case the viscosity maximum gradually shifts in the direction of the less viscous component (indium).

The explanation of this behavior should be sought in those causes which have been pointed out by Belashchenko [581, 582], except that in the present case the formation of metastable states is more likely when temperature is increased rather than lowered.

At high temperatures, the electrical conductivity isotherms show a general broadening of the sharp minima but no shift.

The concentration dependences of the viscosity and the electrical conductivity can be used to draw some general conclusions about the nature of interaction of components in the molten Al—Sb, Ga—Sb, and In—Sb systems. We may regard it as firmly established that aluminum, gallium, and indium antimonides either do not dissociate or dissociate weakly at the melting point. This conclusion is supported particularly strongly by the electrical conductivity data. The conductivity minimum indicates that some electrons form chemical bonds rather than joining the electron gas. When the temperature is increased the compound eventually dissociates and the difference between the electrical conductivity of the compound and of alloys with similar compositions practically disappears.

The conclusions drawn from a study of the viscosity and electrical conductivity are supported by measurements of the magnetic susceptibility.

From these data we may conclude that the maxima in the liquidus curves of the Al—Sb, Ga—Sb, and In—Sb systems should be singular rather than irrational. However, judging by the rounded maxima of the liquidus curves in the phase diagrams of the Ga—Sb and In—Sb systems, gallium and indium antimonides should dissociate considerably at the melting point.

According to Urazov [38, 39], aluminum antimonide also dissociates when it is melted. However, the data on the melting point and on the nature of the liquidus curve near the ordinate representing this compound are very contradictory.

We investigated in detail the phase diagrams of Al—Sb, Ga—Sb, and In—Sb in the regions close to the chemical compounds; this investigation was carried out by the differential thermal analysis method using a Le Chatelier—Saladen pyrometer and the coordinates used in the thermograms were the temperature of the

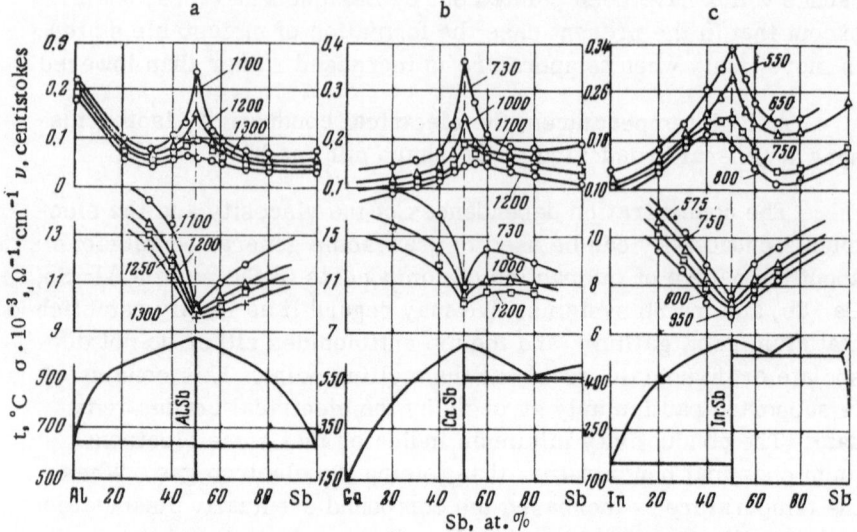

Fig. 117. Viscosity and electrical conductivity isotherms and equilibrium phase diagrams of molten Al−Sb, Ga−Sb, and In−Sb systems.

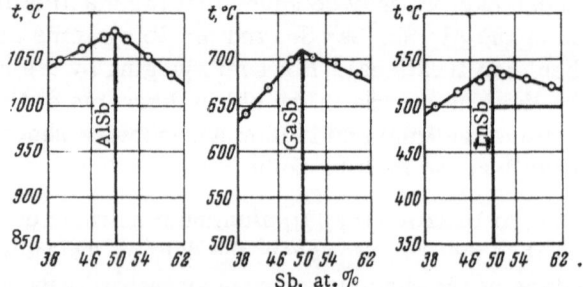

Fig. 118. Liquidus curves of the Al−Sb, Ga−Sb, and In−Sb systems in the immediate vicinity of chemical compounds.

sample and the difference between the temperatures of the sample and a standard.†

Pure silicon was used as a standard. The pyrometer was calibrated using zinc (t_{mp} = 419.4°C), aluminum (t_{mp} = 660°C), and silver (t_{mp} = 960.3°C).

The components of the Ga – Sb and In – Sb systems were placed in Stepanov quartz containers. Al – Sb alloys were placed in corundum crucibles, which were sealed into evacuated quartz ampoules. Only the quartz cap carrying a thermocouple was immersed directly in the melt.

The results obtained were used to plot the liquidus curves of the Al – Sb, Ga – Sb, and In – Sb systems in the immediate vicinity of the chemical compounds (Fig. 118). The experimental points for each of the binary systems were located on almost straight lines, which intersected at the melting point of the chemical compound. Thus, the thermal analyses indicated that the maxima of the liquidus curves of the Al – Sb, Ga – Sb, and In – Sb systems were singular. Consequently, aluminum, gallium, and indium antimonides melt congruently without appreciable dissociation at the melting point and are stable up to certain temperatures in the liquid phase.

The investigated physicochemical properties show a definite correlation.

§ 2.   Physicochemical Analysis of the Mg – B^{IV} Binary Liquid Systems

The physicochemical analysis of the Mg – Sn and Mg – Pb systems in the liquid state is reported in [436-439]; this analysis has been carried out by plotting viscosity – composition and electrical conductivity – composition diagrams. Figure 119 shows the viscosity and electrical conductivity isotherms obtained by Gebhardt et al. [437] and by Roll and Motz [436], respectively, compared with the equilibrium phase diagram of the Mg – Pb system. It is evident from this figure that, at the melting point and at up to 50° above it, the viscosity isotherms have a sharp maximum corresponding to the composition of the compound; at higher temperatures, this

---

† For details, see [583-586]; the apparatus is described in [587].

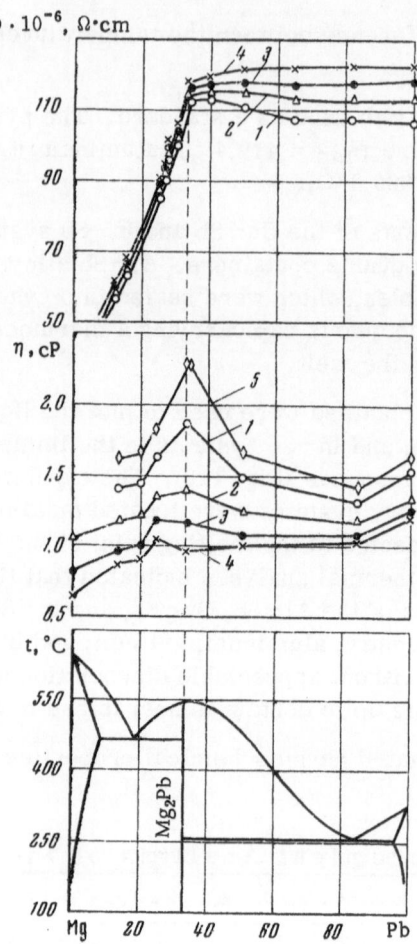

Fig. 119. Viscosity [437] and electrical resistivity [436] isotherms and equilibrium phase diagram of the Mg—Pb system in the molten state. Temperature (°C): 5) 550; 1) 600; 2) 700; 3) 800; 4) 900.

maximum disappears. The electrical resistivity isotherms exhibit a maximum at 600°C; this maximum is less pronounced than the viscosity maximum, and at higher temperatures it broadens out considerably. These extremal points in the property–composition diagrams indicate a strong chemical interaction between the components in the melt at temperatures up to ~50° above the melting point.

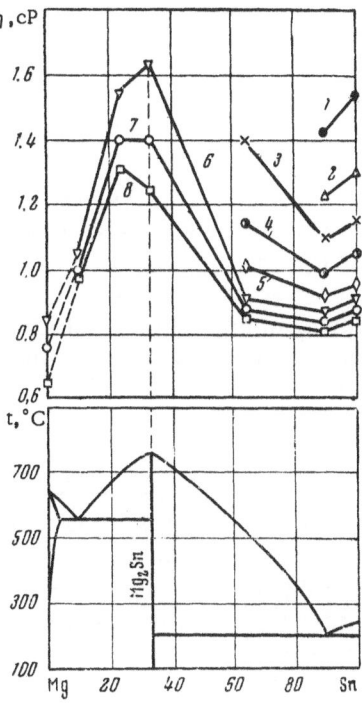

Fig. 120. Viscosity isotherms and equilibrium phase diagram [438] of the Mg—Sn system in the molten state. Temperature (°C): 1) 300; 2) 400; 3) 500; 4) 600; 5) 700; 6)
800; 7) 850; 8) 900.

The thermodynamic properties of the molten Mg—Pb alloys
have been analyzed by Knappwost [435] and Lantratov [588]. They
have shown that liquid magnesium—lead alloys show a strong negative deviation from Raoult's law; this is attributed to the interaction of the components resulting in the formation of Mg$_2$Pb structural groups in the melt. However, this compound is thermally unstable in the molten state and it dissociates strongly even at ~50°
above the melting point, as indicated by the broadening of the extremal points in the isotherms of the properties of the Mg—Pb system (Fig. 119). Properties of Mg—Sn alloys in the molten state
have been investigated only by Gebhardt et al. [438], who have investigated the viscosity over a wide range of temperatures and
concentrations. Figure 120 shows their viscosity isotherms along-

side the equilibrium phase diagram of the Mg—Sn system. At 800°C (i.e., 22° above the melting point of the compound), the viscosity has a maximum corresponding to the composition of the compound; at 850°C, this maximum is strongly broadened and at still higher temperatures it shifts in the direction of pure magnesium. Analysis of the composition dependence of the viscosity of the Mg—Sn system shows that the compound Mg_2Sn is stable at the melting point but it begins to dissociate at temperatures ~20-50° above the melting point. According to Kurnakov's classification [1, 4], the Mg—Sn and Mg—Pb systems should be regarded as irrational. No physicochemical analyses of the Mg—Si and Mg—Ge systems in the liquid state have been published.

§3. Physicochemical Analysis of the A^{II} — Te Binary Liquid Systems [26]

In order to obtain the fullest information on the behavior of the A^{II}Te-type compounds in the molten state and to determine the influence of the nonstoichiometric melt composition on the nature of the interatomic interaction in the melt, we used the physicochemical analysis method.

We prepared six alloys of each system (not counting the compounds), containing 30, 40, 45, 55, 60, and 70 at.% Te.

A study of the electrical conductivity of alloys of the Zn—Te and Cd—Te systems has shown that a deviation from stoichiometry by ±5 at.% has very little effect on the absolute value of the electrical conductivity. Larger deviations (±10-20 at.%) result in a considerable increase in the electrical conductivity in the presence of excess zinc or cadmium, and a smaller increase in the presence of excess tellurium. Figures 121 and 122 give the isotherms of the composition dependences of the electrical conductivity and viscosity of alloys of these two systems alongside the corresponding equilibrium phase diagrams. The electrical conductivity isotherms have strongly pronounced minima and the viscosity isotherms have maxima opposite the compositions corresponding to ZnTe and CdTe. When the temperature is increased, the extremal points lose their singular nature and become smoother. In the Zn—Te system this begins from 1300°C (which is 60° above t_{mp} of this compound); in the Cd—Te system this change starts from 1150-1250°C.

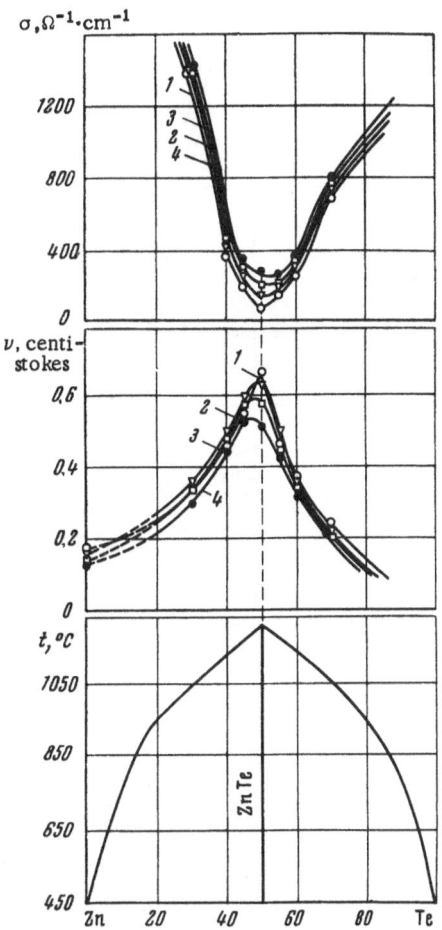

Fig. 121. Viscosity and electrical conductivity isotherms and equilibrium phase diagram of the Zn—Te system in the molten state. Temperature (°C): 1) 1280; 2) 1300; 3) 1350; 4) 1480.

It is very interesting to note that when an alloy departs from the stoichiometric compositions, corresponding to the compounds ZnTe and CdTe, the nature of the intermolecular interaction in the liquid changes little. This is indicated by the observation that the electrical conductivity of all the investigated alloys increases when temperature is increased and the absolute values of the con-

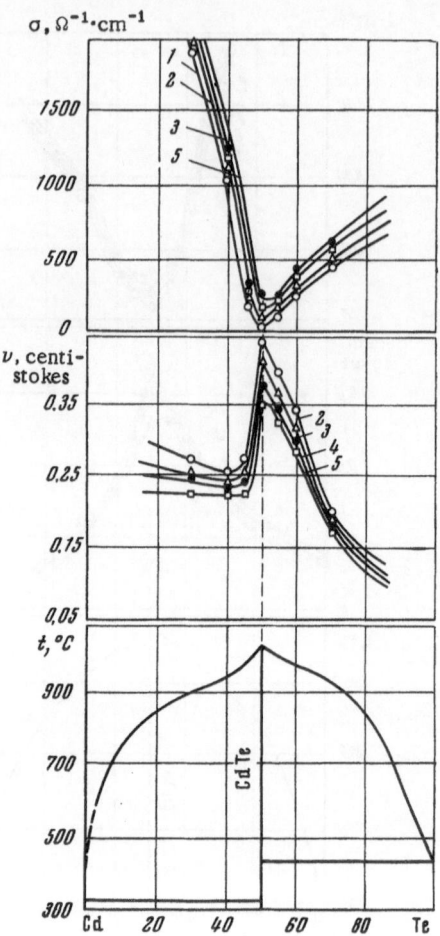

Fig. 122. Viscosity and electrical conductivity isotherms
and equilibrium phase diagram of the Cd−Te system in
the molten state. Temperature (°C): 1) 1050; 2) 1100;
3) 1150; 4) 1200; 5) 1250.

ductivity of all the alloys, with the exception of those with a considerable excess of zinc or cadmium (10-20 at.%), differ little
from the values of the conductivity of the compounds.

All these observations indicate that a considerable fraction
of electrons is used, in a wide range of concentrations, in the formation of chemical bonds and does not take part in conduction.

The nature of the viscosity and electrical conductivity isotherms indicates a strong chemical interaction between zinc (cadmium) and tellurium in the liquid state and the retention of the
properties of solid alloys of stoichiometric composition (corresponding to the ZnTe and CdTe compounds) above the melting
point. This conclusion is in agreement with the results obtained
from the temperature dependences of the various properties.

However, when the temperature is increased considerably,
the extremal points in the isotherms broaden and this is particularly true of the Zn—Te system. It indicates a weakening of the
interaction between the components and, according to Kurnakov's
classification, these systems should be regarded as irrational,
i.e., systems with dissociating chemical compounds.

§4. Physicochemical Analysis of the A^{III} – Te Binary Liquid Systems [28, 29]

The Ga—Te and In—Te systems exhibit the formation (in addition to Ga$_2$Te$_3$ and In$_2$Te$_3$) of congruently melting chemical compounds GaTe and InTe, as well as compounds with latent maxima
[203].

The compound GaTe has a complex layered structure, in
which layers alternate in the following sequence: TeGaGaTe—
TeGaGaTe [402]. Within each packet, consisting of four layers,
the binding is covalent. The packets themselves are bound by van
der Waals forces. The compound InTe has a complex chainlike
structure. We shall compare the behavior of molten Ga$_2$Te$_3$ and
In$_2$Te$_3$ and of alloys whose composition lies in the primary crystallization region of these compounds, with the behavior of GaTe and
InTe in the molten state as well as of alloys whose compositions
are in the primary crystallization region of the latter two compounds.

We must point out that the compositions of the compounds
Ga$_2$Te$_3$ and GaTe and of In$_2$Te$_3$ and InTe differ only by 10 at.%.

We investigated 12 alloys of the Ga—Te system with 33, 40,
45, 50, 53, 55, 57, 58.5, 60, 62.5, 70, 75 at.% Te (they are denoted
by 1-12 in Fig. 123) and 12 alloys of the In—Te system containing
33, 40, 50, 51.5, 53, 55, 56, 57, 58.5, 60, 65, 80 at.% Te (denoted
by 1-12 in Fig. 124).

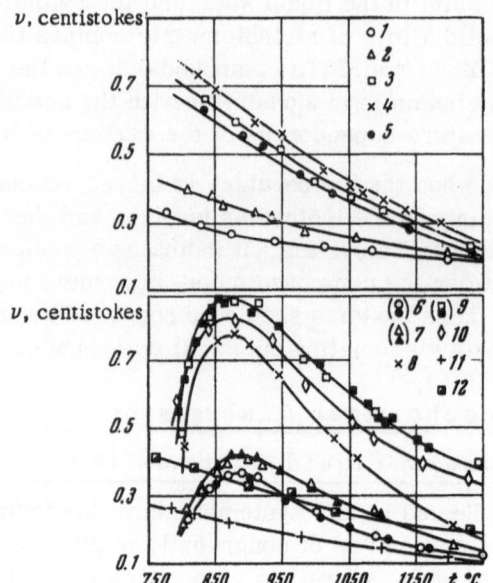

Fig. 123. Temperature dependences of the viscosity of melts of the Ga−Te system. The upper symbols numbered 6, 7, and 9 represent heating; the lower symbols represent cooling.

Figures 123 and 124 give the temperature dependences of the viscosity of the investigated alloys.

It is evident from these figures that the alloys whose compositions are close to the compounds Ga_2Te_3 and In_2Te_3, lying in the region of the primary crystallization of these compounds, have temperature dependences of the viscosity similar to the temperature dependences of the compounds. An increase in the viscosity during heating above the melting point is observed most clearly for Ga_2Te_3 and In_2Te_3; this increase is less when the composition is altered by increasing the gallium, indium, or tellurium concentrations and it disappears completely in alloys with 70-75% Te in the Ga−Te system. The compounds GaTe and InTe and the alloys lying in their primary crystallization region behave quite differently. These alloys have smooth monotonically decreasing temperature dependences of the viscosity (Figs. 123 and 124).

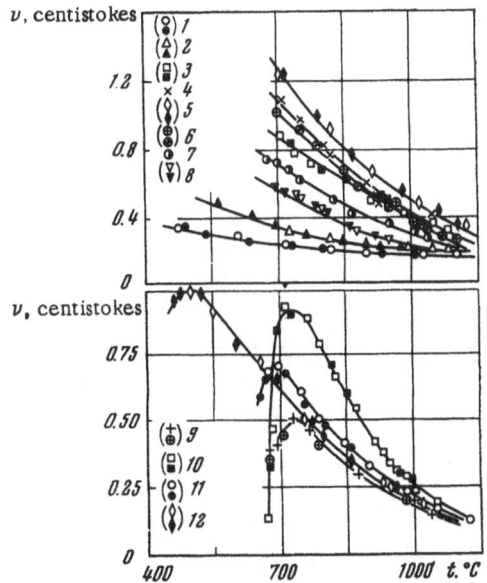

Fig. 124. Temperature dependences of the viscosity of melts of the In—Te system. The upper symbols numbered 1-12 represent heating; the lower symbols represent cooling.

The temperature dependences of the electrical conductivity of all the investigated alloys of the Ga—Te and In—Te systems are similar. When temperature is increased, the conductivity increases as well. The conductivity of alloys whose compositions lie in the range GaTe(InTe)—Ga$_2$Te$_3$(In$_2$Te$_3$), as well as of alloys containing an excess of tellurium (compared with Ga$_2$Te$_3$ or In$_2$Te$_3$) does not vary greatly.

However, alloys with an excess of gallium or indium, compared with GaTe or InTe, exhibit a fairly strong rise of the electrical conductivity.

Investigations of the temperature dependences of the properties of these alloys have been used to plot the concentration dependences of the properties, which together with the equilibrium diagrams of the Ga—Te and In—Te systems, are given in Figs. 125 and 126. It is evident from these figures that the electrical conductivity isotherms of the compounds Ga$_2$Te$_3$, GaTe, and In$_2$Te$_3$ have pronounced minima.

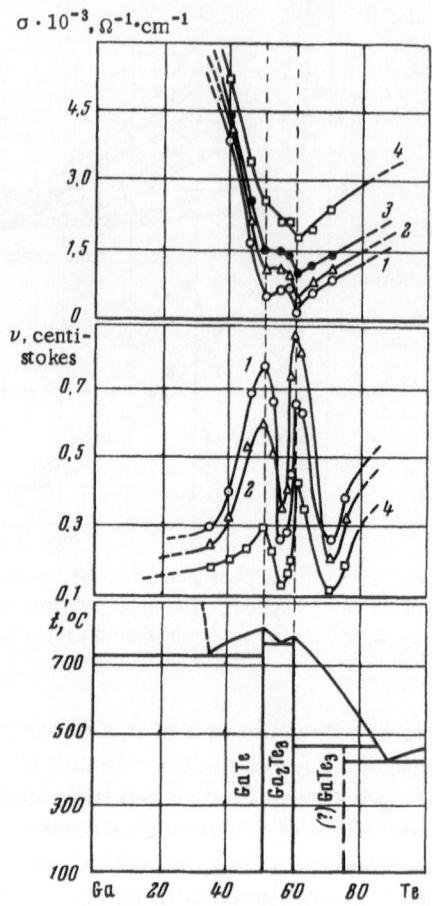

Fig. 125. Viscosity and electrical conductivity isotherms and equilibrium phase diagram of the Ga−Te system in the molten state. Temperature (°C): 1) 825; 2) 900; 3) 1000; 4) 1200.

The viscosity isotherms of the Ga−Te system have two pronounced singular maxima (Fig. 125) corresponding to the compounds Ga_2Te_3 and GaTe. In the In−Te system, the first maximum is shifted somewhat away from the compound InTe in the direction of excess tellurium, but the second corresponds to the compound In_2Te_3 (Fig. 126).

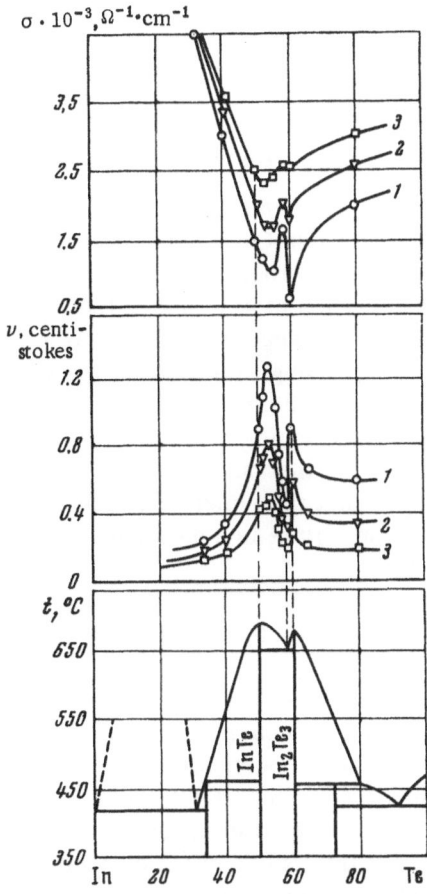

Fig. 126. Viscosity and electrical conductivity isotherms and equilibrium phase diagram of the In−Te system in the molten state. Temperature (°C): 1) 700; 2) 850; 3) 1000.

When the temperature is increased, the extremal points in the isotherms of the properties of the Ga−Te system, corresponding to the compound Ga_2Te_3, retain their singular nature even at temperatures more than 400° higher than the melting point of this compound. In the In−Te system, the maximum in the viscosity isotherms and the minimum in the conductivity isotherms broaden out strongly by the time 1000°C is reached.

In spite of the similarity of the changes in the structure and chemical binding, which take place in molten alloys of various compositions lying in the primary crystallization region of a given compound, the compound itself has distinguishing chemical properties. This is indicated by the singular nature of the extremal points in the isotherms of the properties of the compound and the alloys.

The viscosity minimum observed for eutectic alloys indicates a weak interaction between Ga_2Te_3 and GaTe, In_2Te_3, and InTe, and once again it stresses the individual properties of these compounds. The observation that even at 400° above the melting point the compound Ga_2Te_3 retains a singular maximum in the viscosity isotherms and a corresponding minimum in the conductivity isotherms indicates a considerable thermal stability of this compound. On the other hand, the broadening of the extremal points in the viscosity and conductivity isotherms, observed at 1000°C for the compound In_2Te_3, shows that the range of the thermal stability of this compound in the liquid state is about 300°, above which it begins to dissociate strongly.

§5.   Physicochemical Analysis of the A^{IV} − Te Binary Liquid Systems

A systematic physicochemical analysis of the A^{IV}−Te binary liquid systems has not yet been carried out. There is only one paper, that by Honda and Endo [589], which reports the concentration dependence of the magnetic susceptibility of the Pb−Te and Sn−Te systems, but this paper gives only one isotherm for each system and therefore no conclusions can be drawn about the temperature ranges of the stability of the corresponding compounds in the liquid state.

Moreover, the magnetic susceptibility is a complex property which does not always yield unambiguous information.

Recently, Dancy [590] reported an investigation of the electrical conductivity and thermoelectric power of the Sn−Te system. Dancy's paper appeared simultaneously with papers of the present authors [232, 498], in which we reported our studies of the viscosity and electrical conductivity of the Ge−Te, Sn−Te, and Pb−Te liquid systems. Dancy's results are in good agreement with ours and therefore we shall not discuss his paper in detail.

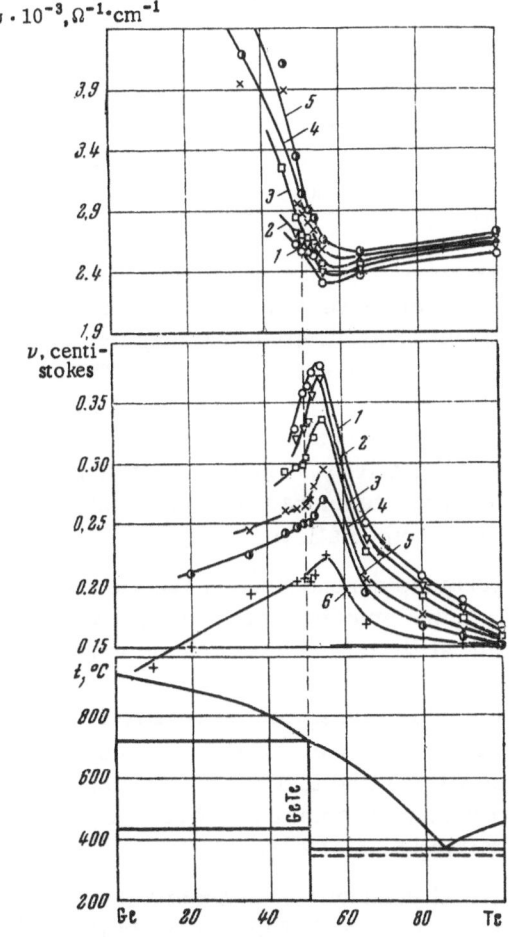

Fig. 127. Viscosity and electrical conductivity isotherms and equilibrium phase diagram of the Ge–Te system in the molten state. Temperature (°C): 1) 730; 2) 750; 3) 800; 4) 850; 5) 900; 6) 1000.

We investigated alloys of various compositions (32 alloys in all). These alloys were prepared by the same method and from the same original materials as the compounds in these systems (cf. Chapter 6, §2). The experimental method is described in Chapter 2. The measurements were carried out up to 1200°C. These measurements yielded the temperature dependences of the

viscosity and electrical conductivity for all the alloys. The temperature dependences of the viscosity of the Ge−Te alloys are smooth curves, and germanium telluride (of stoichiometric composition) is not distinguished by a maximum of the viscosity (compared with the neighboring alloys) over the whole investigated range of temperatures. The temperature coefficient of the viscosity of germanium telluride $(d\nu/dt)$ is somewhat higher than that of neighboring alloys.

The temperature dependences of the electrical conductivity of alloys of this system have a number of characteristic features. First, the electrical conductivity of germanium telluride (of stoichiometric composition) and of alloys of similar compositions increases fairly rapidly when the temperature is increased. Secondly, alloys with higher germanium concentrations have higher values of the electrical conductivity. The alloy with 65 at.% Te and pure tellurium has smaller temperature coefficients of the electrical conductivity than the other alloys.

The temperature dependences of the viscosity of the Sn−Te and Pb−Te alloys are smooth curves, but the values of the viscosity have maxima at tin telluride (of stoichiometric composition) over the whole investigated range of temperatures, while the viscosity of lead telluride is maximal at relatively low temperatures. Since the temperature coefficient of the viscosity of lead telluride is the largest of all, we find that at higher temperatures the viscosity of this compound becomes practically equal to the viscosity of an alloy containing 51.5 at.% Te.

The electrical conductivity of tin and lead tellurides (of stoichiometric compositions) and of alloys of similar compositions increases fairly rapidly, as in the Ge−Te system, when the temperature is increased. Alloys with high concentrations of tellurium and tin (or lead) have high electrical conductivities, which are practically independent of temperature. The conductivity has no minima corresponding to tin telluride (compared with the neighboring alloys) at all the investigated temperatures. The electrical conductivity of lead telluride at relatively low temperatures has lower values than that of other alloys. However, because this compound has the largest value of the temperature coefficient of conductivity $(d\sigma/dt)$ at higher temperatures, its electrical conductivity becomes practically equal to the conductivity of alloys containing excess tellurium.

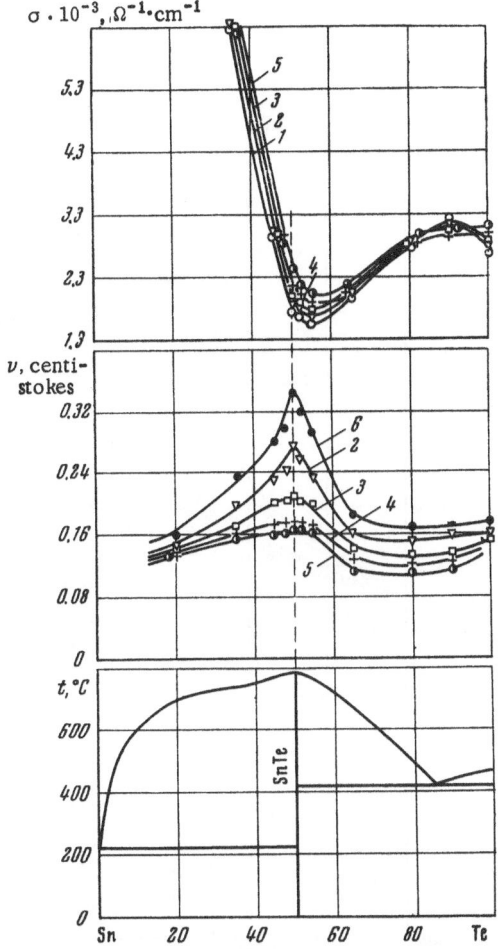

Fig. 128. Viscosity and electrical conductivity isotherms
and equilibrium phase diagram of the Sn—Te system in
the molten state. Temperature (°C): 1) 800; 2) 900; 3)
1000; 4) 1100; 5) 1150; 6) 825.

The temperature dependences of the viscosity and electrical
conductivity have been used to plot the isotherms of these properties, from which one can deduce the general nature of the chemical
interaction in two investigated systems. The results are shown in
Figs. 127-129, in conjunction with the corresponding phase diagrams. It is evident from these figures that the viscosity –

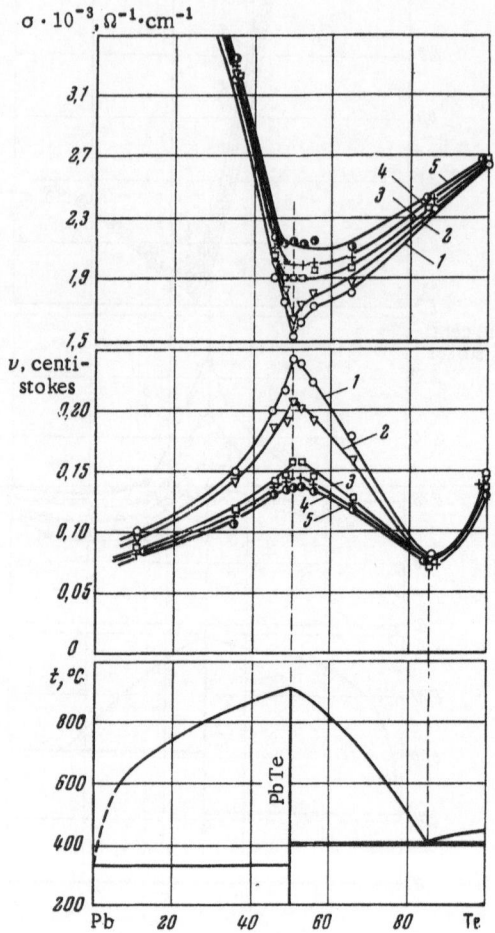

Fig. 129. Viscosity and electrical conductivity isotherms
and equilibrium phase diagram of the Pb−Te system in
the molten state. Temperature (°C): 1) 950; 2) 1000; 3)
1100; 4) 1150; 5) 1200.

composition and electrical conductivity – composition diagrams
have, respectively, maxima and minima near the compounds.

Analysis of the viscosity and electrical conductivity isotherms gives some information on the thermal stability of germanium, tin, and lead tellurides, if we use the basic relationships

of physicochemical analysis, established by Kurnakov [1, 4]. It
follows from Figs. 127 and 128, as well as from the figures given
in Dancy's paper [590], that the electrical conductivity minima and
the viscosity maxima in the isotherms of the Ge−Te and Sn−Te
systems are not singular; this shows that germanium and tin tellurides dissociate even at the melting point. The minima in the
electrical conductivity isotherms of the Ge−Te and Sn−Te systems are shifted in the direction of tellurium, whose electrical
conductivity in the liquid state is much lower than that of germanium or tin. This shift is in full agreement with the relationships
established by Kurnakov for strongly dissociating compounds
formed from components whose electrical conductivities differ
considerably. The maxima in the viscosity isotherms of the Ge−
Te system are broad and shifted in the direction of tellurium,
whose viscosity is higher; when the temperature is increased,
these maxima broaden, which is again in agreement with the relationships governing the physicochemical analysis. A corresponding shift is not observed in the concentration dependences of the
viscosity of the Sn−Te system, which may be due to the small difference between the viscosity of tin and tellurium in the investigated range of temperatures.

The 950°C viscosity and electrical conductivity isotherms of
the Pb−Te system (Fig. 129) have strongly pronounced singular
points corresponding to stoichiometric lead telluride, which indicate the relatively strong thermal stability of this compound at the
melting point and in a certain range of temperatures above it. In
any case, we can definitely say that the degree of dissociation of
lead telluride at the melting point is slight. When the temperature
is increased, the nature of the isotherms changes. The extrema
in the viscosity and conductivity isotherms broaden; hence, we
may conclude that appreciable dissociation of lead telluride begins
at temperatures 50-70° above the melting point.

These experimental data should be borne in mind in the development of synthesis and purification methods based on crystallization. In particular, to retain the stoichiometric composition in
molten germanium and tin tellurides, it is necessary to establish
an equilibrium vapor pressure of tellurium (at a suitable temperature) above the melt, in exactly the same manner as it is done for
A^{III}P and A^{III}As compounds. In the case of lead telluride, it is

necessary to control closely the temperature conditions in order
to remain within the range of the thermal stability of this compound.

§ 6.   Physicochemical Analysis of the $A^V - B^{VI}$ Binary Liquid Systems

The published data on the physicochemical properties of the
A^V –Te binary liquid systems [589, 591] do not give unambiguous
information on the thermal stability of antimony and bismuth tellurides. Thus, Honda and Endo [589] have investigated the concentration dependences of the magnetic susceptibility of the Bi –Te and
Sb –Te systems. For both systems, they have obtained isotherms
with broad minima corresponding to Bi_2Te_3 and Sb_2Te_3, which have
been interpreted as indicating the dissociation of these compounds
above their melting points. However, this conclusion is contra-
dicted by the nature of the liquidus curve of the Sb –Te systems
and by the viscosity data [48], according to which antimony telluride is relatively stable at its melting point and begins to dissociate only after heating to temperatures above this point. On the
other hand, the electrical conductivity isotherms, reported in [591],
show no singular points corresponding to antimony telluride. There
are no published data on the physicochemical properties of the
Bi –Se melts.

We investigated the viscosity and electrical conductivity of
alloys in the Bi –Se, Bi –Te, Sb –Te systems [552]. Alloys of various compositions have been prepared from bismuth of the V-000
grade, antimony of Su-"Extra" grade, tellurium of the TA-1 grade
(99.98% Te), and selenium of 99.92% purity. In our investigations
of the electrical conductivity of the Bi –Te alloys, we used spectroscopically pure tellurium, prepared by double vacuum sublimation of the TA-1 tellurium. The alloys were prepared by direct
fusion of the components in quartz ampoules evacuated to 10^{-3} mm
Hg. The fusion took three to four hours at 80-100° above the liquidus temperature, taken from the corresponding phase diagram.
Cooling was carried out in a furnace at a rate of 30-40° per hour.
The viscosity and electrical conductivity were measured by the
method described in Chapter 2.

In an investigation of the first batch of the Bi –Te alloys, it
was found that the viscosity rises anomalously above ~900-950°C.
It was concluded that the Bi –Te alloys reacted with quartz above

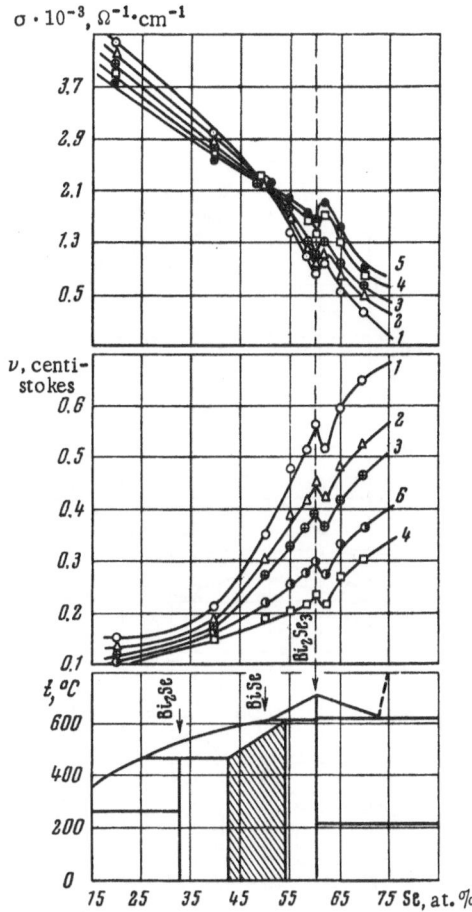

Fig. 130. Viscosity and electrical conductivity isotherms and equilibrium phase diagram of the Bi –Se system in the molten state. Temperature (°C): 1) 710; 2) 760; 3) 800; 4) 960; 5) 1050; 6) 880.

900°C and this was responsible for the increase in the viscosity. In view of this, a second batch of samples of the alloys of this system was prepared, and in this case the viscosity was measured in quartz ampoules coated internally with a mirrorlike layer of pyrocarbon [98]. These measurements showed that if the internal pyrocarbon coating was sufficiently dense, no anomalies were observed in the temperature dependence of the viscosity. In those

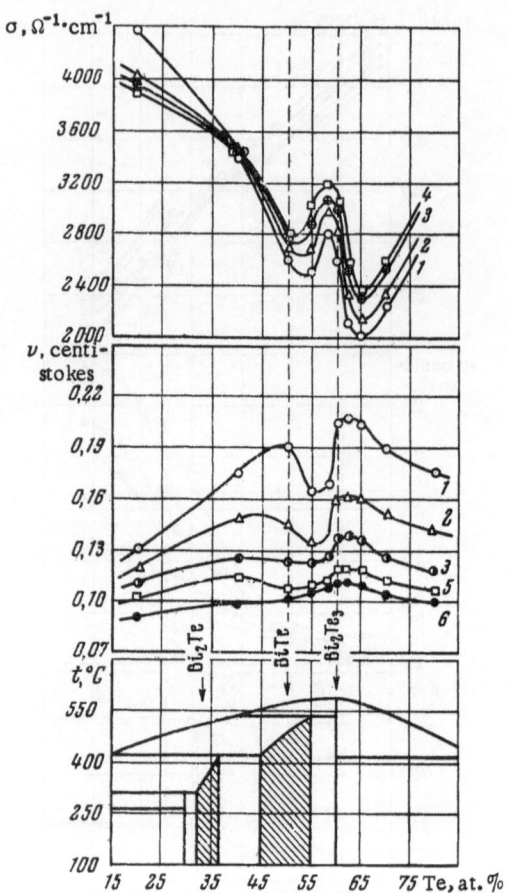

Fig. 131. Viscosity and electrical conductivity isotherms
and equilibrium phase diagram of the Bi−Te system in the
molten state. Temperature (°C): 1) 600; 2) 700; 3) 800;
4) 860; 5) 900; 6) 1100.

cases when the coatings were poor or when they split off from the
substrate (as observed after the experiments), the temperature dependence of the viscosity was found to be of the type observed for
the first batch. The general nature of the changes in the viscosity
measured in carbon-coated and uncoated quartz ampoules supported
the conclusion that the Bi−Te alloys reacted with quartz above a
certain temperature and therefore only the results obtained in

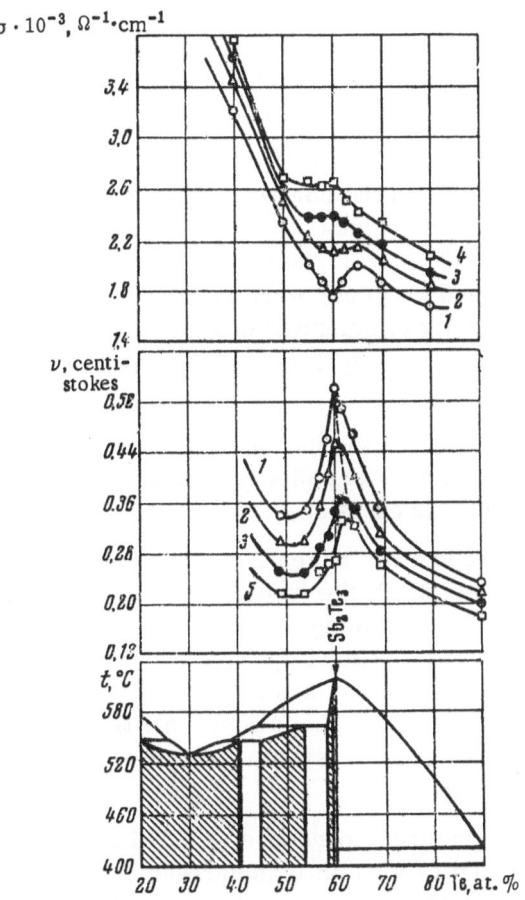

Fig. 132. Viscosity and electrical conductivity isotherms
and equilibrium phase diagram of the Sb —Te system in
the molten state. Temperature (°C): 1) 650; 2) 700; 3)
800; 4) 1000; 5) 900.

carbon-coated ampoules were used. No anomalies were found for
the Bi—Se system and all the measurements were carried out in
quartz ampoules.

The electrical conductivity of the Bi—Se and Sb—Te alloys
was measured in evacuated quartz ampoules using samples of 8
mm diameter and about 10 mm high.

The electrical conductivity of alloys of the Bi—Te system was measured in corundum crucibles, sealed into evacuated quartz ampoules. To maintain the constancy of the moment of inertia of the suspension system, the crucibles were selected so that they did not differ by more than ±0.05 g in weight.

In measurements of the electrical conductivity, the dependence of the volume on the composition was taken into account: it was assumed that the values of the density in the systems $A^V - A_2^V B_3^{VI}$ and $A_2^V B_3^{VI} - B^{VI}$ were additive. The values of the density of bismuth and antimony were taken from a handbook [80]. The values of the density of selenium and tellurium in the liquid state were determined at appropriate temperatures by the pycnometric method.

The results of all the measurements of the viscosity and electrical conductivity were plotted in the form of temperature dependences of these properties.

These dependences indicate that the viscosity of bismuth selenide is higher, at all the investigated temperatures, than the viscosity of alloys containing 58 and 62 at.% Se, whose viscosities are practically equal. The temperature coefficient of the viscosity $(d\nu/dt)$ of bismuth selenide differs little from the corresponding coefficients of two neighboring alloys. All the temperature dependences of the alloys of this system are smooth at all the investigated temperatures.

The temperature dependence of the viscosity of alloys belonging to the Bi—Te system is somewhat different. Bismuth telluride has almost the same viscosity, over the whole investigated range of temperatures, as alloys with 62 and 65 at.% Te. The temperature coefficient of the viscosity of the molten form of this compound differs little from the temperature dependences of the viscosity of the alloys.

The temperature dependences of the electrical conductivity of alloys of the Bi—Se and Bi—Te systems are similar: the electrical conductivity of alloys with 20 at.% Te (or Se) decreases when the temperature is increased, while the conductivity of alloys with 40 and 50 at.% Te changes slightly. The electrical conductivity of other alloys in these systems increases when the temperature is raised. The temperature coefficient of the electrical conductivity

of bismuth selenide is somewhat less than the coefficients of neighboring alloys but in the case of bismuth telluride the coefficients
of the compound and neighboring alloys are very similar.

The electrical conductivity of all the investigated alloys of
the Sb−Te system in the liquid state rises when the temperature
is increased and the temperature coefficient of the conductivity of
antimony telluride (of stoichiometric composition) is, at 620-800°C,
somewhat higher than the coefficients of alloys containing 58 and
62 at.% Te; above this temperature, there is little difference between the temperature coefficients. Consequently, the temperature dependences of the electrical conductivity of antimony telluride and of neighboring alloys intersect at ~720°C. The absolute
values of the conductivity obtained for alloys of this system are
somewhat lower than the data reported in [591]; this is likely to be
due to the higher purity of the original materials used in the present investigation.

The temperature dependences obtained for the viscosity and
electrical conductivity were used to plot the isotherms of these
two properties, which are shown in Figs. 130-132, along with the
equilibrium phase diagrams.

It is evident from Fig. 130 that the viscosity and electrical
conductivity isotherms have pronounced singular extrema (a viscosity maximum and a conductivity minimum) at a composition
corresponding to the compound Bi$_2$Se$_3$. These results indicate that
bismuth selenide is stable over the whole investigated range of
temperatures (up to 960°C), which is in good agreement with the
singular nature of the liquidus curve of the Bi−Se system. We
must stress particularly the full agreement between the concentration dependences of the viscosity and electrical conductivity.

We shall now consider the dependences of the viscosity and
electrical conductivity on the composition when the selenium concentration is increased over 62 at.%. It follows from the curves
given in Fig. 130 that above this selenium concentration the viscosity rises appreciably and the electrical conductivity falls markedly. We may assume that a slight excess of selenium, above the
stoichiometric ratio, may be present in the melt in the atomic
state. This should help to reduce the viscosity because of the presence, in the melt, of structural units smaller than those in the case

of the compound. When the concentration of selenium is increased
further, polymerization may take place, resulting in the formation
of selenium chains, which greatly increase the viscosity.

The changes in the electrical conductivity can be explained
in the same way, since atomic selenium may contribute additional
carriers. The electrical activity of atomic selenium may be
greater at higher temperatures and therefore the temperature coefficient of the electrical conductivity of an alloy with 62 at.% Se
is higher than that of bismuth selenide (of stoichiometric composition). When the temperature is increased, the singular minimum
in the electrical conductivity isotherms is enhanced, but the usual
broadening is not produced.

The viscosity isotherms of the Bi−Te melts (Fig. 131) have
two broad maxima, located near Bi_2Te_3 and near a composition
Bi : Te = 1 : 1. The electrical conductivity isotherms exhibit the
corresponding minima. Thus, there is a definite correlation between the values of the viscosity and electrical conductivity.

The nature of the isotherms of the investigated physicochemical properties suggests the dissociation of bismuth telluride at
its melting point and an enhancement of this process when the melt
temperature is raised. The presence of a viscosity maximum and
an electrical conductivity minimum in the region of the composition Bi : Te = 1 : 1 indicates that although the compound BiTe does
not exist in the solid state, it may exhibit its characteristic properties in the melt at temperatures a little above the melting point.
The observed nature of the concentration dependences of the viscosity and electrical conductivity in the range of compositions near
this compound confirms Shakhparonov's conclusion [592] that there
is no one-to-one correspondence between the form of the phase diagram and the nature of the physicochemical composition−property
diagrams in the liquid state.

The dependences presented in Fig. 132 show a strong correlation between the values of the viscosity and electrical conductivity
and a good agreement with the nature of the liquidus curve for the
Sb−Te system.

§ 7. Investigations of the Viscosity of
Ternary Liquid Systems along the
Lines Intersecting the Ge(Si) – $A^{III}B^V$
Quasi-Binary Lines in Primary
Solvent – Crystallization Region

Since the viscosity is a structure-sensitive property [1, 63], a study of this property yields information on the behavior of a binary compound in the molten state. The viscosity isotherms show a singular maximum corresponding to the ordinate of a stable chemical compound. If this compound dissociates, the maximum gradually broadens as the degree of dissociation increases, and shifts usually in the direction of the more viscous component [1, 4]. In a ternary system, at concentrations corresponding to the primary crystallization region of such a compound, the nature of the concentration dependence of the viscosity should be similar to that observed for a binary system.

From general considerations of physicochemical analysis and from the high sensitivity of the viscosity method to structural changes in the liquid, we may assume that in the case of a ternary system the general behavior should be the same as in the case of binary systems. This means that if a stable binary chemical compound is dissolved in some substance, the dependence of the viscosity on the composition in the range of concentrations corresponding to the primary crystallization region of the solvent should not be a smooth surface but should have a convex fold, corresponding to a line joining the solvent and the compound. The nature of this fold should reflect the behavior of the dissolved compound; when such a compound is stable at any temperature, the fold should be singular, but if the compound dissociates partly, then the fold should gradually broaden at higher temperatures and it should shift in the direction of that binary system which has a higher value of the viscosity in the molten state at a given temperature.

Consequently, the viscosity isotherms should have maxima for lines intersecting a quasi-binary line and the nature and positions of such maxima should depend on the thermal stability of the dissolved compound.

One of the present authors has investigated the dependence of the viscosity on the composition at various temperatures along the lines intersecting quasi-binary lines in the following systems [593-595]:

1. Ge—In—Sb   for   60 and 90%   (at.)   Ge
2. Ge—Ga—Sb   for   50 and 75%   (at.)   Ge
3. Ge—Al—Sb   for   80 and 90%   (at.)   Ge
4. Si—Al—Sb   for   80%   (at.)   Si

All these lines lie in the primary crystallization region of germanium or silicon [114, 115]. Figure 133 shows the positions of the investigated lines and the compositions of the corresponding alloys (shown as dots).

The starting materials were germanium, silicon, and high-purity dopants. Recrystallized aluminum, gallium, and indium antimonides were used to prepare alloys lying along quasi-binary lines Si(Ge) – AlSb, Ge – GaSb, and Ge – InSb.

Samples of the Ge – Ga – Sb and Ge – In – Sb alloys were placed in cylindrical quartz ampoules, of 11 mm internal diameter, which were evacuated down to 10^{-4} mm Hg and sealed. The ratio of the height of the liquid column in the cylinder to the cylinder radius was ~4.5. Samples of Ge–Al–Sb and Si–Al–Sb alloys were placed in cylindrical corundum cans, which were enclosed in graphite cylinders with screw-on covers. The ratio of the height of the liquid column in the can to its radius was also 4.5. The viscosity was calculated from the formulas for weakly viscous liquids.

The results obtained were plotted as temperature dependences of the kinematic viscosity for two lines in each of the investigated systems.

It follows from these dependences that the viscosity of alloys along quasi-binary lines is higher than the viscosity of other alloys. Moreover, the temperature coefficients of the viscosity of these quasi-binary alloys are greater than the temperature coefficients of other alloys. Since the temperature coefficients are negative, when the temperature is increased sufficiently high, we find that the viscosities of the alloys along quasi-binary lines become smaller than the viscosities of alloys enriched with one of the components.

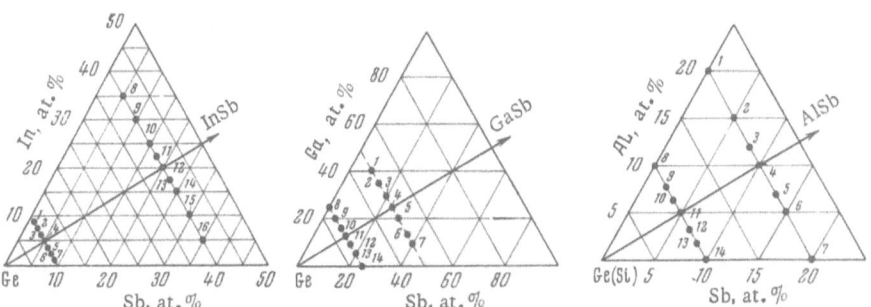

Fig. 133. Positions of lines and compositions of investigated alloys (1-16) in ternary systems based on germanium and silicon.

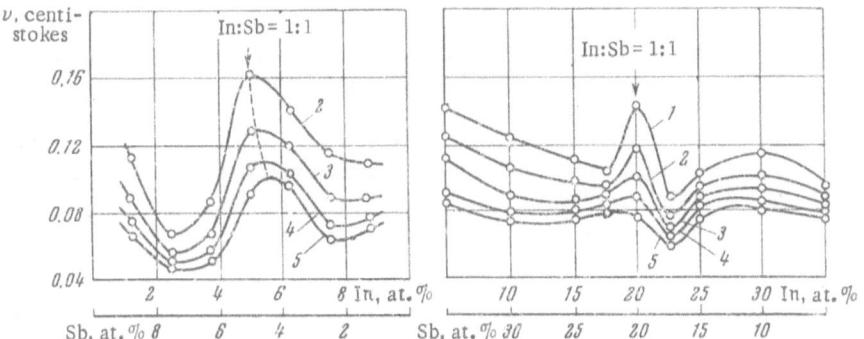

Fig. 134. Viscosity isotherms of melts in the Ge—In—Sb system. Temperature (°C): 1) 900; 2) 950; 3) 1000; 4) 1050; 5) 1100.

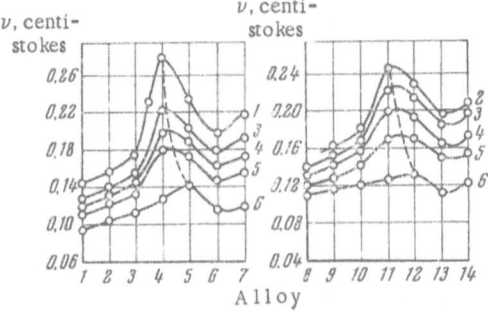

Fig. 135. Viscosity isotherms of melts in the Ge—Ga—Sb system. Temperature (°C): 1) 825; 2) 875; 3) 900; 4) 950; 5) 1000; 6) 1100. For compositions of alloys 1-14, see Fig. 133.

The temperature dependences of the viscosity of alloys of the Ge—Ga—Sb system have one further property. Above ~850°C, the temperature dependences show some deviations from their usual shape. Similar deviations are observed for pure gallium antimonide at somewhat higher temperatures. We have demonstrated in [22] that these deviations may be due to the onset of appreciable dissociation of a chemical compound in the liquid state. It is very likely that the deviations observed in the temperature dependences of the viscosity of the alloys are due to the same cause.

If the results obtained are presented in the form of isotherms (Figs. 134-137), we observe the following features.

In the Ge—In—Sb system the intersection of a 60 at.% Ge line with a quasi-binary Ge—InSb line corresponds to a maximum in the viscosity isotherms; when the temperature is increased, this maximum gradually broadens. For a 90% Ge line (Fig. 134), this maximum is much broader and shifted in the direction of Ge—In.

A basically similar behavior is observed for the Ge—Ga—Sb and Ge—Al—Sb systems for lines corresponding to 50 and 75 at.% Ge, as well as 80 and 90 at.% Ge (Figs. 135 and 136).

These dependences are similar to those observed for binary irrational systems, i.e., systems in which a dissociating chemical compound is formed.

Thus, the general nature of the concentration dependences of the viscosity of Ge—A^{III}—Sb alloys, in the range of concentrations corresponding to the primary crystallization region of germanium-base solid solutions, is characterized by the presence of a convex fold above a quasi-binary line, which broadens when the temperature is increased.

Consequently, on the basis of general relationships governing the physicochemical analysis, we may conclude that antimonides of Group III dissolved in molten germanium dissociate partly into their original components and the degree of dissociation increases when the temperature is increased.

If we compare the dependences obtained for alloys representing lines corresponding to different germanium concentrations, we see that at lower germanium concentrations the broadening and

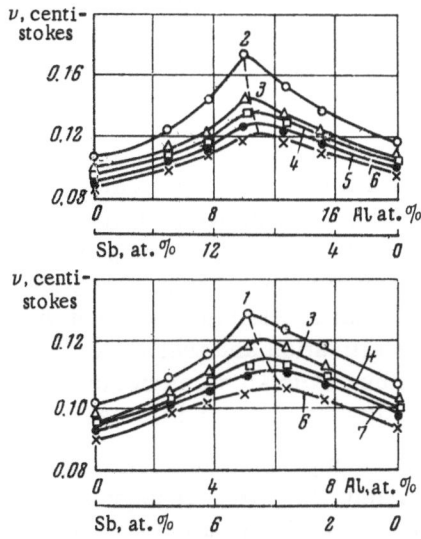

Fig. 136. Viscosity isotherms of melts in the Ge−Al−Sb system. Temperature (°C): 1) 960; 2) 970; 3) 1050; 4) 1100; 5) 1150; 6) 1160; 7) 1250.

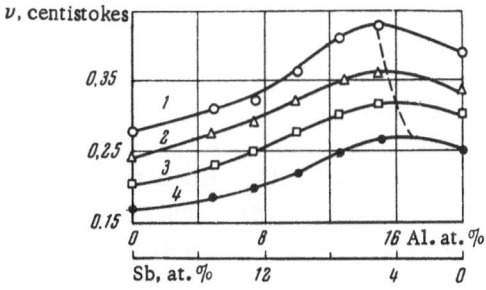

Fig. 137. Viscosity isotherms of melts in the Si−Al−Sb system. Temperature (°C): 1) 1400; 2) 1450; 3) 1500; 4) 1580.

shift of the viscosity maxima from a quasi-binary Ge−A^{III}Sb line occurs at higher temperatures than is the case of alloys with higher germanium concentrations. This may be due to an enhancement of the dissociation of the corresponding compound (A^{III}Sb) when the degree of dilution of the melt is increased. In this re-

spect, the behavior of the $A^{III}Sb$ compounds, which obviously dissociate thermally when dissolved in germanium, is similar to the behavior of a weak binary electrolyte during electrolytic dissociation.

When germanium single crystals are prepared by pulling from a melt doped with Group III antimonides, the Group III element and the antimony should be captured separately, because, as a result of the dissociation of the $A^{III}Sb$ compounds in molten germanium, the quasi-binary lines Ge−InSb, Ge−GaSb, and Ge−AlSb do not reflect the actual equilibrium between the solid and liquid phases during crystallization and therefore they do not possess all the properties of a binary system, as demonstrated by Zemskov et al. [597-603] for the Ge−InSb and Ge−AlSb systems.

Even when the compound dissolved in molten germanium is only partly dissociated, the different degrees of capture of A^{III} and of antimony during the pulling of a crystal should − in accordance with the Le Chatelier principle − shift continuously the $A^{III}Sb \rightleftharpoons A^{III} + Sb$ reaction equilibrium in the direction of dissociation, and the ratio of the A^{III} and antimony concentrations in the pulled crystal should represent the ratio of the effective capture coefficients of A^{III} and antimony. In the Si−Al−Sb system (Fig. 137), the viscosity isotherms corresponding to a line with 80% Si show no maximum at the intersection of this line with the quasi-binary line. The highest values of the viscosity at all temperatures are exhibited by an alloy in which the dopants are in the ratio Al:Sb = 3:1. The temperature coefficients of the viscosity of alloys lying along the quasi-binary line differ little from the coefficients of other alloys. The strong broadening and the shift of the maximum away from the composition corresponding to the quasi-binary line indicates a considerable dissociation of aluminum antimonide dissolved in silicon. This dissociation is evidently due to high temperatures, exceeding the dissociation temperature of even the pure compound; moreover, the dilution effect may be important.

Nevertheless, the presence of a broad maximum in the viscosity isotherms of this system indicates the appearance of some preferential chemical interaction between aluminum and antimony in the melts based on silicon. This is in good agreement with the observation that when silicon-base alloys are melted in the presence of aluminum, antimony escapes less by evaporation than when in the absence of aluminum.

The behavior of aluminum antimonide in the liquid and solid solutions based on germanium and silicon, is similar. This supports our conclusion that the interaction in the solid state is related to the interaction in the liquid state.

In conclusion, we must mention that in some systems such as $Ge-A^{III}Sb$, we cannot neglect the interaction in the liquid state in the explanation of the nature of the solidus isotherms (i.e., the ratio of the activities cannot be assumed to be unity).

Thus, the interaction between donors and acceptors in silicon and germanium is complex, since it is governed by a number of factors mentioned here. In some systems the role of the thermal stability of the compounds in the melts may be important.

§ 8. Some General Problems in the

Physicochemical Analysis of Liquid Systems

Kurnakov has established general relationships governing the dependences of various properties on the composition of liquid binary systems. He has shown that the nature of these dependences is, to some extent, governed by the nature of the chemical interaction in a system which can be described by a phase equilibrium diagram.

However, changes in the nature of the chemical binding during the process of fusion of the components and the intermediate phases was not allowed for by Kurnakov. Regel' and other investigators (including the present authors), have shown that a variety of changes in the nature of the chemical binding is possible at the melting point and these changes may include even a basic transformation from a semiconductor to a metal. Moreover, various types of chemical binding can exist in the liquid state, and, therefore, the temperature dependences of some of these properties may be directly opposite. For example, a substance with the metallic binding should exhibit a fall of the electrical conductivity when the temperature is increased, whereas a substance with a predominantly covalent binding should exhibit a rise of the electrical conductivity. We can also have cases in which a particular system contains components which are bound covalently, as well as components with the metallic binding. Covalent bonds may be retained in intermediate phases, formed by two metals, etc. In

other words, the problem of the influence of the type of binding in
the liquid state on the general nature of the concentration dependence of physicochemical (and particularly electrical) properties
has not yet been solved. This problem has not been considered by
Kurnakov, possibly because the majority of the binary liquid systems investigated by him and his colleagues consisted of organic
compounds. Moreover, the experimental data on changes in the
chemical binding due to fusion have been insufficient at the time
of Kurnakov's investigations and, therefore, the question of the relationship between the composition–property diagrams of liquids
and the phase-equilibrium diagrams has not been considered.
Glazov et al. [604] were first to draw attention to this point. They
pointed out that the relationship between the phase diagrams and
the property–composition dependences should be considered, taking into account those changes in the nature of the chemical binding which are observed during the fusion of the components and
the intermediate phases. We shall now consider our experimental
data for systems with congruently melting chemical compounds,
taking into account such changes in the chemical binding.

1. Relationship between Property–Composition Dependences and Phase Diagrams of Liquid Systems with Congruently Melting Chemical Compounds

We have investigated ten binary systems in which the chemi-
cal compounds ($A^{II}Te$, $A_2^{III}Te_3$, $A^{IV}Te$, and $A_2^V B_3^{VI}$) retain their
predominantly covalent binding above the melting point. However,
the chemical binding of the components of these compounds changes
in various ways at the melting point. Thus, germanium becomes
metallic; zinc, cadmium, gallium, indium, lead, tin, antimony, and
bismuth are metals in the liquid phase. Tellurium becomes metallic at some temperature above the melting point, but selenium remains a semiconductor with a predominantly covalent type of chemical binding between its atoms. Consequently, there is no unique
relationship between the form of the phase diagram and the nature
of the property–composition dependences in the liquid state. In
any case, the "left-hand sides" of these systems (assuming that
the $A^{II}Te$, $A_2^{III}Te_3$, $A^{IV}Te$, or $A_2^V B_3^{VI}$ compounds form the bounda-
ries which divide these systems) should differ from the "right-hand
sides" which consists of two substances retaining (in a certain range
of temperatures) the original nature of the chemical binding. In

this respect, it is worth considering the left-hand side of the Bi−Te system. In this partial system (Bi−Bi$_2$Te$_3$), the compound BiTe (which does not exist in the solid state) exerts its influence on the properties of the liquid phase (Fig. 131). Here, we must agree with the view of Shakhparonov [592], reflected also in the investigations of Samoilov [605] and Khokhlov [606], that the phase diagram can be used only to draw qualitative conclusions about the state of the melt. Even in those cases when the agreement between the phase diagram and the property−composition dependences is apparently good, we must remember that, even when the covalent binding is retained in the melt, the nature of the molten compound differs considerably from its nature in the solid state. This is indicated by numerous experimental data, which show that such short-range order changes take place above the melting point. We can even observe an apparent increase in the chemical interaction due to heating, such as that found from the viscosity of the Ga−Te and In−Te systems. The heating of a molten compound may, in some cases, produce a completely different result from that expected for systems which include chemical compounds. The best example of such behavior is provided by the Sb−Te and Bi−, Se systems.

2. Apparent Singularities in the Property−Composition Dependences of Liquid Systems

Figure 132 shows the isotherms of the concentration dependence of the electrical conductivity of the Sb−Te system in the molten state. It follows from that figure that at high temperatures the nature of the concentration dependence of the conductivity changes considerably so that a minimum is replaced by a maximum corresponding to the chemical compound Sb$_2$Te$_3$. Analysis of the temperature dependences of the electrical conductivity of liquid alloys in the system shows that this behavior is due to a higher value of the temperature coefficient of the electrical conductivity of antimony telluride, compared with alloys containing 58 and 62 at.% Te. This may be because the conductivity is a function not only of the carrier density but also of the carrier mobility and the mobility may be affected in different ways by the excess of one or the other component. It may be that the nature of the carrier scattering depends in a complex manner on the composition at various temperatures, in accordance with the chemical nature of the atoms of the excess component.

Another interesting example is the Bi—Se system. In this system an electrical conductivity minimum in the isotherms (Fig. 130) becomes more pronounced when the temperature is increased. Analysis of the temperature dependences of the electrical conductivity shows that this is due to the smaller value of the temperature coefficient of the conductivity of bismuth selenide than the coefficients of the neighboring alloys, containing an excess of bismuth or selenium. However, the concentration dependence of the viscosity does not indicate any enhancement of the chemical interaction at higher temperatures. Consequently, the deepening of the minimum in the isotherms of the electrical conductivity of the Bi—Se system at higher temperatures is a false clue. Thus, we may conclude that the nature of the electrical conductivity isotherms cannot always be relied upon to give information on the interaction between the components in the melt. It follows that to determine the behavior of a compound in the molten state it is always necessary to investigate several physicochemical properties, particularly the viscosity.

3. Nature of the Viscosity—Composition Diagrams of Eutectic Systems [604]

We have shown that an analysis of the relationship between the property—composition dependences and the form of the phase diagram must allow for the changes in the nature of the chemical binding due to the fusion of the components forming a given system and to the fusion of the intermediate phases. We shall now consider the problem of the concentration dependence of the viscosity in eutectic systems on the basis of modern ideas on the structure of liquid eutectics.

Attempts have been made to solve this problem in a purely empirical manner, by investigating the viscosity in some arbitrarily selected system ignoring modern knowledge of the structure of liquid eutectics and without allowing for the interaction between the components in each particular case.

Three types of viscosity—composition diagram of eutectic systems are discussed in the published papers [73, 74, 159, 161, 162, 164, 166-169, 175, 176, 437, 438, 607-618].

1. Diagrams with a minimum at the eutectic concentration [168, 169, 607, 610-615].

2. Diagrams with a maximum at the eutectic concentration [73, 167, 616, 617].

3. Diagrams without singular points [164, 166, 175, 607-609, 618].

We shall consider the viscosity–composition diagrams of binary systems using modern ideas on the structure of liquid eutectics and analyzing the nature of the intermolecular interaction in eutectic systems on the basis of energy of mixing.

Danilov et al. [178, 619, 620] have shown that x-ray diffraction patterns of many eutectic melts are the result of the superposition of the diffraction patterns of the two components. Hence, they have concluded that complete molecular mixing does not take place in a liquid eutectic and such a eutectic consists of regions enriched with one or the other component.

Bartenev [621, 622] has investigated the heats of mixing of liquid components at the eutectic concentration and has shown that the process of mixing is not favored by energy considerations; hence, he concluded that there is no complete mutual mixing of the components in liquid eutectics.

Danilov's ideas on the structure of liquid eutectics have been confirmed by investigations of many systems using x-ray diffraction [623-628], electron diffraction [629-631], and neutron diffraction [551] methods, by investigations of the nature of the temperature dependence of the electrical conductivity [56, 632], as well as experiments on the centrifuging of liquid eutectics [181, 248]. This view about the structure of liquid eutectics does not contradict Shakhparonov's ideas that a eutectic is a solution with strong concentration fluctuations [396]. However, this may not apply to all cases.

The chemically microinhomogeneous structure of liquid eutectics may be stable if the binding forces between like atoms are stronger than between unlike atoms, i.e., if in a system $A - B$, we have $F_{A-A} > F_{A-B}$ and $F_{B-B} > F_{A-B}$. The chemical microinhomogeneity may also be stable if $F_{A-A} > F_{A-B} \geq F_{B-B}$ or $F_{B-B} > F_{A-B} \geq F_{A-A}$. However, in the latter case, the inhomo-

geneity should be weaker. Thus, the structure of eutectic melts is governed by the nature of the intermolecular interaction.

Consequently, the nature of the viscosity–composition diagrams is governed by the same interaction.

Let us first consider the case when $F_{A-A} > F_{A-B}$ and $F_{B-B} > F_{A-B}$.

Such systems should exhibit minima in their viscosity–composition diagrams and these minima need not, in general, be related to the position of the eutectic point. In fact, the structure of the melt is strongly microinhomogeneous in the chemical sense. In the transition regions lying between microregions enriched with one or the other component, the binding may be mainly of the $A-B$ type and consequently these transition regions may have weak binding, which would tend to make it easier for the liquid to flow and, therefore, should result in a fall of the viscosity. A change resulting in an increase in the concentration of one or the other component should increase the relative number of stronger bonds between like atoms so that the viscosity should increase. The viscosity minimum should correspond to a concentration with the maximum chemical microinhomogeneity.

We shall now consider systems for which $F_{A-A} > F_{A-B}$ but $F_{A-B} > F_{B-B}$. Again, a chemical microinhomogeneity of the melt is assumed.

The viscosity–composition diagrams of such systems should have a slight kink because a change in the concentration resulting in an increase of a component A should increase the number of strong $A-A$ bonds, which should tend to increase the viscosity. A rise in the concentration of a component B should increase the number of weaker $B-B$ bonds and consequently it should reduce the viscosity. The rate of change of the viscosity should be different for changes in the concentration along these two directions, and consequently there should be an inflection in the concentration dependence of the viscosity.

Finally, we shall consider the third case for which $F_{A-B} > F_{A-A}$ and $F_{A-B} > F_{B-B}$.

In this case; the melt should not be chemically microinhomogeneous and the viscosity–composition curves should be smooth without any singularities.

Table 45. Energies of Mixing for Eutectic Alloys in Simple
Eutectic Systems

System	Compo-nent	x_N, at. %	Q_N, kcal/g-atom	T_{mp}, °K	T_{eut}, °K	V'/k
Ag—Bi	Bi	95.3	2.60	544	535	+2280
Ag—Pb	Pb	95.3	1.19	600	576	+1100
Ag—Tl	Tl	97.2	1.03	575	564	+7350
Au—Bi	Bi	81.1	2.60	544	514	+740
Be—Si	Be	33.0	2.80	1557	1363	+3240
Cd—Pb	Pb	72.0	1.19	600	521	+1180
Pb—As	Pb	7.4	1.19	600	561	+8400
Ge—Sb	Sb	17.0	4.75	903	863	+1800
Cd—Bi	Cd	45.0	1.53	594	417	+87
Cd—Sn	Cd	66.5	1.53	594	450	+682
Cd—Tl	Cd	72.8	1.53	594	476	+825
Na—Rb	Na	75.5	0.63	372	267	+501
Pb—Sb	Pb	17.5	1.19	600	525	+7700
Sn—Pb	Sn	26.1	1.69	505	456	+805
Bi—Sn	Bi	43.0	2.60	544	412	+92
Si—Al	Al	11.3	2.50	933	850	—8000
Si—Au	Au	31.0	3.05	1336	643	—5860
Ge—Au	Au	27.2	3.05	1336	629	—8500
Ge—Ag	Ag	25.9	2.69	1233	924	—955
Si—Ag	Ag	15.4	2.69	1233	1103	—35
Ge—Al	Al	30.3	2.50	933	697	—750

In this case, if the interaction between unlike atoms is much
stronger than the interaction between like atoms, "chemical compounds" may form in the melt.† The viscosity–composition curves
may have singularities unrelated to the position of the eutectic point.

We must bear in mind that the chemical microinhomogeneity
of eutectic melts, if present at all, is retained only up to a certain
temperature, above which a true solution is formed, as demonstrated by Regel and Gaibullaev [56, 632]. Since the structure of
eutectic melts changes when the temperature is increased, the nature of the viscosity–composition diagrams should change as well.

---

† This has been confirmed experimentally by Duwez, Willens, and Klement [634], who
have quenched a eutectic Ge–Ag melt with a negative energy of mixing (Table 45)
to obtain an alloy with a crystal structure different from the structure of the components.

When the temperature is sufficiently high, the viscosity isotherms should become smooth curves, not very different from the straight line obtained by the simple addition of the viscosities.

We shall now consider the concentration dependences of the viscosity in real eutectic systems.

To do this, we must know first the nature of the intermolecular interaction in the systems to be investigated.

We shall consider "pure" eutectic systems (Table 45) with a slight mutual solubility of the components in the solid state.

Kamenetskaya et al. [398, 399, 401] and Pines [633] have described a method for the analysis of the nature of the intermolecular interaction of components in binary systems. Starting from a minimum of the free energy of a system, these authors have obtained equations for a two-phase equilibrium which relate the temperatures to the concentration of the coexisting phases:†

$$x^2 V' + kq_A (T_A - T) + kT \ln (1 - x) = 0; \tag{116}$$

$$(1 - x)^2 V' + kq_B (T_B - T) + kT \ln x = 0; \tag{117}$$

$$q_A = \frac{Q_A}{RT_A}; \qquad q_B = \frac{Q_B}{RT_B}. \tag{118}$$

Here, q_A and q_B are the fusion entropies of the components (per molecule), divided by the Boltzmann constant; Q_A and Q_B are the heats of fusion of the components, and V' is the energy of mixing.

Table 45 gives the results of calculations of the energy of mixing for simple eutectic systems.‡

It follows from Table 45 that the majority of the systems considered have positive energies of mixing. The structure of liquid eutectics of these systems is characterized by a strong

---

† Equation (116) represents the left-hand branch of the liquidus curve and Eq. (117) the right-hand branch.

‡ The phase diagrams of the selected systems have been taken from Hansen and Anderko's handbook [203] and the corresponding heats of fusion of the components have been taken from Kubaschewski and Evans's book [113, 424] (cf. Table 45).

chemical microinhomogeneity up to a certain temperature. The viscosity–composition diagrams of these systems may have a minimum or an inflection corresponding to the eutectic concentration provided the general nature of the curvature of the liquidus curves is basically the same over the whole range of concentrations. We can also have a case when the interaction forces between unlike atoms differ little from the corresponding forces between like atoms. In this case, a minimum or an inflection in the viscosity curve may be hardly noticeable and the curve may be practically smooth.

Such systems have a low energy of mixing (for example, BiSn). In this case, the nature of the structure of the melt may be deduced from the energy of mixing.

Six systems in Table 45 have negative energies of mixing. Consequently, the viscosity–composition diagrams of these systems should have inflections (if the liquidus curves are of suitable nature), or they should be smooth curves. Moreover, if the chemical interaction between the components in the liquid state is strong, the viscosity–composition diagrams of these systems may have singularities which need not coincide with the position of the eutectic point.

In the latter case, the kinetic factors may be important and the results obtained by different investigators for the same system may not be identical.

Analysis of the studies of the viscosity of eutectic systems shows that the results obtained fit reasonably well these representations of the possible nature of the viscosity–composition diagrams. In analyzing a given system, we must bear in mind the reliability of the experimental method. Because of the lack of such reliability and in view of the analysis given here, we must regard as erroneous the results reported in [73, 167, 616, 617].

However, the question arises as to the origin of the strong chemical interaction in a simple liquid eutectic system or, in other words, how can we explain why a eutectic transition takes place in systems with a negative energy of mixing when the interaction between unlike atoms is stronger than that between like atoms.

This is possible if the nature of the chemical binding of one or both components changes in the transition from the solid to the liquid state.

We must bear in mind also the relative positions of the components in the periodic table.

If we adopt this approach in dealing with systems with a negative energy of mixing (Table 45), we find that the components of these systems are germanium and silicon. In the solid state, germanium and silicon have structures whose coordination number is low (4), and they have a three-dimensional system of rigid covalent bonds. Therefore, in the solid state, the Si−Si or Ge−Ge forces are considerably stronger than, for example, the Si−Al, Ge−Al, Si−Ag, and similar forces.

The nature of the chemical binding of germanium and silicon changes considerably at their melting points (cf. Chapter 3, §1) and they become much more strongly active in the chemical sense. In the liquid state, the nature of the interaction is opposite to that observed in the solid state, i.e., the interaction between unlike atoms is stronger than that between like atoms. Because of this, the process of separation of atoms of different kind, which is responsible for eutectic transitions in such systems, takes place during crystallization.

Thus, by analyzing the intermolecular interaction and the nature of the viscosity−composition diagrams, we may deduce information about the structure of molten eutectics.

In conclusion, we must mention the characteristic features of the viscosity−composition diagrams of systems with chemical compounds and eutectic transitions. Two cases are possible: when a eutectic transition takes place between a compound and a pure component or between two compounds.

In these cases, the energy of mixing should be calculated regarding the compounds as independent components.

The viscosity−composition dependences of such systems should be the same as for simple binary systems, except that we must take into account the possibility of dissociation of chemical compounds due to fusion. Among such systems are In−Sb, Ga−Te, In−Te, and others which are considered in the present chapter.

Appendix*

---

*The Appendix tabulates the numerical values of various quantities obtained experimentally by the present authors and other investigators.

Table 1. Electrical Conductivity of Semiconductors in Liquid State

Si

t, °C	σ, $\Omega^{-1}\cdot cm^{-1}$
1450	12860
1475	12640
1510	11810
1545	11150
1550	11950
1560	11880
1600	11570

Ge

t, °C	σ, $\Omega^{-1}\cdot cm^{-1}$
945	15200
950	15300
960	15100
970	15100
978	16600
995	16000
1010	16300
1090	15100
1130	13600
1185	12400
1300	11200

Te

t, °C	σ, $\Omega^{-1}\cdot cm^{-1}$
320	259
370	1135
415	1630
440	1755
450	1820
452	1910
470	2020
505	2085
555	2330
585	2350
600	2370
635	2440
650	2460
660	2510
695	2480
705	2575
765	2675
780	2485
790	2520
810	2650
855	2610
920	2650
950	2580
1025	2680

AlSb

t, °C	σ, $\Omega^{-1}\cdot cm^{-1}$
1085	9650
1090	9850
1095	9850
1100	9180
1105	10200
1115	9260
1120	9690
1125	9650
1135	9700
1140	9850
1150	9880
1155	9800
1170	9800
1177	10500
1200	11000
1210	10450
1212	10650
1285	11200
1320	11200

GaSb

t, °C	σ, $\Omega^{-1}\cdot cm^{-1}$
740	10580
760	10440
780	10300
800	10440
820	10880
850	10150
900	9820
1000	9780

InSb

t, °C	σ, $\Omega^{-1}\cdot cm^{-1}$
540	9350
575	9700
580	9430
620	9350
720	9350
800	9450
860	9450
950	9400

GaAs

t, °C	σ, $\Omega^{-1}\cdot cm^{-1}$
1240	6500
1248	8000
1250	7940
1255	8180
1270	7930
1290	7900
1310	8130
1320	7940
1330	7600

InAs

t, °C	σ, $\Omega^{-1}\cdot cm^{-1}$
950	7000
960	6960
970	6760
975	6560
980	6450
985	6940
990	6170
995	6710
1000	6490
1010	5980
1013	6100
1018	6260
1020	6180
1025	6210
1030	6230
1035	6610
1037	5980
1040	8610
1045	6110
1055	6220
1060	5920
1070	6220
1085	5910
1105	5950
1125	5850
1160	6050
1220	6120

ZnTe

t, °C	σ, $\Omega^{-1}\cdot cm^{-1}$
1245	40
1250	55
1267	69
1287	77
1298	84
1313	103
1326	130
1338	165

CdTe

t, °C	σ, $\Omega^{-1}\cdot cm^{-1}$
1050	45
1055	50
1060	65
1080	86
1100	96
1120	104
1150	115
1175	130
1200	150
1225	168

Ga$_2$Te$_3$

t, °C	σ, $\Omega^{-1}\cdot cm^{-1}$
780	11
790	39
795	70
800	39
805	103
835	139
855	45
863	250
875	307
900	300
920	462
940	535
952	560
973	696
1015	1035
1027	1000
1070	1360
1107	1560
1125	1860
1145	1640
1165	1990
1190	2050
1210	2090
1258	2290

In$_2$Te$_3$

t, °C	σ, $\Omega^{-1}\cdot cm^{-1}$
650	5
655	13
660	22
665	25
667	79
670	180
675	310
680	723
685	740
715	800
718	880
730	940
760	985
770	1310
800	1500
845	1930
858	2030
900	2220
905	2200
965	2400
995	2550
1040	2910
1050	2750

Table 1 (continued)

t, °C	σ, $\Omega^{-1}\cdot cm^{-1}$	t, °C	σ, $\Omega^{-1}\cdot cm^{-1}$	t, °C	σ, $\Omega^{-1}\cdot cm^{-1}$	t, °C	σ, $\Omega^{-1}\cdot cm^{-1}$
In₂Te₃		828	11250	785	2690	955	1610
1105	2930	831	10610	795	2700	992	1735
1140	3170	856	9660	815	2650	1010	1600
1150	3110	858	11050	825	2775	1030	1780
1160	3300	893	11200	835	2770	1050	1725
1215	3250	903	11870	855	2700	1070	1830
1235	3420	933	11470	870	2930	1090	1780
1270	3470	934	12400	875	3000	1112	1890
1275	3320	948	12320	885	3090	1130	1825
1310	3370	973	11640	915	3065	1170	2030
		993	12690	920	3090	1180	2035
Mg₂Si		994	12940	925	3040	1220	2030
1102	9900	1013	12050	935	3120	1240	2110
1119	9890	1043	13110	940	2930		
1122	9850	1046	13190	970	3220		
1127	8840	1055	12380	975	3290	**PbSe**	
1142	8980	1088	13120	1020	3400	1100	505
1152	9970	1096	13080	1025	3220	1110	520
1157	10100	1113	12560	1035	3045	1130	525
1162	10200	1148	12420	1040	3225	1140	565
1177	10050	1156	12920	1045	3290	1162	600
1207	10210	1216	12610	1065	3210	1165	605
1217	10220	1226	12400	1085	3360	1180	600
1227	10450			1100	3480	1220	710
1232	9600	**Mg₂Pb**		1135	3480		
1242	10200	550	8290	**SnTe**		**PbS**	
1257	10420	553	8710	800	1840	1120	255
1262	9860	570	9740	810	1855	1123	320
1272	10060	575	8930	820	1885	1130	270
1292	10400	590	8940	830	1920	1140	275
1337	10480	600	8950	840	1905	1148	285
		620	9060	860	1975	1150	280
Mg₂Ge		635	9000	870	1940	1163	305
1130	8260	650	9070	900	2015	1165	290
1145	8450	685	9120	940	2060	1173	310
1155	8440	690	10150	950	2130	1175	310
1170	8610	725	9420	1000	2220	1183	325
1180	8460	735	9510	1040	2250	1190	330
1195	8600	755	10580	1050	2235	1205	350
1208	8710	785	9640	1080	2300	1208	340
1220	8810	807	9840	1090	2310	1213	350
1245	8950	830	10000	1110	2490	1233	385
1250	8730	840	10820	1130	2450		
1285	9100	910	11400	1140	2470	**Bi₂Se₃**	
1335	9300	955	11880	1150	2470	707	755
				1205	2805	710	910
Mg₂Sn		**GeTe**		**PbTe**		715	920
778	10390	745	2590	925	1510	720	980
783	10580	750	2675	930	1558	730	820
793	10840	765	2640	940	1580		
803	10890						

Table 1 (concluded)

t, °C	σ, $\Omega^{-1}\cdot cm^{-1}$	t, °C	σ, $\Omega^{-1}\cdot cm^{-1}$	t, °C	σ, $\Omega^{-1}\cdot cm^{-1}$	t, °C	σ, $\Omega^{-1}\cdot cm^{-1}$
Bi₂Se₃		950	1370	630	2680	Sb₂Te₃	
735	800	955	1390	645	2640	625	1850
740	1040	960	1390	650	2700	650	2060
750	1010	970	1485	655	2955	670	2055
760	1040	990	1430	670	2750	700	2130
780	1030	995	1410	685	2780	715	2215
800	1110	1020	1550	705	3090	750	2310
820	1215	1030	1530	720	2800	765	2390
830	1100	1080	1615	750	2880	800	2440
835	1035	1090	1580	800	3000	835	2590
840	1220	1130	1740	805	3020	850	2480
870	1245	1140	1600	830	3050	900	2530
885	1110			850	3100	915	2710
890	1320	Bi₂Te₃		900	3200	950	2670
910	1190	590	2600	950	3300	970	2755
920	1210	595	2600	990	3460	1000	2720
930	1305	610	2615	1000	3480	1025	2810
940	1385	620	2670			1040	2710

Table 2. Thermoelectric Power of Molten Semiconductors

t, °C	α, μV/deg	t, °C	α, μV/deg	t, °C	α, μV/deg	t, °C	α, μV/deg
Ge		853	+6.0	856	+12	Ga₂Te₃	
961	−1.0	901	+4.5	871	+11	839	−91
962	−3.7	960	+4.0	873	+10	846	−101
966	−3.5			934	+11	880	−58
967	−0.9	GaSb		969	+1	902	−75
973	0.0	706	−61	972	0	903	−74
975	−2.0	721	−3	983	−2	940	−46
988	−0.4	726	−4			965	−58
1009	−5.8	727	−2	AlSb		974	−40
1017	−1.3	727	−48	1092	−63.7	1003	−42
1032	−2.9	728	−51	1092	−66.7	1027	−52
1072	−0.3	731	−62	1148	−50.9	1075	−51
1075	−1.7	743	+7	1184	−22.5	In₂Te₃	
1083	−1.2	763	+4			674	−1.0
		765	+4	Ga₂Te₃		683	+13.0
InSb		766	−59	790	−170	684	+27.5
540	−30.0	805	−39	799	−158	686	+19.0
605	−6.0	810	−41	802	−86	690	+28.0
664	−5.0	843	+12	816	−114	691	+47.0
760	−1.0	852	−35	817	−111	704	+22.8
848	+6.0	854	−31	818	−94	707	+19.0
		854	+12				

Table 2 (continued)

In₂Te₃

t, °C	α, μV/deg	t, °C	α, μV/deg
721	+17.3	803	+20.6
728	+21.0	803	+20.4
745	+15.2	819	+21.4
749	+19.6	833	+19.0
758	+19.0	839	+18.7
760	+17.0	847	+22.0
766	+18.2	856	+19.0
770	+11.5	856	+14.5
801	+17.8	877	+16.4
801	+14.7	880	+20.5
804	+17.4	882	+18.0
831	+12.2	890	+17.0
833	+13.6	919	+14.8
840	+18.0	920	+18.9
852	+17.4	922	+15.9
870	+17.5	940	+15.3
947	+17.7	974	+13.8
		981	+14.9

CuI

t, °C	α, μV/deg
606	+0.47
606	+0.48
615	+0.44
622	+0.44
631	+0.65
633	+0.65
633	+0.72
634	+0.62
635	+0.69
639	+0.48
641	+0.48
676	+0.48
680	+0.45
692	+0.46
718	+1.02
723	+0.39
724	+0.39
725	+0.39
759	+0.38
767	+0.80

GeTe

t, °C	α, μV/deg
770	+20.0
777	+22.1
779	+19.5
798	+20.4

SnTe

t, °C	α, μV/deg
801	+41.3
808	+27.3
817	+25.1
822	+30.2
828	+32.8
833	+18.2
867	+18.26
879	+20.5
884	+17.3
925	+14.0
926	+15.0
960	+8.5

PbTe

t, °C	α, μV/deg
913	+1.3
923	−3.7
932	−4.7
950	−5.5
954	−8.0
974	−6.7
975	−10.2
978	−11.2
994	−7 5
1008	−10.2

PbSe

t, °C	α, μV/deg	t, °C	α, μV/deg
1086	−63.0	720	+2.2
1105	−79.0	723	+2.7
1106	−71.0	736	+0.2
1140	−74.0	768	0.0
1142	−67.0	783	+0.3
1188	−63.0	799	+1.4
1201	−62.0	808	+6.6
		835	+3.4
		857	+0.8

PbS

t, °C	α, μV/deg
1146	−217
1149	−168
1156	−254
1168	−174
1189	−251
1209	−195

Bi₂Se₃

t, °C	α, μV/deg
721	−13.0
730	−45.3
782	−32.2
803	−20.4
828	−20.0
848	−12.9
873	−11.7
877	−10.3
911	−0.5
976	+7.4

Bi₂Te₃

t, °C	α, μV/deg
587	−12.8
599	0.0
619	+2.3
620	−1.6
638	+0.4
644	−0.2
660	+2.9
664	−0.2
665	−0.2
674	−0.4
688	+2.5
691	+2.6
695	+3.5

Sb₂Te₃

t, °C	α, μV/deg
623	+10.3
626	+14.5
627	+25.5
630	+10.4
640	+10.0
656	+12.0
658	+17.0
659	+18.9
663	+13.3
666	+9.8
697	+16.1
699	+10.2
710	+13.0
717	+3.7
723	+9.5
747	+11.3
754	+10.3
754	+9.0
758	+10.0
760	+2.6
766	+11.0
770	+8.4
826	+8.0
874	+7.0
897	+9.8
937	+9.0

Table 3. Temperature Dependences of Thermoelectric Power of the $A^{III}B^V$, $A^{IV}B^{VI}$, $A_2^V B_3^{VI}$ Compounds, $\alpha = a + bt$ (α is the thermoelectric power in $\mu V/deg$; t is the temperature in °C; $\Delta\alpha$ is the rms error)

Compound	a	b	$\Delta\alpha$	Compound	a	b	$\Delta\alpha$
AlSb	−533	0.4288	10	PbSe	−174	0.092	7
InSb	−28.2	0 0369	2.7	PbS	−292	0.079	10
GeTe	44.8	−0.0307	1.7	Bi_2Te_3	−7.4	0.0121	1.9
SnTe	84.6	−0.0767	2 2	Bi_3Se_3	−188	0.204	4
PbTe	65.2	−0.0752	1.9	Sb_2Te_3	27.8	−0.0231	3.6

Table 4 Density of Semiconductors in Solid and Liquid States

t, °C	d, g/cm³	t, °C	d, g/cm³	t, °C	d, g/cm³	t, °C	d, g/cm³
Si		1570	2.465	Lucas et al.		122	0.2102
		1594	2.463	[128, 129]		131	0.2105
Solid		1633	2.445			138	0.2108
25	2.3280	Ge		1000	5.46	156	0.2109
100	2 3278	Solid		1100	5.41	168	0.2114
200	2.3275	200	5 3544	1200	5.36	168	0.2115
300	2 3272	240	5.3520	1300	5.31	170	0.2121
400	2.3268	280	5.3444	1400	5.265	182	0.2122
500	2.3264	320	5.3423	1500	5.215	183	0.2121
600	2.3260	360	5.3385	1600	5.17	196	0.2127
700	2.3210	400	5.3332			213	0.2134
800	2.3160	440	5.3275	Present authors			
900	2.3110	480	5.3260	975	5.49	Campbell	
1000	2.306	520	5 3224	975	5 49	et al. [200]	
1100	2.301	560	5.3171	980	5.49	20.4	4.7924
1200	2.296	600	5.3117	1000	5.48	90.0	4.7641
1300	2.291	640	5.3064	1005	5.47	159.5	4.7152
1400	2.286	680	5.3010	1025	5.45	188.5	4.6954
		720	5 3010	1025	5.47	197.5	4.6940
Liquid		760	5.2956	1030	5.46	205.5	4.6919
Lucas et al.		800	5.2900	1055	5.43		
[128, 129]		840	5 2849	1060	5.45		
1410	2.52	880	5.2840	1075	5.41	Present authors	
1450	2.51			1090	5.43	18	4.816
1500	2.49	Liquid		1095	5.425	137	4.739
1550	2.47	Klemm et al. [132]		1110	5.38	155	4.726
1600	2.45	960	5.570	1120	5.41	171	4.718
1650	2.44	960	5.574	Se		180	4.726
		960	5 570	Solid		196	4.687
Present authors		980	5.565	Bonnier et al.[230]•		211	4.675
1455	2.515	1000	5.557	25	0.2077		
1473	2.50	1050	5.532	42	0.2093		
1500	2.49	1100	5.505	100	0.2099		
1530	2.485						

• Data for specific volume.

Table 4 (continued)

t, °C	d, g/cm³	t, °C	d, g/cm³	t, °C	d, g/cm³	t, °C	d, g/cm³
Liquid		186	0.1617	690	5.63	Liquid	
Campbell et al.		253	0.1625	740	5.59	Sauerwald [635]*	
[200]		271	0.1626	788	5.57		
		295	0.1629	800	5.54		
225	3.9705	295	0.1628	848	5.52	1091	0.2134
241	3.9540	306	0.1628	884	5.49	1092	0.2134
244.5	3.9394	329	0.1631	977	5.41	1096	0.2147
257	3.9276	337	0.1633	985	5.41	1132	0.2108
277	3.8904	337	0.1632	1000	5.38	1135	0.2123
		356	0.1634	AlSb		1135	0.2113
Lucas et al. [229]		378	0.1637	Solid		1136	0.2113
250	3.97	394	0.1638	Sauerwald [635]*		1194	0.2157
300	3.91	398	0.1638				
350	3.85	420	0.1642	885	0.2398	Present authors	
400	3.79	428	0.1642	911	0.2387	1080	4.72
				990	0.2368	1090	4.74
Dobinskii et al.		Present authors		1012	0.2259	1100	4.72
[228]		130	6.213	1017	0.2237	1120	4.69
228	3.972	165	6.193	1017	0.2213	1150	4.64
235	3.962	192	6.187			1155	4.52
246	3.946	225	6.180	Present authors			
252	3.936	245	6.173	100	4.258	GaSb	
263	3.923	275	6.164	160	4.252	Solid	
274	3.902	290	6.160	172	4.252	112	5.672
283	3.885	335	6.153	200	4.250	120	5.671
294	3.868	370	6.137	212	4.250	172	5.666
303	3.853	400	6.137	224	4.248	200	5.663
315	3.834	450	6.115	240	4.248	208	5.660
328	3.814			268	4.246	248	5.658
337	3.800	Liquid		316	4.243	268	5.651
345	3.784	Regel' et al. [16]		360	4.239	292	5.652
		350	5.717	404	4.236	304	5.650
Present authors		400	5.733	412	4.236	340	5.642
230	3.97	450	5.738	456	4.233	368	5.640
287	3.90	500	5.737	508	4.229	388	5.635
356	3.82	550	5.733	528	4.228	410	5.637
425	3.76	600	5.729	552	4.227	420	5.631
503	3.67	650	5.722	600	4.224	458	5.630
526	3.64	700	5.716	644	4.221	476	5.625
575	3.59	750	5.710	670	4.221	500	5.622
631	3.54	800	5.703	680	4.219	520	5.623
666	3.50	850	5.696	760	4.214	552	5.616
738	3.42	900	5.686	788	4.216	584	5.615
819	3.35			800	4.211	600	5.609
		Present authors		810	4.214	624	5.607
Te		492	5.77	836	4.209	648	5.608
Solid		527	5.76	860	4.209	656	5.603
Bonnier et al.[201]*		564	5.76	892	4.207	680	5.599
25	0.1602	584	5.71	896	4.204	692	5.589
110	0.1610	596	5.70	920	4.202	692	5.601
		610	5.71	928	4.204		
				972	4.202		

*Data for specific volume.

Table 4 (continued)

t, °C	d, g/cm³	t, °C	d, g/cm³	t, °C	d, g/cm³	t, °C	d, g/cm³
Liquid		292	5.784			784	5.513
		316	5.782	652	5.230	852	5.508
720	6.03	348	5.778	668	5.228	854	5.513
720	6.01	360	5.778	688	5.226	880	5.510
740	6.04	388	5.775	714	5.223	908	5.504
750	6.01	408	5.774	716	5.223		
755	5.99	440	5.769	764	5.218	Liquid	
775	5.97	444	5.771	796	5.215		
790	5.99	468	5.769	800	5.216	980	5.85
800	6.02	480	5.768	824	5.211	1020	5.78
805	5.95			848	5.209	1030	5.81
810	5.99	Liquid		872	5.205	1050	5.75
820	5.98			880	5.209	1085	5.73
830	6.00	550	6.43	896	5.204		
850	5.97	580	6.42	936	5.198	CuI	
850	5.97	594	6.43	956	5.196	Solid	
850	5.93	620	6.41	968	5.194		
870	5.95	630	6.42	996	5.192	100	5.648
870	5.90	655	6.40	1004	5.190	105	5.643
880	5.98	666	6.41	1064	5.184	120	5.640
895	5.93	690	6.38	1088	5.182	160	5.628
900	5.94	710	6.36			180	5.616
910	5.89	720	6.33			220	5.600
920	5.92	755	6.31	Liquid		242	5.591
940	5.87	770	6.31			264	5.582
950	5.87	780	6.28	1245	5.71	284	5.517
955	5.91	820	6.27	1320	5.63	305	5.565
975	5.86					325	5.536
985	5.86	GaAs		InAs		350	5.520
990	5.88	Solid		Solid		355	5.506
1000	5.88					358	5.511
1005	5.85	120	5.275	106	5.575	365	5.495
1035	5.85	160	5.274	112	5.574	407	5.483
1040	5.84	180	5.273	170	5.569	410	5.466
1045	5.86	216	5.269	194	5.568	450	5.450
1050	5.83	252	5.267	210	5.566	468	5.440
1060	5.82	282	5.264	252	5.562	468	5.450
1090	5.84	296	5.264	280	5.559	490	5.434
		340	5.259	290	5.558	505	5.416
InSb		348	5.258	328	5.555		
Solid		372	5.256	352	5.552	Liquid	
		388	5.255	404	5.549		
104	5.803	400	5.254	440	5.545	613	4.820
120	5.800	432	5.251	448	5.545	630	4.821
148	5.799	436	5.250	490	5.541	640	4.780
152	5.797	464	5.248	520	5.543	645	4.780
172	5.796	488	5.245	532	5.537	664	4.770
196	5.795	524	5.243	556	5.537	665	4.769
212	5.792	568	5.238	612	5.529	683	4.749
220	5.790	580	5.237	680	5.524	717	4.720
268	5.786	612	5.234	694	5.525	720	4.720
272	5.787	616	5.233	744	5.518	754	4.685
				772	5.519	755	4.680
						763	4.679

Table 4 (continued)

t, °C	d, g/cm³	t, °C	d, g/cm³	t, °C	d, g/cm³	t, °C	d, g/cm³
Liquid		1017	5.094	605	5.787	185	5.412
		1020	5.088	605	5.781	210	5.403
770	4.670	1070	5.076	612	5.773	225	5.405
787	4.659	1120	5.059	622	5.781	250	5.400
792	4.660	1130	5.063	640	5.780	252	5.395
807	4.647	1140	5.054			275	5.409
810	4.578	1175	5.034	Liquid		285	5.389
824	4.608	1225	5.044			295	5.391
825	4.610	1225	5.021	680	5.559	330	5.383
				700	5.564	345	5.392
Ga₂Te₃		In₂Te₃		720	5.564	365	5.379
Solid		Solid		735	5.567	370	5.392
100	5.477	112	5.883	755	5.578	375	5.375
100	5.486	117	5.895	770	5.576	387	5.371
140	5.472	140	5.863	790	5.566	405	5.382
160	5.467	145	5.874	806	5.578	412	5.387
203	5.461	147	5.888	807	5.566	420	5.372
220	5.459	180	5.848	830	5.581	422	5.385
267	5.446	182	5.865	860	5.578	430	5.371
285	5.447	187	5.878	875	5.566	440	5.377
305	5.434	215	5.856	878	5.566	455	5.360
347	5.425	217	5.872	880	5.581	465	5.372
387	5.417	222	5.845	912	5.567	480	5.378
407	5.428	250	5.848	913	5.578	490	5.375
417	5.409	262	5.836	930	5.566	495	5.365
467	5.419	282	5.851	965	5.559	500	5.361
475	5.398	285	5.837	968	5.568	510	5.357
510	5.390	317	5.841	1000	5.546	520	5.364
530	5.408	325	5.830	1015	5.538	522	5.351
557	5.379	332	5.825	1020	5.489	550	5.369
582	5.373	350	5.830	1050	5.528	555	5.358
593	5.398	382	5.812	1070	5.519	557	5.363
637	5.365	390	5.822	1070	5.567	560	5.349
655	5.390	408	5.809	1085	5.502	565	5.350
673	5.354	420	5.805	1102	5.519	587	5.336
717	5.349	435	5.809	1115	5.491	600	5.344
720	5.377	452	5.797	1120	5.479	610	5.351
740	5.346	470	5.797	1123	5.499	610	5.342
760	5.377	470	5.796	1140	5.545	615	5.359
760	5.346	477	5.792	1145	5.485	625	5.357
		500	5.790	1175	5.464	660	5.335
Liquid		510	5.790	1176	5.457	675	5.326
807	5.094	520	5.790	1205	5.424	680	5.341
860	5.094	530	5.787	1230	5.431	692	5.344
864	5.088	550	5.784			700	5.328
913	5.094	552	5.785	GaTe		710	5.324
915	5.084	557	5.789	Solid		715	5.337
920	5.088	570	5.784	150	5.412	727	5.311
965	5.094	570	5.778	150	5.433	727	5.339
968	5.088	587	5.783	150	5.426	750	5.322
1015	5.084	592	5.773	165	5.420	752	5.338

Table 4 (continued)

t, °C	d, g/cm³
760	5.306
760	5.331
777	5.317
780	5.328
795	5.315
795	5.335
805	5.324
805	5.333
Liquid	
835	5.16
865	5.14
870	5.13
885	5.11
895	5.07
900	5.07
905	5.07
920	5.06
925	5.06
930	5.05
960	5.02
960	5.03
980	5.00
InTe	
Solid	
135	5.588
150	6.007
170	5.996
175	5.988
200	5.986
220	5.979
230	5.974
242	5.971
245	5.968
260	5.966
285	5.965
295	5.950
310	5.949
335	5.948
345	5.944
360	5.938
390	5.932
412	5.933
425	5.923
455	5.912
480	5.920
490	5.911
510	5.904
515	5.916
520	5.929

t, °C	d, g/cm³
550	5.907
575	5.895
582	5.901
615	5.896
650	5.885
650	5.890
Liquid	
715	5.64
775	5.62
825	5.59
865	5.58
875	5.57
885	5.61
905	5.55
915	5.53
920	5.55
925	5.58
940	5.54
995	5.48
Mg₂Si	
Solid	
142	1.874
150	1.874
155	1.871
167	1.870
180	1.869
185	1.870
210	1.865
217	1.866
235	1.864
240	1.862
250	1.862
270	1.858
275	1.859
295	1.858
300	1.855
315	1.856
330	1.855
345	1.853
360	1.849
420	1.846
430	1.845
480	1.842
507	1.839
540	1.837
570	1.836
600	1.835
630	1.835

t, °C	d, g/cm³
660	1.835
690	1.835
720	1.835
Liquid	
1130	2.25
1140	2.23
1150	2.20
1160	2.15
1170	2.06
Mg₂Sn	
Solid	
142	3.527
160	3.525
170	3.519
190	3.517
195	3.512
200	3.514
210	3.510
217	3.508
225	3.509
242	3.505
260	3.504
265	3.502
295	3.500
300	3.497
310	3.496
330	3.494
335	3.494
345	3.491
352	3.489
357	3.491
365	3.490
380	3.488
395	3.486
395	3.482
412	3.483
422	3.483
425	3.480
430	3.481
448	3.478
465	3.475
468	3.480
480	3.470
482	3.476
485	3.473
500	3.475
507	3.470
515	3.468

t, °C	d, g/cm³
522	3.464
535	3.471
535	3.466
550	3.469
565	3.465
575	3.468
595	3.463
600	3.466
620	3.462
625	3.464
640	3.461
650	3.462
667	3.461
675	3.459
692	3.459
697	3.458
700	3.458
715	3.456
727	3.457
745	3.455
Liquid	
800	3.51
850	3.49
900	3.45
950	3.42
1090	3.40
Mg₂Pb	
Solid	
142	5.079
150	5.076
160	5.085
175	5.065
192	5.062
200	5.056
210	5.053
225	5.049
235	5.044
242	5.044
250	5.042
260	5.039
270	5.041
285	5.035
295	5.039
310	5.034

Table 4 (continued)

t, °C	d, g/cm³	t, °C	d, g/cm³	t, °C	d, g/cm³	t, °C	d, g/cm³
Liquid		524	6.058	484	6.251	268	8.046
		560	6.041	504	6.256	292	8.035
600	5.17	564	6.039	508	6.242	304	8.021
650	5.15	612	6.019	532	6.232	328	8.014
700	5.13	640	6.013	568	6.216	360	7.994
750	5.05	648	6.001	572	6.216	384	7.983
800	5.02	668	6.993	592	6.207	412	7.968
880	4.94	670	6.008	620	6.192	420	7.963
GeTe		684	5.982	626	6.190	456	7.946
Solid		Liquid		656	6.179	472	7.937
108	6.179			688	6.165	504	7.925
114	6.179	816	5.52	700	6.163	512	7.918
140	6.168	844	5.50	716	6.155	552	7.897
146	6.167	872	5.48	744	6.153	600	7.879
148	6.169	880	5.45	746	6.147	637	7.860
150	6.164	892	5.46	764	6.148	644	7.847
176	6.158	920	5.44	768	6.145	675	7.841
176	6.159	926	5.44			684	7.816
200	6.153	944	5.43	Liquid		728	7.813
204	6.155	960	5.42			732	7.802
210	6.151	988	5.42	808	5.87	764	7.797
220	6.148	996	5.39	815	5.87	768	7.779
236	6.143	1008	5.39	844	5.83	800	7.764
248	6.144	1045	5.35	848	5.84	812	7.769
256	6.140	1048	5.38	850	5.85	824	7.749
276	6.133	1085	5.36	880	5.83	844	7.754
296	6.132	1088	5.36	884	5.81	860	7.745
300	6.129	1145	5.35	890	5.83	876	7.736
312	6.125			920	5.78	880	7.719
316	6.127	SnTe		924	5.81	888	7.731
328	6.124	Solid		930	5.80		
338	6.122			950	5.79	Liquid	
340	6.120	100	6.427	956	5.77		
344	6.119	112	6.422	964	5.76	930	7.42
352	6.118	136	6.412	992	5.74	955	7.41
352	6.117	144	6.406	1008	5.72	960	7.40
360	6.117	176	6.391	1032	5.70	975	7.37
376	6.118	192	6.386	1040	5.68	985	7.34
386	6.116	208	6.376	1080	6.67	1000	7.33
388	6.112	220	6.374			1020	7.32
408	6.108	248	6.355	PbTe		1025	7.31
424	6.098	280	6 342	Solid		1060	7.28
432	6.098	288	6.377			1080	7.26
440	6.094	324	6.320	104	8.135	1090	7.25
456	6.084	340	6.318	112	8.128	1110	7.22
460	6.085	352	6.306	136	8.122	1125	7.20
476	6.078	384	6.294	160	8.103	1140	7.24
480	6 076	400	6.289	172	8.102	1160	7.20
492	6.071	412	6.282	188	8.089	1199	7.17
520	6.058	444	6.266	220	8.069		
		456	6.268	256	8.052		

Table 4 (continued)

t, °C	d, g/cm³	t, °C	d, g/cm³	t, °C	d, g/cm³	t, °C	d, g/cm³
PbSe		740	7.752	410	7.309	160	7.453
Solid		765	7.740	440	7 300	184	7.446
		775	7.720	450	7 300	188	7.448
100	8.053	810	7.715	470	7.298	224	7.429
120	8.050	825	7.700	500	7.295	260	7.414
150	8.025	865	7 700	513	7.289	268	7.417
165	8.015	875	7.678	515	7.268	290	7.410
200	8.005	875	7.690	520	7.263	316	7.404
205	7.995	880	7.700	547	7.268	352	7.391
215	7.998	910	7.682	570	7.253	364	7.391
242	7.950	925	7.675	585	7.245	370	7.377
243	7.955	945	7.670	615	7.245	396	7.383
250	7.958	980	7 648	620	7.227	416	7.370
260	7.953	1000	7.642	622	7.252	436	7.370
280	7.951	1020	7.625	650	7.218	520	7.336
305	7.945	1025	7.615	655	7.225	564	7.326
315	7.945	1030	7.600	675	7.225	564	7.304
320	7.941	1040	7.602	720	7.205		
345	7.910	1050	7.592	725	7.190	**Liquid**	
350	7.918	1060	7.598	755	7.175		
355	7.910	1060	7.598	770	7.188	724	6.86
355	7.921	1075	7 580	800	7.185	760	6.84
365	7.918			815	7.150	770	6.86
400	7.900			825	7 175	820	6.84
410	7.890	**Liquid**		830	7.150	870	6.81
417	7.885	1100	7.08	930	7.130	885	6 75
430	7.881	1120	7.07	965	7.121	950	6.75
435	7.880	1140	7.05	980	7.125	965	6.70
450	7.865	1150	7.04	1025	7.105	1010	6.70
455	7.868	1160	7.03	1035	7.105	1015	6.65
460	7.868	1165	7.03	1085	7.091	1090	6.62
470	7.865	1190	7.01	1115	7.072	1100	6.57
505	7.851	1210	7.00				
510	7.850	1210	6.99	**Liquid**		**Bi₂Te₃**	
512	7.850			1120	6.42	**Solid**	
515	7.845			1140	6.44	90	7.692
520	7.845	**PbS**		1155	6.43	112	7.688
540	7.846	**Solid**		1160	6.44	125	7.678
555	7.812			1165	6.44	140	7.675
562	7.800	100	7.450	1170	6.40	162	7.666
563	7.818	110	7.445	1170	6.40	176	7.662
570	7.827	150	7.408	1220	6.37	180	7.661
590	7.805	160	7.403	1240	6.32	204	7.651
605	7.803	205	7.400	1250	6.34	208	7.651
615	7.800	230	7.383			210	7.652
620	7.800	255	7.375	**Bi₂Se₃**		248	7.639
625	7.795	310	7.350	**Solid**		268	7.634
640	7.798	315	7.350	96	7.469	300	7.621
675	7.789	325	7 350	120	7.467	312	7.619
710	7.767	365	7 341	120	7.468	332	7.619
715	7.752	385	7.325	152	7.453	352	7.608
720	7.753	392	7.332				

Table 4 (concluded)

t, °C	d, g/cm³	t, °C	d, g/cm³	t, °G	d, g/cm³	t, °C	d, g/cm³
354	7.605	666	7.18	146	6 396	544	6.304
372	7.609	668	7.18	152	6.397	576	6.297
400	7.595	685	7.23	184	6.389		
408	7.587	700	7.16	220	6.379	Liquid	
412	7.599	708	7.15	264	6.364	630	6.10
440	5.579	740	7.12	268	6.366	660	6.08
448	7.573	750	7.15	306	6.356	720	6.05
460	7.585	768	7.09	312	6.349	755	6.02
484	7.554	788	7.07	352	6.344	780	6.06
490	7.564	826	7.04	390	6.339	790	6.03
500	7.564	835	7.08	400	6.333	840	5.97
520	7.535	864	7.03	426	6.329	860	6.00
536	7.546	892	6.99	432	6.328	910	5.94
542	7.525	910	7.03	462	6.319	920	5.94
		1020	6.94	468	6.319	930	5.93
				504	6.310	965	5.93
		Sb₂Te₃		504	6.317	970	5.92
				536	6.303	1015	5.90
Liquid		Solid				1020	5.80
630	7.25	114	6.403				
632	7.22	118	6.406				

Table 5. Temperature Dependences of the Density of the Mg_2B^{IV}, $A^{IV}B^{VI}$, and $A_2^{V}B_3^{VI}$ Compounds: $d = a + bt$ (d is the density in g/cm³; t is the temperature in °C; Δd is the rms error)

Com-pound	Solid state			Liquid state		
	a	$-b \times 10^3$	Δd	a	$-b \times 10^3$	Δd
Mg₂Si	—	—	—	7.699	4.799	0.0279
Mg₂Sn	—	—	—	3.976	0.580	0.0074
Mg₂Pb	—	—	—	5.703	0.858	0.0207
GeTe	6.224	0.312	0.0075	5.869	0.459	0.0115
SnTe	6.467	0.429	0.0057	6.514	0.786	0.0144
PbTe	8.192	0.538	0.0029	8.181	0.843	0.0148
PbSe	8.068	0.441	0.0062	7.696	0.575	0.0065
PbS	7.465	0.352	0.0079	7.529	0.958	0.0234
Bi₂Se₃	7.502	0.308	0.0038	7.521	0.819	0.0091
Bi₂Te₃	7.803	0.537	0.0548	7.772	0.819	0.0135
Sb₂Te₃	6.426	0.230	0.0027	6.448	0.546	0.0141

Table 6. Viscosity of Molten Semiconductors (in centistokes)

t, °C	ν, cS
Si	
1420	0.348
1430	0.328
1440	0.315
1447	0.300
1460	0.285
1480	0.275
1497	0.265
1540	0.250
1560	0.245
1600	0.233
1630	0.230
Ge	
950	0.135
975	0.125
1000	0.119
1025	0.114
1050	0.108
1080	0.1065
1120	0.102
1150	0.100
1210	0.096
Se	
Khalilov [249]	
220	5.56
230	3.69
240	2.49
260	1.53
280	0.948
300	0.612
320	0.398
340	0.276
360	0.182
Dobinski et al. [250]	
234	3.18
254.8	1.64
281.7	0.784
289.0	0.662
296.4	0.565
303.4	0.487
318.1	0.349
337	0.239
345	0.203
Present authors	
230	6.63
250	2.67
270	2.12
320	0.582
370	0.184
445	0.0883
525	0.035
Te	
460	0.357
470	0.301
485	0.296
500	0.282
505	0.290
555	0.283
610	0.222
670	0.171
730	0.160
770	0.156
785	0.154
790	0.159
850	0.156
860	0.154
900	0.148
925	0.138
960	0.142
980	0.149
1080	0.137
1100	0.137
AlSb	
1080	0.250
1090	0.230
1100	0.215
1120	0.198
1150	0.180
1200	0.156
1225	0.146
1235	0.143
1245	0.142
1260	0.123
1300	0.094
1350	0.074
GaSb	
730	0.368
740	0.350
760	0.332
780	0.321
835	0.310
900	0.296
940	0.280
1000	0.260
1050	0.224
1100	0.200
1180	0.159
InSb	
540	0.363
550	0.351
565	0.340
575	0.331
600	0.320
630	0.302
650	0.292
660	0.286
670	0.280
700	0.254
750	0.218
775	0.188
800	0.174
GaAs	
1285	0.320
1290	0.297
1295	0.280
1305	0.266
1310	0.260
1325	0.254
1335	0.250
1350	0.247
1365	0.246
InAs	
995	0.174
950	0.166
960	0.160
970	0.155
980	0.150
1000	0.142
1010	0.139
1050	0.127
1100	0.117
1175	0.107
1250	0.104
1270	0.104
1300	0.104
CuI	
605	0.432
650	0.375
700	0.350
750	0.330
800	0.300
900	0.255
1000	0.220
GaTe	
830	0.730
840	0.693
895	0.889
935	0.571
995	0.442
1035	0.385
1050	0.372
1080	0.352
1120	0.339
1185	0.314
InTe	
705	0.880
730	0.838
770	0.700
815	0.706
830	0.620
860	0.640
930	0.530
940	0.520
985	0.470
1085	0.336
ZnTe	
1262	0.868
1280	0.782
1317	0.636
1332	0.598
1352	0.560
1380	0.543
CdTe	
1070	0.435
1097	0.413
1120	0.397
1140	0.380
1160	0.365
1170	0.360
1190	0.351
1212	0.340
1235	0.330
Ga_2Te_3	
790	0.546
800	0.690
810	0.755
825	0.820
835	0.867
845	0.897
855	0.877

Table 6 (concluded)

t, °C	ν, cS	t, °C	ν, cS	t, °C	ν, cS	t, °C	ν, cS
890	0.838	1170	0.283	850	0.262	**PbS**	
900	0.838	1180	0.281	895	0 246	1105	0.319
925	0.782	1200	0.278	900	0.250	1120	0.281
935	0 782	1210	0.262	970	0.215	1125	0.292
970	0.695	1230	0.249	1035	0.188	1140	0.264
995	0.682	**Mg₂Sn**		1080	0.180	1150	0.252
1000	0.647	780	0.520	1135	0.169	1160	0.233
1040	0.607	800	0.460	**SnTe**		1170	0.258
1070	0.576	810	0.445	830	0.348	1180	0.234
1080	0.557	820	0.430	865	0.326	1190	0.205
1085	0.517	850	0.395	915	0.268	1210	0.237
1100	0.503	870	0.375	965	0.216	1240	0.252
1130	0.449	890	0.365	1010	0.202	**Bi₂Se₃**	
1160	0.428	895	0.355	1050	0.198	720	0.540
1245	0 426	900	0.362	1090	0.189	750	0.484
In₂Te₃		930	0.350	1130	0.184	780	0.425
675	0.323	940	0.340	1170	0.182	810	0.355
680	0.501	960	0.330	1210	0.175	840	0.342
690	0.834	970	0.330	**PbTe**		870	0.314
700	0.851	1000	0.322	940	0.243	900	0.270
710	0.905	**Mg₂Pb**		970	0.216	930	0.253
715	0.844	560	*0.560	980	0.217	960	0.233
720	0.915	565	0.530	1000	0.203	**Bi₂Te₃**	
740	0.903	580	0.510	1020	0.190	600	0.198
765	0.864	590	0.480	1030	0.192	630	0.180
790	0.773	600	0.470	1060	0.173	650	0.169
820	0.707	610	0.460	1060	0.171	660	0.158
825	0.660	630	0.445	1090	0.168	690	0.149
850	0.594	650	0.420	1100	0.157	700	0.147
870	0.537	660	0.400	1120	0.158	720	0.140
915	0.404	670	0.390	1140	0.142	750	0 138
935	0.365	680	0.375	1180	0.134	755	0.137
950	0.328	700	0.360	1190	0.146	780	0.134
980	0.290	720	0.340	1230	0 134	815	0.128
985	0.300	740	0.325	**PbSe**		850	0.123
1015	0.213	750	0.322	1060	0.240	900	0.114
1050	0.197	800	0.280	1080	0.220	950	0 119
Mg₂Si		810	0.275	1095	0.182	1000	0.117
1120	0.299	820	0.270	1100	0.205	1050	0.116
1140	0.284	870	0.240	1110	0.193	1100	0.115
1160	0.249	900	0.222	1125	0.194	1100	0.111
1190	0.233	950	0.200	1130	0.182	**Sb₂Te₃**	
1220	0.222	**GeTe**		1140	0.167	650	0.513
1260	0.195	725	0.375	1150	0.162	750	0.425
1320	0.180	730	0.363	1160	0.176	820	0.344
Mg₂Ge		740	0.344	1170	0.158	890	0.267
1140	0.311	755	0 324	1190	0.152	970	0.216
1150	0.299	775	0.323	1200	0.164	1080	0.206
		800	0.295	1210	0.153		
				1240	0.158		

Table 7. Temperature Dependences of the Viscosity of the
Mg_2B^{IV}, $A^{IV}B^{VI}$, and $A_2^V B_3^{VI}$ Compounds: $\log \nu = A + B/T$
(ν is the viscosity, in centistokes; T is the temperature in °K;
$\Delta \nu$ is the rms error)

Compound	$-A$	B	$\Delta \nu$	Compound	$-A$	B	$\Delta \nu$
Mg_2Si	2.297	2452	0.0170	PbTe	1.869	1496	0.0106
Mg_2Ge	1.882	1933	0.0092	PbSe	2.854	2962	0.0334
Mg_2Sn	1.466	1207	0.0186	PbS	0.595	−13.64	0.0519
Mg_2Pb	1.585	1104	0.0090	Bi_2Se_3	1.751	1274	0.0068
GeTe	1.618	1177	0.0126	Bi_2Te_3	1.679	819.6	0.0298
SnTe	1.681	1319	0.0351				

Literature Cited

1. N. S. Kurnakov, Introduction to Physicochemical Analysis, 3rd ed. Onti-Khimteoret, Leningrad (1936).
2. Ya. I. Frenkel', Kinetic Theory of Liquids (Collection of Selected Papers). Izd. AN SSSR, Moscow (1959), Vol. 3.
3. A. R. Regel', Investigations of Electronic Conductivity of Liquids. Doctoral Dissertation. Leningrad (1956).
4. N. S. Kurnakov, Selected Works, Vol. 1. Izd. AN SSSR, Moscow (1960).
5. A. R. Regel', Collection: Structure and Physical Properties of Matter in the Liquid State. Kiev State University (1954).
6. A. R. Regel', Collection: Theoretical and Experimental Problems in Semiconductors and Semiconductor Metallurgy. Izd. AN SSSR, Moscow (1955).
7. A. R. Regel', Uch. zap. Leningr. gos. ped. in-ta im. A. I. Gertsena, 197: 187 (1958).
8. A. F. Ioffe and A. R. Regel', Progress in Semiconductors, 4: 237 (1960).
9. A. P. Zhuze and A. R. Regel', Proc. Intern. Conf. on Semiconductor Physics, Prague, 1960 . Prague (1961), p. 929.
10. A. I. Blum, N. P. Mokrovskii, and A. R. Regel', Zh. Tekhn. Fiz., 21: 237 (1951).
11. A. I. Blum, N. P. Mokrovskii, and A. R. Regel', Izv. AN SSSR, ser. fiz., 16: 139, 155 (1952).
12. N. P. Mokrovskii and A. R. Regel', Zh. Tekhn. Fiz., 22: 1281 (1952).
13. N. P. Mokrovskii and A. R. Regel', Zh. Tekhn. Fiz., 23: 779 (1953).
14. A. I. Blum and A. R. Regel', Zh. Tekhn. Fiz., 23: 783 (1953).
15. A. I. Blum and A. R. Regel', Zh. Tekhn. Fiz., 23: 964 (1953).
16. N. P. Mokrovskii and A. R. Regel', Zh. Tekhn. Fiz., 25: 2093 (1955).
17. A. R. Regel', Collection: Structure and Properties of Liquid Metals. A. A. Baikov Metallurgy Institute (1959), p. 3.
18. M. S. Ablova, O. D. Elpat'evskaya, and A. R. Regel', Zh. Tekhn. Fiz., 26: 1366 (1956).
19. V. M. Glazov, Investigations Relating to Physicochemical Analysis of Semiconductors in the Liquid State. Doctoral Dissertation. MISIS (1966).
20. V. M. Glazov, Izv. AN SSSR, OTN, No. 6: 111 (1960).

21. V. M. Glazov and D. A. Petrov, Izv. AN SSSR, OTN, No. 2, p. 15 (1958).

22. V. M. Glazov and D. A. Petrov, Izv. AN SSSR, OTN, No. 4, p. 125 (1958).

23. D. A. Petrov and V. M. Glazov, Dokl. AN SSSR, 120: 293 (1958).

24. S. N. Chizhevskaya, Investigations of Physicochemical Properties of Semiconducting Compounds with ZnS-type Lattice near the Melting Point and in the Liquid State. Dissertation for Candidate's Degree. A. A. Baikov Metallurgy Institute (1963).

25. V. M. Glazov and S. N. Chizhevskaya, Zh. neorg. khim., 9: 759 (1964).

26. V. M. Glazov and S. N. Chizhevskaya, Dokl. AN SSSR, 154: 193 (1964).

27. V. M. Glazov and S. N. Chizhevskaya, Izv. AN SSSR, OTN, No. 3, p. 154 (1961).

28. S. N. Chizhevskaya and V. M. Glazov, Dokl. AN SSSR, 145: 115 (1962).

29. S. N. Chizhevskaya and V. M. Glazov, Zh. neorg. khim., Vol. 7, No. 8 (1962).

30. A. N. Krestovnikov et al., Physical Chemistry of Metallurgical Processes and Systems, Trudy MISiS, No. 41, p. 135 (1965).

31. A. F. Ioffe, Jubilee Collection of the Academy of Sciences of the USSR on the Occasion of the Thirtieth Anniversary of the Great October Socialist Revolution. Izd. AN SSSR (1947), Part I, p. 305.

32. A. F. Ioffe, Physics of Semiconductors. Izd. AN SSSR, Moscow (1957).

33. Ya. I. Frenkel', Selected Works. Izd. AN SSSR (1959).

34. I. Z. Fisher, Statistical Theory of Liquids. Izd. AN BSSR, Minsk (1962).

35. V. M. Glazov, A. A. Vertman, and E. G. Shvidkovskii, Izv. AN SSSR, OTN, Met. i topl., No. 3, p. 104 (1961).

36. I. B. Cadoff and E. Miller, Thermoelectric Materials and Devices. Reinhold, New York (1960).

37. J. W. Garner, Engl. Electric. J., 18: 16 (1963).

38. G. G. Urazov, Zh. Russk. fiz.-khim. ob-va, 51: 461 (1919).

39. G. G. Urazov, Izv. in-ta fiz.-khim. analiza, 1: 58 (1921).

40. W. Guertler and A. Bergman, Z. Metallkunde, 25: 81 (1933).

41. D. A. Petrov et al., Collection: Problems in Metallurgy and Physics of Semiconductors. Izd. AN SSSR, Moscow (1957), p. 70.

42. G. N. Nikolaenko, Collection: Problems in Metallurgy and Physics of Semiconductors. Izd. AN SSSR, Moscow (1957), p. 80.

43. V. M. Glazov, Zavodsk. lab., 24: 824 (1958).

44. N. Kh. Abrikosov, L. V. Poretskaya, and I. P. Ivanova, Zh. neorg. khim., 4: 2525 (1959).

45. E. Miller, K. Komarek, and J. Cadoff, Trans. Met. Soc. AIME, 215: 882 (1959).

46. G. Offergeld and J. Van Cakenberghe, J. Phys. Chem. Solids, 11: 310 (1959).

47. J. McHugh and W. Tiller, Trans. Met. Soc. AIME, 218: 187 (1960).

48. L. V. Poretskaya, N. Kh. Abrikosov, and V. M. Glazov, Zh. neorg. khim., 8: 1196 (1963).

49. N. Kh. Abrikosov and L. E. Glukhikh, Zh. neorg. khim., 8: 1792 (1963).

50. L. E. Shelimova and N. Kh. Abrikosov, Zh. neorg. khim., 9: 1879 (1964).

51. L. E. Shelimova, N. Kh. Abrikosov, and V. V. Zhdanova, Zh. neorg. khim., 10: 1200 (1965).

52. A. A. Popov, Collection: Problems in Metallography and Heat Treatment. Mashgiz, Moscow (1956).

53. P. W. Bridgman, Phys. Rev., 27 : 61 (1921).

54. H. Perlitz, Phil. Mag., 2 : 1148 (1926).

55. N. F. Mott, Proc. Roy. Soc. (London), A146 : 465 (1934).

56. F. Gaibullaev, Uch. zap. Namangansk. Gos. Ped. in-ta, No. 2, p. 177 (1957).

57. P. W. Selwood, Magnetochemistry, 2nd ed. Interscience, New York (1956).

58. A. A. Vertman and A. M. Samarin, Zavodsk. lab., 24 : 309 (1958).

59. A. N. Krestovnikov and V. N. Vigdorovich, Chémical Thermodynamics. Metallurgizdat, Moscow (1964).

60. G. Ulick, Chemical Thermodynamics [Russian translation]. Goskhimtekhizdat, Leningrad (1933).

61. V. G. Livshits, Physical Properties of Metals and Alloys. Metallurgizdat, Moscow (1960).

62. A. I. Bachinskii, Vremennik ob-va im. Ledentsova, prilozhenie, No. 3 (1913).

63. A. I. Bachinskii, Selected Works. Izd. AN SSSR, Moscow (1960).

64. E. Stewart, Phys. Rev., 9 : 1575 (1931).

65. J. Bernal, Trans. Faraday Soc., 33 : 27 (1937).

66. G. M. Panchenkov, Theory of Viscosity of Liquids. Gostoptekhizdat, Moscow (1947).

67. S. D. Ravikovich, Collection: Structure and Physical Properties of Matter in the Liquid State. Kiev State University (1954), p. 87.

68. A. Z. Golik and D. N. Karlikov, Dokl. AN SSSR, 114 : 361 (1957).

69. V. M. Glazov, Izv. AN SSSR, OTN, Met. i topl., No. 5, p. 190 (1960).

70. E. Hatschek, Viscosity of Liquids [Russian translation]. ONTI, Moscow (1935).

71. G. Barr, A Monograph of Viscometry. London (1931).

72. M. P. Volarovich, Izv. sektora fiz.-khim. analiza, 8 : 125 (1936).

73. S. V. Sergeev, Physicochemical Properties of Liquid Metals. Oborongiz, Moscow (1952).

74. E. G. Shvidkovskii, Some Problems in Viscosity of Molten Metals. Gostekhteorizdat, Moscow (1955).

75. O. Meyer, Ann. Physik, 43 : 1 (1891).

76. A. R. Regel', Zh. tekhn. fiz., 18 : 1511 (1948).

77. D. A. Petrov and V. M. Glazov, Zavodsk. lab., 24 : 34 (1958).

78. G. I. Goryaga, Vestn. MGU, seriya fiz.-mat. i est. nauk, No. 1, p. 79 (1956).

79. G. I. Goryaga and É. P. Belozerova, Vestn. MGU, seriya mat., fiz., khim., No. 1, p. 133 (1958).

80. Liquid Metals Handbook. New York (1950), p. 31.

81. O. P. Astakhov and V. V. Lobankov, Izmerit. tekhn., No. 9, p. 15 (1965).

82. V. M. Glazov, V. A. Evseev, and V. G. Pavlov, Zavodsk. lab., 32 : 290 (1966).

83. V. M. Glazov, A. N. Krestovnikov, and V. A. Evssev, Zavodsk. lab., 32 : 110 (1966).

84. A. I. Blum and G. P. Ryabtsova, Fiz. tverd. tela, 1 : 761 (1959).

85. Ya. I. Dutchak and V. Ya. Prokhorenko, Fiz. tverd. tela, 6 : 3172 (1964).

86. M. Shtenbek and P. I. Baranskii, Zh. tekhn. fiz., 26 : 1373 (1956);

87. V. A. Osipova and V. I. Fedorova, Teplofiz. vysokikh temperatur, 3 : 228 (1965).

88.   M. A. Vukalovich et al., Izv. AN SSSR, Neorgan. materialy, 2: 844 (1966).
89.   N. B. Vargaftik, Thermal Properties of Matter. Gosénergoizdat, Moscow (1956).
90.   M. Garber, W. G. Henry, and H. G. Hoeve, Can. J. Phys., 38:1595 (1960).
91.   G. Busch and R. Kern, Helv. Phys. Acta, 32: 24 (1959).
92.   V. M. Glazov, N. N. Glagoleva, and S. B. Evgen'ev, Izv. AN SSSR, Neorgan. materialy, 2:418 (1966).
93.   V. M. Glazov, N. N. Glagoleva, and L. A. Romantseva, Izv. AN SSSR, Neorgan. materialy, 2:1737 (1966).
94.   A. A. Vertman and E. S. Filippov, Collection: Properties of Metal in Solid and Liquid States (on the Occasion of the 80th Birthday of Academician I. P. Bardin). "Nauka," Moscow (1964), p. 100.
95.   E. G. Shvidkovskii, Uch. zap. MGU, No. 74, p. 135 (1944).
96.   E. G. Shvidkovskii, Vestn. MGU, seriya fiz.-mat. i est. nauk, No. 12, p. 54 (1950).
97.   A. M. Butov, L. S. Priss, and E. G. Shvidkovskii, Zh. tekhn. fiz., 21:1319 (1951).
98.   L. Ya. Krol', A. Ya. Nashel'skii, and L. D. Khlystovskaya, Zavodsk. lab., 27:177 (1961).
99.   G. B. Bokii, Introduction to Crystallochemistry. Moscow State University (1954).
100. W. Hume-Rothery, Atomic Theory for Students of Metallurgy, 3rd ed. Institute of Metals, London (1962).
101. Ya. K. Syrkin and M. E. Dyatkina, Chemical Binding and Structure of Molecules. Goskhimizdat, Moscow (1946).
102. N. N. Sirota, N. M. Olekhnovich, and A. U. Sheleg, Dokl. AN BSSR, 4:144 (1960).
103. N. N. Sirota and A. U. Sheleg, Dokl. AN SSSR, 135:1176 (1960).
104. I. I. Shafranovskii, Diamonds. Izd. AN SSSR, Moscow (1953).
105. M. E. Straumanis and E. Z. Aka, J. Appl. Phys., 23: 330 (1952).
106. A. Smakula and J. Kalnajs, Phys. Rev., 99:1733 (1955).
107. A. S. Cooper, Acta Cryst., 15:578 (1962).
108. D. N. Dutta, Phys. Status Solidi, 2:984 (1962).
109. N. A. Goryunova, Chemistry of Diamond-like Semiconductors. Chapman and Hall, London (1965).
110. Ya. A. Ugai, Introduction to Chemistry of Semiconductors. Izd. "Vysshaya shkola," Moscow (1965).
111. R. A. Smith, Semiconductors. Cambridge University Press (1961).
112. N. B. Hannay (editor), Semiconductors. Reinhold, New York (1959).
113. O. Kubaschewski and E. Ll. Evans, Metallurgische Thermochemie. Berlin (1959) (see also [424]).
114. V. M. Glazov and S. N. Chizhevskaya, Dokl. AN SSSR, 129:869 (1959).
115. V. M. Glazov and Liu Chên Yüan, Zh. neorg. khim., 7:582 (1962).
116. E. S. Greiner, J. Metals, 4:1044 (1952).
117. L. V. Shipanova and P. V. Gel'd, Izv. VUZov, tsvetnaya metallurgiya, No. 6, p. 111 (1962).
118. P. V. Gul'tyaev and A. V. Petrov, Fiz. tverd. tela, 1: 368 (1959).
119. N. N. Sirota and E. M. Gololobov, Dokl. AN BSSR, 3:368 (1959).

120. B. F. Ormont, Zh. neorg. khim., 3:1281 (1958).

121. B. F. Ormont, Zh. neorg. khim., 5:255 (1960).

122. P. V. Gel'd et al., Surface Phenomena in Metallurgical Processes, Proc. Inter-
university Conference. Metallurgizdat, Moscow (1963), p. 224.

123. B. A. Baum, P. V. Gel'd, and S. I. Suchil'nikov, Izv. AN SSSR (met. i gornoe
delo), No. 2, p. 149 (1964).

124. C. A. Domenicali, J. Appl. Physics, 28:749 (1957).

125. R. W. Keyes, Phys. Rev., 84:367 (1951).

126. D. K. Hamilton and K. G. Seidensticker, J. Appl. Phys., 34:2697 (1963).

127. V. M. Glazov and S. N. Chizhevskaya, Fiz. tverd. tela, 6:1684 (1964).

128. L. D. Lucas, Les Mém. Scient. de Rev. de métallurgie, 61: 1 (1964).

129. L. D. Lucas and G. Urbain, Compt. Rend., 255: 2414 (1962).

130. P. V. Gel'd and Yu. M. Gertman, Izv. AN SSSR, OTN, Met. i topl., No. 6,
p. 134 (1960); Fiz. Metallov i Metallovedenie, 12: 47 (1961).

131. H. V. Wartenberg, Naturwissenschaften, 36: 373 (1949).

132. W. Klemm et al., Montash. Chem., 83:629 (1952).

133. A. Koniger and G. Nagel, Giesserei Techn.-Wiss. Beih. Giessereiwesen und
Metallkunde, 13:57 (1961).

134. R. A. Logan and W. L. Bond, J. Appl. Phys., 30: 322 (1959).

135. N. A. Vatolin and O. V. Esin, Fiz. metallov i metallovedenie, 16:936 (1963).

136. H. Hendus, Z. Naturforsch., 2A:505 (1947).

137. V. K. Grigorovich, Collection: Structure and Properties of Liquid Metals.
A. A. Baikov Metallurgy Institute of the Academy of Sciences of the USSR
(1960), p. 81.

138. V. N. Kondrat'ev, Structure of Atoms and Molecules. Fizmatgiz, Moscow
(1959).

139. J. N. Hodgson, Phil. Mag., 6:509 (1961).

140. A. Abraham, J. Tauc, and B. Velicky, Phys. Status Solidi, 3:767 (1963).

141. V. M. Glazov and S. N. Chizhevskaya, Izv. AN SSSR, Neorgan. materialy,
1:307 (1965).

142. G. Busch and J. Tieche, Helv. Phys. Acta, 35:273 (1962).

143. G. Busch and J. Tieche, Phys. Kondens. Materie, 1:78 (1963).

144. A. R. Regel', Ukr. fiz. zhurnal, 7:833 (1962).

145. G. Busch and O. Vogt, Helv. Phys. Acta, 27:241 (1954).

146. V. M. Glazov, Izv. AN SSSR, OTN, Met. i topl., No. 5, p. 110 (1962).

147. B. M. Turovskii and A. P. Lyubimov, Izv. VUZov, chernaya metallurgiya,
No. 1, p. 24 (1960).

148. S. Glasstone, K. J. Laidler, and H. Eyring, Theory of Rate Processes. McGraw-
Hill, New York (1941).

149. J. Bernal and R. Fowler, J. Chem. Phys., 1:515 (1933).

150. O. Pfannenschmid, Z. Naturforsch., 15A:603 (1960).

151. Ya. I. Dutchak, A. G. Mikolaichuk, and N. I. Klym, Fiz. Metallov i
Metallovedenie, 14:548 (1962).

152. Ya. I. Dutchak, O. G. Mikolaichuk, and N. M. Klym, Visnyk L'dys'k. un-tu,
ser. fiz., No. 1(8), p. 138 (1962).

153. Ya. I. Dutchak, Kristallografiya, 6:124 (1961).

154.   R. Ozerov, Usp. fiz. nauk, 42:161 (1950).
155.   C. Gammertsfelder, J. Chem. Phys., 9:450 (1941).
156.   Ya. I. Dutchak, Fiz. Metallov i Metallovedeniye, 11:290 (1961).
157.   E. Gebhardt and M. Becker, Z. Metallkunde, 42:111 (1951).
158.   E. Gebhardt and S. Dörner, Z. Metallkunde, 42:353 (1951).
159.   E. Gebhardt and G. Wörwag, Z. Metallkunde, 42:358 (1951).
160.   E. Gebhardt and G. Wörwag, Z. Metallkunde, 43:106 (1952).
161.   E. Gebhardt, M. Becker, and K. Kostlin, Z. Metallkunde, 47:684 (1956).
162.   E. Gebhardt, M. Becker, and E. Trägner, Z. Metallkunde, 44:379 (1953).
163.   E. Gebhardt, M. Becker, and S. Schäfer, Z. Metallkunde, 43:292 (1952).
164.   A. Bienias and F. Sauerwald, Z. Anorg. Allgem. Chem., 161:51 (1927).
165.   K. Gering and F. Sauerwald, Z. Anorg. Allgem. Chem., 223:204 (1935).
166.   W. Radecker and F. Sauerwald, Z. Anorg. Allgem. Chem., 203:156 (1931).
167.   É. V. Polyak and S. V. Sergeev, Dokl. AN SSSR, 30:136 (1941).
168.   W. R. D. Jones and W. L. Bartlett, J. Inst. Metals, 81:145 (1952).
169.   T. P. Yao and V. Kondic, J. Inst. Metals, 81: 17 (1952).
170.   E. Gebhardt, M. Becker, and S. Dörner, Z. Metallkunde, 44: 510 (1953).
171.   E. Gebhardt, M. Becker, and S. Dörner, Z. Metallkunde, 44:573 (1953).
172.   E. Gebhardt, M. Becker, and S. Dörner, Z. Metallkunde, 45: 83 (1954).
173.   Yu. D. Chistyakov, Investigation of Physicochemical Properties of Aluminum
       and Its Alloys in the Liquid State. Dissertation for Candidate's Degree. M. I.
       Kalinin Institute for Nonferrous Metals and Gold, Moscow (1956).
174.   V. M. Glazov and Yu. D. Chistyakov, Izv. AN SSSR, OTN, No. 7, p. 141 (1958).
175.   F. Sauerwald and K. Töpler, Z. Anorg. Allgem. Chem., 157:117 (1926).
176.   E. Gebhardt and K. Köstlin, Z. Metallkunde, 48: 601 (1957).
177.   J. Budde et al., Z. Physik. Chem., 218:100 (1961).
178.   V. I. Danilov, Structure and Crystallization of Liquids. Izd. AN UkrSSR, Kiev
       (1956).
179.   A. I. Bublik and A. G. Buntar', Fiz. Metallov i Metallovedenie, 5:53 (1957).
180.   E. G. Shvidkovskii and G. I. Goryaga, Vestn. MGU, seriya fiz.-mat. i est. nauk,
       No. 9, p. 63 (1953).
181.   K. P. Bunin, Izv. AN SSSR, OTN, No. 2, p. 305 (1946).
182.   D. Turnbull and R. E. Cech, J. Appl. Phys., 21: 804 (1950).
183.   D. Turnbull, J. Appl. Phys., 21:1022 (1950).
184.   Yu. I. Petrov, Fiz. tverd. tela, 6: 2160 (1964).
185.   Yu. I. Petrov, Fiz. Metallov i Metallovedenie, 19:219 (1965).
186.   J. Fehlig and E. Scheil, Z. Metallkunde, 53:593 (1962).
187.   S. Roginsky, L. Sena, and J. Zeldowitsch, Physikalische Z. d. Sowjetunion,
       1: 630 (1932).
188.   G. I. Goryaga and E. G. Shvidkovskii, Vestn. MGU, seriya fiz.-mat. i est.
       nauk, No. 6, p. 33 (1956).
189.   T. A. Kontorova, Fiz. tverd. tela, 1:1761 (1959).
190.   T. A. Kontorova, Izv. AN SSSR, OTN, met. i topl., No. 3, p. 157 (1961).
191.   Ya. I. Frenkel', Zh. eksperim. i teor. fiz., 6: 902 (1936).
192.   Ya. I. Frenkel', Izv. AN SSSR, ser. fiz., No. 3, p. 287 (1937).

193. D. M. Chizhikov and V. P. Schastlivyi, Selenium and Selenides. Izd. "Nauka," Moscow (1964).

194. A. Nussbaum, Am. Scientist, 50:312 (1962).

195. E. Mooser and W. C. Pearson, J. Phys. Chem. Solids, 7:65 (1958).

196. B. V. Nekrasov, General Chemistry. Goskhimizdat, Moscow (1960).

197. K. W. Plessner, Proc. Phys. Soc. (London), 64:671 (1951).

198. T. Stubb, Recent Advances in Selenium Physics: Symposium on Solid and Liquid State of Selenium. Pergamon Press, London (1964), p. 53.

199. T. P. Smorodina, Fiz. tverd. tela, 2:883 (1960).

200. A. N. Campbell and S. Epstein, J. Am. Chem. Soc., 64:2679 (1942).

201. E. Bonnier, P. Hicter, and S. Aléonard, Compt. Rend., 258:166 (1964).

202. V. N. Lange and A. R. Regel', Fiz. tverd. tela, 1:559 (1959).

203. M. Hansen and K. Anderko, Constitution of Binary Alloys, 2nd edition. McGraw-Hill, New York (1958).

204. A. A. Kudryavtsev and G. P. Ustyugov, Zh. neorg. khim., 4:2421 (1961).

205. W. E. Spear and H. P. D. Lanyon, Proc. Intern. Conf. on Semiconductor Physics, Prague, 1960. Prague (1961), p. 987.

206. F. Eckart, Ann. Physik, 17:84 (1956).

207. H. W. Henkels, J. Appl. Phys., 22:1265 (1951).

208. P. T. Kozyrev, Zh. tekhn. fiz., 27 (1957).

209. V. N. Lange and A. R. Regel', Fiz. tverd. tela, 1:562 (1959).

210. J. Stuke, Phys. Status Solidi, 6:441 (1964).

211. E. Z. Meilikhov, Fiz. tverd. tela, 7:1743 (1965).

212. A. S. Epstein, H. Fritzsche, and K. Lark-Horovitz, Phys. Rev., 107:412 (1957).

213. F. Eckart, Ann. Physik, 14:233 (1954).

214. G. B. Abdullaev et al., Fiz. tverd. tela, 6:1018 (1964).

215. G. Borelius et al., Arkiv. Mat. Astron. Fysik, 30A:1 (1944).

216. H. W. Henkels, J. Appl. Phys., 21:725 (1950).

217. B. Lizell, J. Chem. Phys., 20:672 (1952).

218. H. W. Henkels and J. Maczuk, J. Appl. Phys., 24:1056 (1953).

219. S. I. Mekhtieva, G. M. Aliev, and D. Sh. Abdinov, Izv. AN AzerbSSR, ser. fiz.-mat. i tekhn. nauk, No. 4, p. 101 (1964).

220. Kh. G. Barkhinkhoev and G. M. Aliev, Izv. AN AzerbSSR, ser. fiz.-mat. i tekhn. nauk, No. 3, p. 95 (1963).

221. M. Cutler and C. E. Mallon, J. Chem. Phys., 37:2677 (1962).

222. G. Busch and O. Vogt, Helv. Phys. Acta, 30:224 (1957).

223. C. Massen, H. Weijts, and J. A. Poulis, Trans. Faraday Soc., 60:317 (1964).

224. H. Richter and F. Herre, Naturwissenschaften, 44:31 (1957).

225. H. Richter and F. Herre, Z. Naturforsch., 13A, 874 (1958).

226. H. Krebs, Z. Anorg. Allgem. Chem., 265:156 (1951).

227. A. Eisenberg and A. Tobolsky, J. Polymer Sci., 46:19 (1960).

228. S. Dobinski and J. Wesolowski, Bull. Acad. Polon. Sci., No. 8-9A, p. 446 (1936).

229. L. D. Lucas and G. Urbain, Compt. Rend., 258:6403 (1964).

230. E. Bonnier, P. Hicter, and S. Aléonard, Compt. Rend., 258:4967 (1964).

231. J. C. Perron, Compt. Rend., 258:4698 (1964).

232. V. M. Glazov, A. N. Kresovnikov, and N. N. Glagoleva, Dokl. AN SSSR, 162:94 (1965).

233.   C. A. Kraus and E. W. Johnson, J. Phys. Chem., 32:1281 (1928).
234.   V. A. Johnson, Phys. Rev., 98:1567 (1955).
235.   A. Epstein and H. Fritzsche, Phys. Rev., 93:922 (1954).
236.   A. Epstein and H. Fritzsche, Bull. Am. Phys. Soc., 29:28 (1954).
237.   G. Busch and J. Tieche, Rept. Intern. Conf. Phys. Semicond., Exeter, 1962. Inst. Phys. and Phys. Soc., London (1962), p. 237.
238.   S. Takeuchi and H. Endo, J. Japan. Inst. Metals, 26:148 (1962).
239.   H. Endo and S. Takeuchi, Trans. Japan Inst. Metals, 2:243 (1961).
240.   P. W. Kendall and N. E. Cusack, Phil. Mag., 5:100 (1960).
241.   N. E. Cusack and P. W. Kendall, Phil. Mag., 6:419 (1961).
242.   N. E. Cusack, P. W. Kendall, and A. S. Marwaha, Phil. Mag., 7:1745 (1962).
243.   J. N. Hodgson, Phil. Mag., 8:735 (1963).
244.   R. Buschert, J. G. Geib, and K. Lark-Horovitz, Bull. Am. Phys. Soc., 30:18 (1955).
245.   R. Buschert, J. G. Geib, and K. Lark-Horovitz, Phys. Rev., 98:1157 (1955).
246.   W. Klemm et al., Monatsh. Chem., 83:629 (1952).
247.   D. L. Ball, J. Chem. and Eng. Data, 8:61 (1963).
248.   A. A. Vertman, A. M. Samarin, and Ya. M. Yakobson, Izv. AN SSSR, OTN, Met. i topl., No. 3, p. 17 (1960).
249.   Kh. M. Khalilov, Izv. AN AzerbSSR, ser. fiz.-mat. i tekhn. nauk, No. 6, p. 67 (1959).
250.   S. Dobinski and J. Wesolowski, Bull. Acad. Polon. Sci., No. 1-2A, pp. 7-14 (1937).
251.   J. N. Stranski, R. Kraischew, and L. Krastanow, Z. Krist., 88:325 (1934).
252.   C. C. Urazovskii and B. D. Lyuft, Zh. fiz. khim., 22:409 (1948).
253.   O. G. Folberth, J. Phys. Chem. Solids, 7:295 (1958).
254.   J. van den Boomgaard and K. Schol, Collection: New Semiconducting Materials [Russian translation]. Metallurgizdat, Moscow (1964).
255.   D. Richman, J. Phys. Chem. Solids, 24:1131 (1963).
256.   H. Welker, Physica, 20:893 (1954).
257.   B. F. Ormont et al., Izv. AN SSSR, ser. fiz., 21:133 (1957).
258.   G. V. Ozolin'sh et al., Kristallografiya, 7:850 (1962); 8:272 (1963).
259.   S. A. Semiletov, Kristallografiya, 8:923 (1963).
260.   D. de Nobel, Philips Res. Rep., 14:361 (1959); 14:430 (1959).
261.   M. R. Lorenz, J. Phys. Chem. Solids, 23:939 (1962).
262.   I. V. Korneeva, A. V. Belyaev, and A. V. Novoselova, Zh. neorg. khim., 5:3 (1960).
263.   K. Nagel and C. Wagner, Ž. Physik. Chem., 25B:71 (1934).
264.   R. J. Maurer, J. Chem. Phys., 13:321 (1945).
265.   H. Vine and R. J. Maurer, Z. Physik. Chem., 198:147 (1951).
266.   K. Weiss, Z. Physik. Chem., 12:68 (1957).
267.   R. N. Kurdyumova and S. A. Semiletov, Kristallografiya, 7:370 (1962).
268.   E. G. Grochowski et al., J. Phys. Chem. Solids, 25:551 (1964).
269.   J. Dale, Nature, 197:242 (1963).
270.   S. A. Semiletov and V. A. Vlasov, Kristallografiya, 8:877 (1963).
271.   P. J. Holmes, I. C. Jennings, and J. E. Parrott, J. Phys. Chem. Solids, 23:1 (1962).

272. S. A. Semiletov and M. Rozsival, Kristallografiya, 2:287 (1957).

273. S. A. Semiletov and L. I. Man, Kristallografiya, 5:314 (1960).

274. S. A. Semiletov, Trudy in-ta kristallografii, 11:121 (1955).

275. S. A. Semiletov, Kristallografiya, 1:306 (1956).

276. E. Krucheanu, Preparation and Properties of Some Semiconducting $A^{II}B^{VI}$ Compounds. Dissertation for Candidate's Degree. Moscow (1960).

277. Ya. S. Pashinkin et al., Kristallografiya, 5:261 (1960).

278. M. A. Rumsh, F. T. Novik, and G. M. Zimkina, Kristallografiya, 7:873 (1962).

279. K. V. Shalimova et al., Kristallografiya, 9:741 (1964).

280. E. Krucheanu, D. Nikulesku, and A. Vanku, Kristallografiya, 9:537 (1964).

281. F. J. M. Kendall, Phys. Letters, 8:237 (1964).

282. S. Miyake, S. Hoshino, and T. Takenaka, J. Phys. Soc. Japan, 7:19 (1952).

283. J. B. Wagner and C. Wagner, J. Chem. Phys., 26:1599 (1957).

284. W. Biermann and H. J. Oel, Z. Physik. Chem., 17:163 (1958).

285. W. Jost, H. J. Oel, and G. Schniedermann, Z. Physik. Chem., 17:175 (1958).

286. J. Krug and L. Sieg, Z. Naturforsch., 7A:369 (1952).

287. H. Hahn, Z. Angew. Chem., 64:203 (1952).

288. J. C. Woolley, B. R. Pamplin, and P. J. Holmes, Less-Common Metals, 1:362 (1959).

289. P. C. Newman and J. A. Cundall, Nature, 200:876 (1963).

290. A. I. Zaslavskii and V. M. Sergeeva, Fiz. tverd. tela, 2:2872 (1960).

291. P. C. Newman, Phys. Chem. Solids, 23:19 (1962).

292. G. Harbeke and G. Lautz, Naturwissenschaften, 45:283 (1958).

293. G. Harbeke and G. Lautz, Z. Naturforsch., 13A:771 (1958).

294. G. Giesecke and H. Pflister, Acta Cryst., 11:369 (1958).

295. V. A. Kotovich and V. A. Frank-Kamenetskii, Uch. zap. LGU im. Zhdanova, ser. geolog., No. 8, p. 135 (1957).

296. N. A. Goryunova, V. A. Kotovich, and V. A. Frank-Kamenetskii, Dokl. AN SSSR, 103:659 (1955).

297. V. N. Vigdorovich et al., Nauchnye trudy Giredmeta, 10:264 (1963).

298. M. Kobayashi, Z. Metallkunde, 2:65 (1912).

299. N. A. Goryunova et al., Zh. tekhn. fiz., 25:1675 (1955).

300. C. Hilsum and A. C. Rose-Innes, Semiconducting III–V Compounds. Pergamon Press, Oxford (1961).

301. J. P. Suchet, Physical Chemistry of Semiconductors. Dunod, Paris (1962).

302. C. H. L. Goodman, Nature, 187:590 (1960).

303. E. Mooser and W. B. Pearson, Nature, 190:406 (1961).

304. C. H. L. Goodman, Nature, 192:355 (1961).

305. E. Mooser and W. B. Pearson, Nature, 192:355 (1961).

306. Ya. K. Syrkin, Usp. khim., 31:397 (1962).

307. S. S. Batsanov, Zh. fiz. khim., 37:1418 (1963).

308. Ya. K. Syrkin, Zh. fiz. khim., 37:1422 (1963).

309. V. Gol'dshmidt, Usp. fiz. nauk, 9A:811 (1929).

310. N. N. Sirota and N. M. Olekhnovich, Dokl. AN SSSR, 136:660 (1961).

311. N. N. Sirota and N. M. Olekhnovich, Dokl. AN SSSR, 136:879 (1961).

312. N. N. Sirota and E. M. Gololobov, Dokl. AN SSSR, 138:162 (1961).

313. N. N. Sirota and E. M. Gololobov, Dokl. AN SSSR, 143:156 (1962).

314. N. N. Sirota and N. M. Olekhnovich, Dokl. AN SSSR, 143:370 (1962).

315. N. N. Sirota and E. M. Gololobov, Dokl. AN SSSR, 144: 398 (1962).

316. L. Pauling, The Nature of the Chemical Bond, 3rd edition. Cornell University Press, Ithaca (1960).

317. C. H. L. Goodman, J. Electronics, 1:115 (1955).

318. W. Gordy and W. Thomas, J. Chem. Phys., 24:439 (1956).

319. A. L. Allred and A. L. Hensley, J. Inorg. and Nucl. Chem., 17:43 (1961).

320. S. S. Batsanov, Electronegativity of Elements and Chemical Binding. Izd. SO AN SSSR, Novosibirsk (1962).

321. H. Pflister, Z. Naturforsch., 10A:79 (1955).

322. G. Wolff et al., Halbleiter und Phosphoren. Berlin (1958).

323. G. A. Wolff and J. G. Broder, Acta Cryst., 12: 313 (1959).

324. W. G. Spitzer and H. Y. Fan, Phys. Rev., 99:1893 (1955).

325. G. S. Picus et al., J. Phys. Chem. Solids, 8:282 (1959).

326. M. Hass and B. W. Henvis, J. Phys. Chem. Solids, 23:1099 (1962).

327. R. A. Barinskii and E. G. Nadzhakov, Dokl. AN SSSR, 129:1279 (1959).

328. R. A. Barinskii and E. G. Nadzhakov, Izv. AN SSSR, ser. fiz., 24: 407 (1960).

329. Ya. A. Ugai and É. P. Domashevskaya, Dokl. AN SSSR, 156:430 (1964).

330. N. N. Sirota and E. M. Gololobov, Dokl. AN SSSR, 156:1075 (1964).

331. N. N. Sirota, Chemical Bonds in Semiconductors and Solids. Consultants Bureau, New York (1967).

332. E. Mooser and W. B. Pearson, Collection: Semiconducting Materials — Problems of Chemical Binding [Russian translation]. IL (1960), p. 134.

333. V. P. Zhuze, V. M. Sergeeva, and A. I. Shelykh, Fiz. tverd. tela, 2:2858 (1960).

334. W. B. Pearson, Collection: Semiconducting Materials — Problems of Chemical Binding [Russian translation]. IL (1960), p. 249.

335. E. M. Shustorovich, Nature of Chemical Binding. Izd. AN SSSR, Moscow (1963).

336. G. A. Wolff, L. Tolman, and J. C. Clark, Phys. Rev., 98:1179 (1955).

337. B. F. Ormont, Dokl. AN SSSR, 124:129 (1959).

338. B. F. Ormont, Zh. neorg. khim., 4:2174 (1959).

339. B. F. Ormont, Dokl. AN SSSR, 106: 687 (1956).

340. B. F. Ormont, Zh. fiz. khim., 31:509 (1957).

341. M. Matyas, Czechoslov. J. Phys., B12: 838 (1962).

342. H. Welker, Z. Naturforsch., 7A:744 (1952).

343. V. P. Zhuze, Dokl. AN SSSR, 99:711 (1954).

344. V. P. Zhuze, Zh. tekhn. fiz., 25: 2079 (1955).

345. N. N. Sirota, Nauchnye trudy Moskovskogo instituta tsvetnykh metallov i zolota, No. 1, p. 117 (1957).

346. N. N. Sirota and L. I. Berger, Inzh.-fiz. zh., 2:104 (1959).

347. C. H. L. Goodman, J. Electronics, 1:115 (1955).

348. O. G. Folberth, Z. Naturforsch., 13A:856 (1958).

349. O. G. Folberth and H. Welker, J. Phys. Chem. Solids, 8:14 (1959).

350. O. G. Folberth, Z. Naturforsch., 15A:425 (1960).

351. P. P. Otopkov and A. M. Evseev, Zh. fiz. khim., 34:815 (1960).

352.   D. P. Belotskii, Problems in Metallurgy and Physics of Semiconductors. Proc.
       Fourth Conf. on Semiconducting Materials. Izd. AN SSSR, Moscow (1961), p. 18.
353.   P. Manca, J. Phys. Chem. Solids, 20 : 268 (1961).
354.   V. A. Presnov, Fiz. tverd. tela, 4 : 548 (1962).
355.   K. A. Sharifov, Dokl. AN AzerbSSR, 19 : 23 (1963).
356.   K. A. Sharifov, Zh. fiz. khim., 31 : 488 (1965).
357.   S. A. Semenkovich, Dokl. AN SSSR, 158 : 442 (1964).
358.   S. A. Semenkovich, Chemical Bonds in Semiconductors and Solids.
       Consultants Bureau, New York (1967), p. 83.
359.   V. A. Petrusevich and V. M. Sergeeva, Fiz. tverd. tela, 2 : 2881 (1960).
360.   V. P. Zhuze et al., Proc. Intern. Conf. on Semiconductor Physics, Prague, 1960.
       Prague (1961), p. 881.
361.   V. P. Zhuze, V. M. Sergeeva, and A. I. Shelykh, Fiz. tverd. tela, 2 : 2858 (1960).
362.   G. Harbeke and G. Lautz, Optik, 14 : 547 (1957).
363.   T. Hirosi, S. Hironichi, and H. Kazuyoshi, J. Phys. Soc. Japan, 16 : 1038 (1961).
364.   B. T. Kolomiets and A. A. Mal'kova, Zh. tekhn. fiz., 28 : 1662 (1958).
365.   H. G. Fozner, Wiss. Z. d. Martin Luther Univ. Halle-Wittenberg. Math.-
       Naturwiss. Reihe, 13 : 731 (1964).
366.   S. I. Radautsan, Izv. Mold. fil. AN SSSR, No. 3(69), p. 49 (1960).
367.   W. Schottky and M. Bever, Acta Met., 6 : 320 (1958).
368.   A. V. Nikol'skaya, V. A. Geiderikh, and Ya. I. Gerasimov, Dokl. AN SSSR,
       130 : 1074 (1960).
369.   H. B. Guthier, Z. Naturforsch., 16A : 268 (1961).
370.   H. Hahn and F. Burow, Z. Angew. Chem., 68 : 382 (1956).
371.   L. J. Vieland, Acta Met., 11 : 137 (1963).
372.   V. M. Glazov and G. L. Malyutina, Zh. neorg. khim., No. 8, p. 1921 (1963).
373.   S. N. Gadzhiev and K. A. Shafirov, Dokl. AN SSSR, 136 : 1339 (1961).
374.   O. J. Kleppa, J. Am. Chem. Soc., 77 : 897 (1955).
375.   A. Schneider and K. Klotz, Naturwissenschaften, 46 : 141 (1959).
376.   A. S. Abbasov et al., Dokl. AN SSSR, 156 : 1399 (1964).
377.   T. Renner, Collection: New Semiconducting Materials [Russian translation].
       Metallurgizdat, Moscow (1964), p. 51.
378.   D. Richman and E. F. Hockings, J. Electrochem. Soc. Japan, 112 : 461 (1965).
379.   N. H. Nachtrieb and N. Clement, J. Phys. Chem., 62 : 876 (1958).
380.   A. S. Abbasov et al., Dokl. AN SSSR, 156 : 1140 (1964).
381.   A. S. Abbasov, A. V. Nikol'skaya, and Ya. I. Gerasimov, Dokl. AN SSSR,
       147 : 835 (1962).
382.   O. G. Folberth, Collection: New Semiconducting Materials [Russian translation].
       Metallurgizdat, Moscow (1964), p. 5.
383.   A. Jayaraman, W. Klement, and G. C. Kennedy, Phys. Rev., 130 : 540 (1963).
384.   J. Terpilowski and W. Trzebiatowski, Bull. Acad. Polon. Sci. Classe III,
       8 : 95 (1960).
385.   Yu. I. Pashintsev and N. N. Sirota, Dokl. AN BSSR, 3 : 38 (1959).
386.   N. N. Sirota and E. M. Gololobov, Chemical Bonds in Semiconductors and
       Solids. Consultants Bureau, New York (1967), p. 69.
387.   V. M. Glazov and S. N. Chizhevskaya, Fiz. tverd. tela, 4 : 1841 (1962).

388. G. Busch and S. Yuan, Helv. Phys. Acta, 32:465 (1959).

389. H. Krebs, H. Weyand, and M. Haucke, Z. Angew. Chemie, 70: 468 (1958); Physics and Chemistry of Metallic Solutions and Intermetallic Compounds, London, H. M. S. O., 1959, Vol. 2, 4C/1-4C/13, Discussion, 41/1-41/12.

390. A. I. Belyaev, E. A. Zhemchuzhina, and L. A. Firsanova, Physical Chemistry of Molten Salts. Metallurgizdat, Moscow (1957).

391. S. S. Batsanov and V. I. Pakhomov, Kristallografiya, 2:183 (1957).

392. J. W. Mellor, Comprehensive Treatise on Inorganic and Theoretical Chemistry. Wiley, New York (1952), Vol. 3, p. 201.

393. K. Baedeker, Ann. Physik, 29:566 (1909).

394. C. Tubandt, E. Rindtorff, and W. Jost, Z. Anorg. Allgem. Chem., 165:195 (1927).

395. J. Nölting, Z. Physik. Chem. (Neue Folge), 19:118 (1959).

396. M. I. Shakhparonov, Structure and Properties of Liquid Metals. A. A. Baikov Metallurgy Institute (1959), p. 103.

397. A. I. Ioffe, Semiconductor Thermoelements. Izd. AN SSSR, Moscow (1960).

398. D. S. Kamenetskaya, Investigation of the Influence of Intermolecular Interaction on Phase Equilibrium. Dissertation for Candidate's Degree. Dnepropetrovsk (1946).

399. V. I. Danilov and D. S. Kamenetskaya, Zh. fiz. khim., 22:69 (1948).

400. V. P. Zhuze and A. I. Shelykh, Fiz. tverd. tela, 7:1175 (1965).

401. D. S. Kamenetskaya, Collection: Problems of Metallography and Physics of Metals. Metallurgizdat, Moscow (1949), p. 113.

402. F. Fielding, G. Fischer, and E. Mooser, J. Phys. Chem. Solids, 8:434 (1959).

403. Ya. A. Ugai, Zh. neorg. khim., 3:678 (1958).

404. Ya. A. Ugai, Investigations of Semiconducting Phases Based on Antimony, Arsenic, and Phosphorus. Author's Abstract of Doctoral Dissertation. N. S. Kurnakov Institute of General and Inorganic Chemistry, Moscow (1965).

405. N. F. Mott and H. Jones, The Theory of the Properties of Metals and Alloys, Clarendon Press, Oxford (1936).

406. U. Winkler, Helv. Phys. Acta, 28:633 (1955).

407. B. F. Ormont, Structures of Inorganic Substances. GTTI, Moscow (1950).

408. W. Klemm and H. Westlinning, Z. Anorg. Allgem. Chem., 245: 365 (1940).

409. A. Sacklowski, Ann. Physik, 77: 264 (1925).

410. L. Pauling, J. Am. Chem. Soc., 45:2777 (1923).

411. J. Friauf, J. Am. Chem. Soc., 48:1906 (1926).

412. G. Brauer and J. Tiesler, Z. Anorg. Chem., 262:319 (1950).

413. G. Busch and U. Winkler, Helv. Phys. Acta, 26:578 (1953).

414. E. A. Owen and G. D. Preston, Proc. Phys. Soc. (London), 36:541 (1924).

415. E. Zintl and H. Keiser, Z. Anorg. Chem., 211:113 (1933).

416. A. F. Ioffe, Zh. Tekhn. Fiz., 27:1153 (1957).

417. W. Klemm and H. Klein, Z. Anorg. Allgem. Chem., 248:167 (1941).

418. F. Hund, Z. Anorg. Allgem. Chem., 263:102 (1950).

419. A. Neuhaus, Forschr. Mineralog., 32: 37 (1954).

420. H. Krebs, Z. Anorg. Allgem. Chem.,278: 82 (1955); Acta Cryst. 9: 95 (1956).

421. H. Welker, Ergebnisse der exakten Naturwissenschaften. Berlin (1955).

422. Status of the Theory of Chemical Structure in Organic Chemistry. Paper presented by a commission of the Division of Chemical Sciences of the Academy of Sciences of the USSR. Izd. AN SSSR, Moscow (1954).

423. B. I. Boltaks, Zh. tekhn. fiz., 20:180 (1950); Dokl. AN SSSR, 64:487 (1949).

424. O. Kubaschewski and E. Ll. Evans, Metallurgical Thermochemistry, 4th edition. Pergamon, New York (1958).

425. R. G. Morris, R. D. Redin, and G. C. Danielson, Phys. Rev., 109:1909 (1958).

426. R. F. Blunt, H. P. Frederikse, and W. R. Hosler, Phys. Rev., 100:663 (1955).

427. W. D. Lawson et al., J. Electronics, 1:203 (1955).

428. W. D. Robertson and H. H. Uhlig, Trans. AIME, 180:345 (1949).

429. R. D. Redin, R. G. Morris, and G. C. Danielson, Phys. Rev., 109:1916 (1958).

430. T. Dŭlamită, D. Drimer, and E. Fokt, Acad. R. P. R., 4:511 (1959).

431. G. Busch and M. Moldovanova, Helv. Phys. Acta, 35:500 (1962).

432. N. A. Vul'onkov, K. A. Bol'shakov, P. I. Fedorov, and M. S. Tsirlin, Author's Certificate No. 150495, October 11, 1962.

433. V. M. Glazov and N. N. Glagoleva, Izv. AN SSSR, Neorgan. materialy, 1:1079 (1965).

434. A. Knappwost, Z. Elektrochem., 56:594 (1952).

435. A. Knappwost, Z. Physik. Chem., 21:358 (1959).

436. A. Roll and H. Motz, Z. Metallkunde, 48:435 (1957).

437. E. Gebhardt, M. Becker, and E. Tragner, Z. Metallkunde, 46:90 (1955).

438. E. Gebhardt, M. Becker, and H. Sebastian, Z. Metallkunde, 46:669 (1955).

439. A. Knappwost and G. Horz, Z. Physik. Chem., 22:139 (1959).

440. F. Sauerwald, Z. Metallforsch., 2:188 (1947).

441. O. Kubaschewski and R. Hörnle, Z. Metallkunde, 42:129 (1951).

442. V. M. Glazov, N. N. Glagoleva, and L. A. Romantseva, Izv. AN SSSR. Neorgan. materialy, 2:1953 (1966).

443. W. Klemm and G. Frischmuth, Z. Anorg. Chem., 218:249 (1934).

444. R. F. Brebrick, J. Phys. Chem. Solids, 24:27 (1963).

445. J. Umeda, M. Jeong, and T. Okada, Japan J. Appl. Phys., 1:277 (1962).

446. E. Miller, K. Komarek, and J. Cadoff, Trans. Met. Soc. AIME, 218:382 (1960).

447. R. F. Brebrick and R. S. Allgaier, J. Chem. Phys., 32:1826 (1960).

448. R. F. Brebrick and E. Gubner, J. Chem. Phys., 36:1283 (1962).

449. A. E. Goldberg and G. R. Mitchell, J. Chem. Phys., 22:220 (1954).

450. R. F. Brebrick and E. Gubner, J. Chem. Phys., 36:170 (1962).

451. J. Bloem and F. A. Kröger, Z. Physik Chem. (Neue Folge), 7:1 (1956).

452. C. Hirayama, J. Phys. Chem., 66:1563 (1962).

453. W. Johnston and D. Sestrich, J. Inorg. and Nucl. Chem., 68:428 (1964).

454. L. Kufman, Trans. Met. Soc. AIME, 224:1006 (1962).

455. W. Scanlon, Solid State Physics, 9:83 (1959).

456. R. Nozato and K. Igaki, Bull. Naniva Univ. (Japan), 3:125, 135 (1955).

457. K. Schubert and H. Fricke, Z. Metallkunde, 44:457 (1953).

458. K. Schubert and H. Fricke, Z. Naturforsch., 6a:781 (1951).

459. I. M. Tsidil'kovskii, Thermomagnetic Effects in Semiconductors. Infosearch, London (1962).

460. I. M. Tsidil'kovskii, Dokl. AN SSSR, 102:737 (1955).

461. E. D. Devyatkova and I. A. Smirnov, Fiz. tverd. tela, 2(8) (1960).
462. A. Sagar and R. C. Miller, Bull. Am. Phys. Soc., 7: 1203 (1962).
463. D. H. Damon, C. R. Martin, and R. C. Miller, J. Appl. Phys., 34: 3083 (1963).
464. Y. Moriguchi and Y. Koga, J. Phys. Soc. Japan, 12: 100 (1957).
465. T. Okada, J. Phys. Chem. Solids, 8: 428 (1959).
466. R. S. Allgaier and P. O. Scheie, Bull. Am. Phys. Soc., 6: 436 (1961).
467. R. F. Brebrick and A. J. Strauss, Bull. Am. Phys. Soc., 7: 203 (1962).
468. R. F. Brebrick and A. J. Strauss, Phys. Rev., 131: 104 (1963).
469. J. A. Kafalas, R. F. Brebrick, and A. J. Strauss, Appl. Phys. Letters, 4: 93 (1964).
470. N. V. Kolomoets, E. Ya. Lev, and L. M. Sysoeva, Fiz. tverd. tela, 5: 2871 (1963).
471. N. V. Kolomoets, E. Ya. Lev, and L. M. Sysoeva, Fiz. tverd. tela, 6: 706 (1964).
472. V. M. Glazov, N. N. Glagoleva, and K. V. Panoyan, Abstracts of Papers
      Presented at the Second All-Union Conf. on Problems of Chemical Binding in
      Semiconductors. Izd. AN BSSR, Minsk (1963), p. 39.
473. V. M. Glazov, N. N. Glagoleva, and K. V. Panoyan, Chemical Bonds in
      Semiconductors and Solids. Consultants Bureau, New York (1967), p. 107.
474. A. N. Krestovnikov et al., Izv. AN SSSR, Neorgan. materialy, 2(5) (1966).
475. J. Sherman, Chem. Rev., 11: 93 (1932).
476. G. B. Bokii, Crystallochemistry. Moscow State University (1960).
477. C. Hirayama, J. Chem. Eng. Data, 9: 65 (1964).
478. R. W. Fritts, Thermoelectric Materials and Devices (1960).
479. O. M. Konovalov, Semiconducting Materials. Khar'kov State University (1963).
480. R. F. Brebrick, J. Chem. Phys., 41: 1140 (1964).
481. B. M. Kulwick, Doctoral Dissertation. University of Michigan (1963), p. 58.
482. R. F. Brebrick and A. J. Strauss, J. Chem. Phys., 41: 197 (1964).
483. É. V. Britske and A. F. Kapustinskii, Thermal Constants of Inorganic Sub-
      stances. Izd. AN SSSR, Moscow (1949).
484. F. D. Rossini et al., Selected Values of Chemical and Thermodynamic
      Properties, Circular of the Nat. Bur. Standards, Washington (1952), p. 1.
485. G. Bauer and F. Sauerwald, Wiss. Z. Martin Luther Univ. Halle-Wittenberg,
      Math.-Naturwiss. Reihe, 10: 1029 (1961).
486. J. H. McAteer and H. Seltz, J. Am. Chem. Soc., 58: 2081 (1936).
487. R. Colin and J. Drowart, J. Phys. Chem., 68: 428 (1964).
488. E. Putley, Proc. Phys. Soc. (London), B65: 993 (1952).
489. K. Shogenji and S. Uchiyama, J. Phys. Soc. Japan, 12: 252 (1957).
490. R. A. Smith, Physica, 20: 910 (1954).
491. P. M. Starik, Fiz. tverd. tela, 7: 2246 (1965).
492. V. F. Ormont, Problems of Metallurgy and Physics of Semiconductors. Izd.
      AN SSSR, Moscow (1961), p. 5.
493. E. Putley, Proc. Phys. Soc. (London), 65B: 388, 736 (1951).
494. R. F. Brebrick and W. W. Scanlon, Phys. Rev., 96: 598 (1954).
495. D. G. Avery, Proc. Phys. Soc. (London), 67B: 2 (1954).
496. M. Moldovanova, S. Dimitrova, and S. Decheva, Fiz. tverd. tela, 6: 3717
      (1964).
497. R. Allgaier and W. W. Scanlon, Phys. Rev., 111: 1029 (1958).

498. A. N. Krestovnikov, V. M. Glazov, and N. N. Glagoleva, Collection: Physical Chemistry of Metallurgical Processes and Systems. Trudy MISiS, No. 41, p. 232 (1965).

499. W. W. Scanlon, Phys. Rev., 109:47 (1958).

500. V. M. Glazov, A. N. Krestovnikov, and V. A. Evseev, Dokl. AN SSSR, 169:868 (1966).

501. A. N. Krestovnikov et al., Izv. VUZov, Tsvetnaya metallurgiya, No. 2, p. 79 (1967).

502. Ya. G. Dorfman, Diamagnetism and Chemical Binding. Fizmatgiz, Moscow (1961).

503. V. M. Glazov, N. N. Glagoleva, and S. B. Evgen'ev, Izv. AN SSSR, Neorgan. materialy, 2:418 (1966).

504. N. Kh. Abrikosov and F. V. Bankina, Zh. neorg. khim., 3:661 (1958).

505. N. Kh. Abrikosov, V. F. Bankina, and K. F. Kharitonovich, Zh. neorg. khim., 5:2011 (1960).

506. S. B. Satterthwaite and R. W. Ure, Phys. Rev., 108:1164 (1957).

507. M. J. Smith, Appl. Phys. Letters, 1:79 (1962).

508. S. A. Semiletov, Trudy in-ta kristallografii, 10:76 (1954).

509. N. Parravano, Gazz. Chim. Ital., 43:201 (1913).

510. P. W. Lange, Naturwissenschaften, 27:133 (1939).

511. L. S. Ramsdell, Am. Miner., 15:119 (1930).

512. E. Dönges, Z. Anorg. Chem., 265:56 (1951).

513. J. A. Bland and S. J. Basinski, Can. J. Phys., 39:1040 (1961).

514. J. R. Wiese and L. Muldawer, J. Phys. Chem. Solids, 15:13 (1960).

515. K. Shubert and K. Anderko, Naturwissenschaften, 40:269 (1953).

516. C. Frondell, Am. Miner., 24:185 (1939).

517. D. Harker, Z. Krist., 89:175 (1934).

518. F. Roessler, Z. Anorg. Allgem. Chem., 9:546 (1895).

519. S. Nakajima, J. Phys. Chem. Solids, 24:479 (1963).

520. V. G. Kuznetsov and K. K. Palkina, Zh. neorg. khim., 8:1204 (1963).

521. S. A. Semiletov, Kristallografiya, 1:403 (1956).

522. J. R. Drabble and C. H. L. Goodman, J. Phys. Chem. Solids, 5:142 (1958).

523. P. P. Konorov, Investigation of Electrical Properties of Bismuth Sulfide, Selenide, and Telluride. Candidate's Dissertation. Leningrad State University (1951).

524. H. Krebs, Z. Elektrochem., 61:925 (1957).

525. S. V. Airapetyants and B. A. Efimova, Zh. tekhn. fiz., 28:1768 (1958).

526. E. K. Kudinov, Fiz. tverd. tela, 3:317 (1961).

527. J. Lagrenaudie, J. Phys. Radium Suppl., 18:39A (1957).

528. J. P. Suchet, J. Phys. Chem. Solids, 12:174 (1959).

529. J. P. Suchet, J. Phys. Chem. Solids, 16:265 (1960).

530. E. L. Shtrum, Ternary Semiconducting Compounds of ABX_2-Type. Candidate's Dissertation. Institute for Semiconductors, Academy of Sciences of the USSR, Leningrad (1960).

531. B. W. Howlett, S. Misra, and M. B. Bever, Trans. Met. Soc. AIME, 230:1367 (1964).

532.   G. F. Bolling, J. Chem. Phys., 33: 305 (1960).

533.   J. P. McHugh and W. A. Tiller, Trans. Met. Soc. AIME, 15: 651 (1959).

534.   Ya. I. Gerasimov and A. V. Nikol'skaya, Collection: Problems in Metallurgy and Physics of Semiconductors. Izd. AN SSSR, Moscow (1961), p. 79.

535.   S. Shigetoni and S. Mori, J. Phys. Soc. Japan, 11: 915 (1956).

536.   J. R. Drabble and R. Wolfe, Proc. Phys. Soc. (London), 13: 69, 1101 (1956).

537.   H. J. Goldsmid, Proc. Phys. Soc. (London), 71: 635 (1958).

538.   H. J. Goldsmid, Proc. Phys. Soc. (London), 72: 17 (1958).

539.   J. Black et al., J. Phys. Chem. Solids, 2: 240 (1957).

540.   R. Mansfield and W. Williams, Proc. Phys. Soc. (London), 72: 733 (1958).

541.   T. Harman et al., J. Phys. Chem. Solids, 2: 181 (1957).

542.   J. G. Austin, Proc. Phys. Soc. (London), 72: 545 (1958).

543.   L. Aimsworth, Proc. Phys. Soc. (London), B69: 606 (1956).

544.   J. Austin and A. Sheard, J. Electronics Control, 3: 236 (1957).

545.   R. Sehr and L. Testardi, J. Phys. Chem. Solids, 23: 1219 (1962).

546.   P. P. Konorov, Zh. tekhn. fiz., 26: 1394 (1956).

547.   P. P. Konorov, Zh. tekhn. fiz., 26: 1400 (1956).

548.   J. R. Drabble, R. D. Groves, and R. Wolfe, Proc. Phys. Soc. (London), 71: 430 (1958).

549.   J. R. Drabble, Proc. Phys. Soc. (London), 72: 380 (1958).

550.   R. M. Vlasova and L. S. Stil'bans, Zh. tekhn. fiz., 25: 569 (1955).

551.   P. C. Sharrah, J. I. Petz, and R. F. Kruh, J. Chem. Phys., 32: 241 (1960).

552.   V. M. Glazov, A. N. Krestovnikov, and N. N. Glagoleva, Izv. AN SSSR, Neorgan. materialy, 2: 453 (1966).

553.   L. Ball, J. Chem. and Eng. Data, 10: 37 (1965).

554.   V. M. Glazov et al., Izv. AN SSSR, Neorgan. materialy, 2: 1477 (1966).

555.   V. M. Glazov, A. N. Krestovnikov, and N. N. Glagoleva, Dokl. AN SSSR, 161: 629 (1965).

556.   V. M. Glazov et al., Proc. All-Union Conf. on Physical Properties of $A^{III}B^V$ and $A^{III}B^{VI}$ Semiconductors. Izd. AN AzSSR, Baku (1966).

557.   A. A. Bochvar and G. M. Kuznetsov, Dokl. AN SSSR, 98: 227 (1954).

558.   H. T. Hall, J. Phys. Chem., 59: 1144 (1955).

559.   A. Jayaraman, R. C. Newton, and G. C. Kennedy, Nature, 191: 1288 (1961).

560.   A. Jayaraman, W. Klement, and G. C. Kennedy, Phys. Rev., 130: 540 (1963).

561.   A. Jayaraman et al., J. Phys. Chem. Solids, 24: 7 (1963).

562.   P. L. Smith and J. E. Martin, Nature, 196: 762 (1962).

563.   E. P. Bundy, J. Chem. Phys., 41: 3809 (1964).

564.   D. L. Ball, Inorg. Chemistry, 1: 805 (1962).

565.   N. A. Tikhomirova and S. N. Stishov, Zh. eksperim. i teor. fiz., 43: 2321 (1962).

566.   V. V. Evdokimova, Usp. fiz. nauk, 88: 93 (1966).

567.   V. I. Fedorov, Experimental Investigation of Thermal Conductivity of Liquid Semiconductors. Author's Abstract of Candidate's Dissertation. Moscow Power Institute (1966).

568.   V. M. Glazov and V. N. Vigdorovich, Zh. fiz. khim., 33: 2164 (1959).

569.   S. Arrhenius. Z. Physik. Chem., 1: 285 (1887).

570.   J. Kendall and K. P. Monroe, J. Am. Chem. Soc., 39:1787 (1917).

571.   J. Kendall and A. H. Wright, J. Am. Chem. Soc., 39:1787 (1917).

572.   D. A. Pospekhov, Zh. fiz. khim., 30:228 (1956).

573.   A. N. Sakhanov and N. A. Ryakhovskii, Zh. Russk. khimich. ob-va, 47:118 (1915).

574.   R. Kremann, F. Jugl, and R. Meingast, Monatsh. Chem., 35: 1365 (1914).

575.   V. A. Kireev, Course of Physical Chemistry. Goskhimizdat, Moscow (1955).

576.   V. M. Glazov and S. N. Chizhevskaya, Zavodsk. lab., 26:720 (1960).

577.   G. I. Goryaga and I. A. Nosyreva, Vestn. MGU, seriya mat., fiz., khim., No. 6, p. 59 (1958).

578.   A. A. Bochvar, Metallography. Metallurgizdat, Moscow (1956).

579.   A. A. Bochvar and I. I. Novikov, Izv. AN SSSR, OTN, No. 2, p. 217 (1952).

580.   A. A. Bochvar and I. I. Novikov, Technology of Nonferrous Metals, Trudy Mintsvetmetzol. VNITOM, No. 23, p. 5 (1952).

581.   D. K. Belashchenko, Zh. fiz. khim., 31:2269 (1957).

582.   D. K. Belashchenko, Dokl. AN SSSR, 117:98 (1958).

583.   N. V. Ageev and D. S. Shoikhet, Thermal Analysis of Metals and Alloys. KUBUCh, Leningrad (1936).

584.   L. G. Berg, A. V. Nikolaev, and E. Ya. Rode, Thermography. Izd. AN SSSR, Moscow (1944).

585.   M. V. Zakharov, Trudy Moskovsk. in-ta tsvetnykh metallov i zolota im. M. I. Kalinina, No. 8, p. 78 (1940).

586.   V. M. Glavov, M. V. Zakharov, and M. V. Stepanova, Metallovedenie i obrabotka metallov, No. 3, p. 23 (1957).

587.   V. M. Glazov, M. V. Mal'tsev, and Yu. D. Chistyakov, Izv. AN SSSR, OTN, No. 4, p. 131 (1956).

588.   M. F. Lantratov, Zh. neorg. khim., Vol. 4 (1959).

589.   K. Honda and H. Endo, J. Inst. Metals, 37:29 (1927).

590.   E. A. Dancy, Trans. Metal. Soc. AIME, 233:270 (1965).

591.   G. F. Gubskaya and I. V. Evfimovskii, Zh. neorg. khim., 7:615 (1962).

592.   M. I. Shakhparonov, Izv. AN SSSR, OTN, Met. i topl., No. 3, p. 118 (1961).

593.   V. M. Glazov, Zh. neorg. khim., 4:933 (1961).

594.   V. M. Glazov, Izv. AN SSSR, OTN, Met. i topl., No. 1, p. 89 (1962).

595.   V. M. Glazov, A. N. Krestovnikov, and G. L. Malyutina, Izv. AN SSSR, Neorgan. materialy, 2:1692 (1966).

596.   B. G. Zhurkin et al., Izv. AN SSSR, OTN, Met. i topl., No. 4, p. 156 (1959).

597.   V. S. Zemskov, A. D. Suchkova, and Wang Kuei Hua, Izv. AN SSSR, OTN, Met. i topl., No. 6, p. 149 (1961).

598.   V. S. Zemskov et al., Proc. Conf. on Impact Ionization and Tunnel Effect in Semiconductors, October 11-15, 1960. Izd. AN AzerbSSR, Baku (1963).

599.   V. S. Zemskov, A. D. Suchkova, and B. G. Zhurkin, Zh. fiz. khim., 36:1914 (1962).

600.   V. S. Zemskov and A. D. Belaya, Fiz. tverd. tela, 5:1601 (1963).

601.   V. S. Zemskov, A. D. Belaya, and T. E. Puris, Fiz. tverd. tela, 5:1100 (1963).

602.   V. S. Zemskov, A. D. Belaya, and G. N. Podkorytova, Fiz. tverd. tela, 6:2552 (1964).

603.   A. D. Belaya, Experimental Investigation of Heterogeneous Equilibrium be-
        tween Liquid and Solid Phases during Crystallization of Germanium, Doped
        with Group III, IV, and V Elements. Candidate's Dissertation. A. A. Baikov
        Metallurgy Institute, Moscow (1965).

604.   V. M. Glazov and A. A. Vertman, Collection: Investigation of Nonferrous
        Metal Alloys. Izd. AN SSSR, Moscow (1963), No. 4, p. 85.

605.   O. Ya. Samoilov, Structure of Aqueous Solutions of Electrolytes and the
        Hydration of Ions. Consultants Bureau, New York (1965).

606.   S. F. Khokhlov, Izv. AN SSSR, OTN. Met. i topl., No. 6, p. 80 (1960).

607.   M. Plüss, Z. Anorg. Allgem. Chem., 93:1 (1915).

608.   F. Sauerwald, Z. Anorg. Allgem. Chem., 135: 255 (1924).

609.   A. Bienias and F. Sauerwald, Z. Anorg. Allgem. Chem., 203:15 (1931).

610.   W. Jones and W. Bartlett, J. Inst. Metals, 83:59 (1954/1955).

611.   W. Jones and J. B. Davies, J. Inst. Metals, 86:164 (1957).

612.   H. Y. Fisher and A. Phillips, J. Inst. Metals, 6:1060 (1954).

613.   A. M. Korol'kov, Casting Properties of Metals and Alloys. Consultants
        Bureau, New York (1963).

614.   A. A. Vertman and A. M. Samarin, Izv. AN SSSR, OTN. Met. i topl.,
        No. 4, p. 95 (1960).

615.   B. M. Turovskii and A. P. Lyubimov, Izv. VUZov, Chernaya metallurgiya,
        No. 2, p. 24 (1960).

616.   É. V. Polyak and S. V. Sergeev, Dokl. AN SSSR, 33:244 (1941).

617.   É. V. Polyak and S. V. Sergeev, Dokl. AN SSSR, 57:677 (1947).

618.   F. Sauerwald, Z. Metallkunde, 41:97 (1950).

619.   V. I. Danilov and I. V. Radchenko, Zh. eksperim. i teor. fiz., 7:1158 (1937).

620.   A. I. Danilova and V. I. Danilov, Collection: Problems in Metallography and
        Physics of Metals. Metallurgizdat, Moscow (1951), Vol. 2, p. 31.

621.   G. M. Bartenev, Zh. tekhn. fiz., 17:1325 (1947).

622.   G. M. Bartenev, Collection: Structure and Properties of Liquid Metals.
        A. A. Baikov Metallurgy Institute, Moscow (1959), p. 93.

623.   A. S. Lashko, Dokl. AN UkrSSR, No. 1, p. 30 (1957).

624.   A. S. Lashko, Collection: Problems in Physics of Metals and Metallography.
        Izd. AN UkrSSR, Kiev (1957), No. 8, p. 182.

625.   A. S. Lashko and A. V. Romanova, Ukr. Fiz. Zh., 3:375 (1958).

626.   A. S. Lashko, Zh. fiz. khim., 33:1730 (1959).

627.   A. F. Skryshevskii, Collection: Problems in Physics of Metals and Metallography.
        Izd. AN UkrSSR, Kiev (1957), No. 8, p. 187.

628.   Ya.I. Dutchak and N. M. Bondar', Ukr. fiz. zh., 4:402 (1959).

629.   A. I. Bublik and A. G. Buntar', Kristallografiya, 3:32 (1958).

630.   A. I. Bublik, A. G. Buntar', and N. P. Gaevaya, Trudy fiz. otdel. fiz.-mat.
        fak-ta. Khar'kovsk. Gos. un-ta im. A. M. Gor'kogo, 7:251 (1958).

631.   N. V. Alekseev and Ya. I. Gerasimov, Dokl. AN SSSR, 121:488 (1958).

632.   F. Gaibullaev and A. R. Regel', Zh. tekhn. fiz., 27:1997 (1957).

633.   B. Ya. Pines, Zh. fiz. khim., 23:625 (1949).

634.   P. Duwez, R. H. Willens, and W. Klement, J. Appl. Phys., 31:1136 (1960).

635.   F. Sauerwald, Z. Metallkunde, 14:457 (1922).

Index